Basic Engineering Mathematics

Third Edition

John Bird BSc(Hons), CMath, CEng, FIMA, MIEE, FIIE(Elec), FCollP

Newnes

OXFORD   AMSTERDAM   BOSTON   LONDON   NEW YORK   PARIS
SAN DIEGO   SAN FRANCISCO   SINGAPORE   SYDNEY   TOKYO

Newnes
An imprint of Elsevier Science
Linacre House, Jordan Hill, Oxford OX2 8DP
200 Wheeler Rd, Burlington, MA 01803

First published 1999
Second edition 2000
Reprinted 2001
Third edition 2002

British Library Cataloguing in Publication Data
A catalogue record for this book is available from the British Library

ISBN 0 7506 5775 8

For information on all Newnes publications
visit our website at www.newnespress.com

Typeset by Laserwords Private Limited, Chennai, India
Printed and bound in Great Britain

Contents

Preface

Basic Engineering Mathematics, 3rd edition introduces and then consolidates basic mathematical principles and promotes awareness of mathematical concepts for students needing a broad base for further vocational studies. In this third edition, new material has been added on the introduction to differential and integral calculus.
The text covers:

(i) the Applied Mathematics content of the GNVQ mandatory unit '**Applied Science and Mathematics for Engineering**' at Intermediate level (i.e. GNVQ 2)

(ii) the mandatory '**Mathematics for Engineering**' at Advanced level (i.e. GNVQ 3) in Engineering

(iii) the optional '**Applied Mathematics for Engineering**' at Advanced level (i.e. GNVQ 3) in Engineering

(iv) the Mathematics content of '**Applied Science and Mathematics for Technicians**' for Edexcel/BTEC First Certificate

(v) the mandatory '**Mathematics for Technicians**' for National Certificate and National Diploma in Engineering

(vi) **Mathematics 1** for City & Guilds Technician Certificate in Telecommunications and Electronics Engineering

(vii) **basic mathematics** for a wide range of introductory/access/foundation mathematics courses

(viii) **GCSE revision** and for similar mathematics courses in English-speaking countries world-wide.

Basic Engineering Mathematics, 3rd edition provides a lead into *Engineering Mathematics*.
Each topic considered in the text is presented in a way that assumes in the reader little previous knowledge of that topic.
Theory is introduced in each chapter by a brief outline of essential theory, definitions, formulae, laws and procedures. However these are kept to a minimum, for problem solving is extensively used to establish and exemplify the theory. It is intended that readers will gain real understanding through seeing problems solved and then solving similar problems themselves.
This textbook contains some **575 worked problems**, followed by over **1000 further problems** (all with answers – at the end of the book). The further problems are contained within some **118 Exercises**; each Exercise follows on directly from the relevant section of work. **260 line diagrams** enhance the understanding of the theory. Where at all possible the problems mirror practical situations found in engineering and science.
At regular intervals throughout the text are fifteen **Assignments** to check understanding. For example, Assignment 1 covers material contained in Chapters 1 and 2, Assignment 2 covers the material contained in Chapters 3 and 4, and so on. These Assignments do not have answers given since it is envisaged that lecturers could set the Assignments for students to attempt as part of their course structure. Lecturers' may obtain a complimentary set of solutions of the Assignments in an **Instructor's Manual** available from the publishers via the internet – see below.
At the end of the book a list of relevant **formulae** contained within the text is included for convenience of reference.
'Learning by Example' is at the heart of *Early Engineering Mathematics*

John Bird
University of Portsmouth

Instructur's Manual

Full worked solutions and mark scheme for all the Assignments are contained in this Manual which is available to lecturers only. To obtain a password please e-mail *J.Blackford@Elsevier.com* with the following details: course title, number of students, your job title and work postal address.
To download the Instructor's Manual visit *http://www.newnepress.com* and enter the book title in the search box, or use the following direct URL: *http://www.bh.com/manuals/0750657758/*

1

Basic arithmetic

1.1 Arithmetic operations

Whole numbers are called **integers**. $+3$, $+5$, $+72$ are called positive integers; -13, -6, -51 are called negative integers. Between positive and negative integers is the number 0 which is neither positive nor negative.

The four basic arithmetic operators are: add $(+)$, subtract $(-)$, multiply $(\times)$ and divide $(\div)$

For addition and subtraction, when **unlike signs** are together in a calculation, the overall sign is **negative**. Thus, adding minus 4 to 3 is $3+-4$ and becomes $3-4 = -1$. **Like signs** together give an overall **positive sign**. Thus subtracting minus 4 from 3 is $3 - -4$ and becomes $3 + 4 = 7$. For multiplication and division, when the numbers have **unlike signs**, the answer is **negative**, but when the numbers have **like signs** the answer is **positive**. Thus $3 \times -4 = -12$, whereas $-3 \times -4 = +12$. Similarly

$$\frac{4}{-3} = -\frac{4}{3} \quad \text{and} \quad \frac{-4}{-3} = +\frac{4}{3}$$

Problem 1. Add 27, -74, 81 and -19

This problem is written as $27 - 74 + 81 - 19$

Adding the positive integers:

$$\begin{array}{r} 27 \\ 81 \\ \hline \end{array}$$

Sum of positive integers is: $\quad\underline{108}$

Adding the negative integers:

$$\begin{array}{r} 74 \\ 19 \\ \hline \end{array}$$

Sum of negative integers is: $\quad\underline{93}$

Taking the sum of the negative integers from the sum of the positive integers gives:

$$\begin{array}{r} 108 \\ -93 \\ \hline 15 \\ \hline \end{array}$$

Thus $\mathbf{27 - 74 + 81 - 19 = 15}$

Problem 2. Subtract 89 from 123

This is written mathematically as $123 - 89$

$$\begin{array}{r} 123 \\ -89 \\ \hline 34 \\ \hline \end{array}$$

Thus $\mathbf{123 - 89 = 34}$

Problem 3. Subtract -74 from 377

This problem is written as $377 - -74$. Like signs together give an overall positive sign, hence

$$377 - -74 = 377 + 74 \qquad \begin{array}{r} 377 \\ +74 \\ \hline 451 \\ \hline \end{array}$$

Thus $\mathbf{377 - -74 = 451}$

Problem 4. Subtract 243 from 126

The problem is $126 - 243$. When the second number is larger than the first, take the smaller number from the larger and make the result negative.

Thus $126 - 243 = -(243 - 126)$

$$\begin{array}{r} 243 \\ -126 \\ \hline 117 \end{array}$$

Thus **$126 - 243 = -117$**

Problem 5.   Subtract 318 from -269

$-269 - 318$. The sum of the negative integers is

$$\begin{array}{r} 269 \\ +318 \\ \hline 587 \end{array}$$

Thus **$-269 - 318 = -587$**

Problem 6.   Multiply 74 by 13

This is written as 74×13

$$\begin{array}{r} 74 \\ 13 \\ \hline 222 \\ 740 \\ \hline 962 \end{array}$$

$\leftarrow 74 \times 3$
$\leftarrow 74 \times 10$

Adding:

Thus **$74 \times 13 = 962$**

Problem 7.   Multiply by 178 by -46

When the numbers have different signs, the result will be negative. (With this in mind, the problem can now be solved by multiplying 178 by 46)

$$\begin{array}{r} 178 \\ 46 \\ \hline 1068 \\ 7120 \\ \hline 8188 \end{array}$$

Thus $178 \times 46 = 8188$ and **$178 \times (-46) = -8188$**

Problem 8.   Divide l043 by 7

When dividing by the numbers 1 to 12, it is usual to use a method called **short division**.

$$7 \overline{) 10^34^63} \quad 1\ 4\ 9$$

Step 1. 7 into 10 goes 1, remainder 3. Put 1 above the 0 of 1043 and carry the 3 remainder to the next digit on the right, making it 34;

Step 2. 7 into 34 goes 4, remainder 6. Put 4 above the 4 of 1043 and carry the 6 remainder to the next digit on the right, making it 63;

Step 3. 7 into 63 goes 9, remainder 0. Put 9 above the 3 of 1043.

Thus **$1043 \div 7 = 149$**

Problem 9.   Divide 378 by 14

When dividing by numbers which are larger than 12, it is usual to use a method called **long division**.

$$14 \overline{) 378} \quad 27$$

(2) $2 \times 14 \rightarrow$ 28

98
(4) $7 \times 14 \rightarrow$ 98

00

(1) 14 into 37 goes twice. Put 2 above the 7 of 378.
(3) Subtract. Bring down the 8. 14 into 98 goes 7 times. Put 7 above the 8 of 378.
(5) Subtract.

Thus **$378 \div 14 = 27$**

Problem 10.   Divide 5669 by 46

This problem may be written as $\dfrac{5669}{46}$ or $5669 \div 46$ or $5669/46$

Using the long division method shown in Problem 9 gives:

$$46 \overline{) 5669} \quad 123$$

$$\begin{array}{r} 46 \\ \hline 106 \\ 92 \\ \hline 149 \\ 138 \\ \hline 11 \end{array}$$

As there are no more digits to bring down,

$$5669 \div 46 = 123, \quad \text{remainder } 11 \quad \text{or} \quad 123\frac{11}{46}$$

Now try the following exercise

Exercise 1   Further problems on arithmetic operations (Answers on page 252)

In Problems 1 to 21, determine the values of the expressions given:

1. $67 - 82 + 34$

2. $124 - 273 + 481 - 398$

3. $927 - 114 + 182 - 183 - 247$

4. $2417 - 487 + 2424 - 1778 - 4712$

5. $-38419 - 2177 + 2440 - 799 + 2834$

6. $2715 - 18250 + 11471 - 1509 + 113274$

7. $73 - 57$

8. $813 - (-674)$

9. $647 - 872$

10. $3151 - (-2763)$

11. $4872 - 4683$

12. $-23148 - 47724$

13. $38441 - 53774$

14. (a) 261×7 (b) 462×9

15. (a) 783×11 (b) 73×24

16. (a) 27×38 (b) 77×29

17. (a) 448×23 (b) $143 \times (-31)$

18. (a) $288 \div 6$ (b) $979 \div 11$

19. (a) $\dfrac{1813}{7}$ (b) $\dfrac{896}{16}$

20. (a) $\dfrac{21432}{47}$ (b) $15904 \div 56$

21. (a) $\dfrac{88738}{187}$ (b) $46857 \div 79$

1.2 Highest common factors and lowest common multiples

When two or more numbers are multiplied together, the individual numbers are called **factors**. Thus a factor is a number which divides into another number exactly. The **highest common factor (HCF)** is the largest number which divides into two or more numbers exactly.

A **multiple** is a number which contains another number an exact number of times. The smallest number which is exactly divisible by each of two or more numbers is called the **lowest common multiple (LCM)**.

Problem 11. Determine the HCF of the numbers 12, 30 and 42

Each number is expressed in terms of its lowest factors. This is achieved by repeatedly dividing by the prime numbers 2, 3, 5, 7, 11, 13 ... (where possible) in turn. Thus

$$12 = 2 \times 2 \times 3$$
$$30 = 2 \quad\times 3 \times 5$$
$$42 = 2 \quad\times 3 \times 7$$

The factors which are common to each of the numbers are 2 in column 1 and 3 in column 3, shown by the broken lines. Hence the **HCF is 2×3, i.e. 6**. That is, 6 is the largest number which will divide into 12, 30 and 42.

Problem 12. Determine the HCF of the numbers 30, 105, 210 and 1155

Using the method shown in Problem 11:

$$30 = 2 \times 3 \times 5$$
$$105 = \quad 3 \times 5 \times 7$$
$$210 = 2 \times 3 \times 5 \times 7$$
$$1155 = \quad 3 \times 5 \times 7 \times 11$$

The factors which are common to each of the numbers are 3 in column 2 and 5 in column 3. Hence **the HCF is $3 \times 5 = 15$**

Problem 13. Determine the LCM of the numbers 12, 42 and 90

The LCM is obtained by finding the lowest factors of each of the numbers, as shown in Problems 11 and 12 above, and then selecting the largest group of any of the factors present. Thus

$$12 = 2 \times 2 \times 3$$
$$42 = 2 \quad\times 3 \qquad\times 7$$
$$90 = 2 \quad\times 3 \times 3 \times 5$$

The largest group of any of the factors present are shown by the broken lines and are 2×2 in 12, 3×3 in 90, 5 in 90 and 7 in 42.

Hence **the LCM is $2 \times 2 \times 3 \times 3 \times 5 \times 7 = 1260$**, and is the smallest number which 12, 42 and 90 will all divide into exactly.

Problem 14. Determine the LCM of the numbers 150, 210, 735 and 1365

Using the method shown in Problem 13 above:

$$150 = \boxed{2} \times \boxed{3} \times \boxed{5 \times 5}$$

$$210 = 2 \times 3 \times 5 \times 7$$

$$735 = \qquad 3 \times 5 \times \boxed{7 \times 7}$$

$$1365 = \qquad 3 \times 5 \times 7 \qquad \times \boxed{13}$$

The LCM is $2 \times 3 \times 5 \times 5 \times 7 \times 7 \times 13 = 95550$

Now try the following exercise

Exercise 2 Further problems on highest common factors and lowest common multiples (Answers on page 252)

In Problems 1 to 6 find (a) the HCF and (b) the LCM of the numbers given:

1. 6, 10, 14   2. 12, 30, 45

3. 10, 15, 70, 105   4. 90, 105, 300

5. 196, 210, 910, 462   6. 196, 350, 770

1.3 Order of precedence and brackets

When a particular arithmetic operation is to be performed first, the numbers and the operator(s) are placed in brackets. Thus 3 times the result of 6 minus 2 is written as $3 \times (6-2)$. In arithmetic operations, the order in which operations are performed are:

(i) to determine the values of operations contained in brackets;

(ii) multiplication and division (the word 'of' also means multiply); and

(iii) addition and subtraction.

This **order of precedence** can be remembered by the word **BODMAS**, standing for **B**rackets, **O**f, **D**ivision, **M**ultiplication, **A**ddition and **S**ubtraction, taken in that order.
The basic laws governing the use of brackets and operators are shown by the following examples:

(i) $2 + 3 = 3 + 2$, i.e. the order of numbers when adding does not matter;

(ii) $2 \times 3 = 3 \times 2$, i.e. the order of numbers when multiplying does not matter;

(iii) $2 + (3+4) = (2+3) + 4$, i.e. the use of brackets when adding does not affect the result;

(iv) $2 \times (3 \times 4) = (2 \times 3) \times 4$, i.e. the use of brackets when multiplying does not affect the result;

(v) $2 \times (3+4) = 2(3+4) = 2 \times 3 + 2 \times 4$, i.e. a number placed outside of a bracket indicates that the whole contents of the bracket must be multiplied by that number;

(vi) $(2 + 3)(4 + 5) = (5)(9) = 45$, i.e. adjacent brackets indicate multiplication;

(vii) $2[3 + (4 \times 5)] = 2[3 + 20] = 2 \times 23 = 46$, i.e. when an expression contains inner and outer brackets, the inner brackets are removed first.

Problem 15. Find the value of $6 + 4 \div (5 - 3)$

The order of precedence of operations is remembered by the word BODMAS.

$$\text{Thus } 6 + 4 \div (5 - 3) = 6 + 4 \div 2 \qquad \text{(Brackets)}$$

$$= 6 + 2 \qquad \text{(Division)}$$

$$= 8 \qquad \text{(Addition)}$$

Problem 16. Determine the value of

$$13 - 2 \times 3 + 14 \div (2 + 5)$$

$$13 - 2 \times 3 + 14 \div (2 + 5) = 13 - 2 \times 3 + 14 \div 7 \quad \text{(B)}$$

$$= 13 - 2 \times 3 + 2 \quad \text{(D)}$$

$$= 13 - 6 + 2 \quad \text{(M)}$$

$$= 15 - 6 \quad \text{(A)}$$

$$= 9 \quad \text{(S)}$$

Problem 17. Evaluate

$$16 \div (2 + 6) + 18[3 + (4 \times 6) - 21]$$

$$16 \div (2 + 6) + 18[3 + (4 \times 6) - 21]$$

$$= 16 \div (2 + 6) + 18[3 + 24 - 21] \quad \text{(B)}$$

$$= 16 \div 8 + 18 \times 6 \quad \text{(B)}$$

$$= 2 + 18 \times 6 \quad \text{(D)}$$

$$= 2 + 108 \qquad (M)$$

$$= \mathbf{110} \qquad (A)$$

Problem 18. Find the value of

$$23 - 4(2 \times 7) + \frac{(144 \div 4)}{(14 - 8)}$$

$$23 - 4(2 \times 7) + \frac{(144 \div 4)}{(14 - 8)} = 23 - 4 \times 14 + \frac{36}{6} \quad (B)$$

$$= 23 - 4 \times 14 + 6 \quad (D)$$

$$= 23 - 56 + 6 \quad (M)$$

$$= 29 - 56 \quad (A)$$

$$= \mathbf{-27} \quad (S)$$

Now try the following exercise

Exercise 3   Further problems on order of precedence and brackets (Answers on page 252)

Simplify the expressions given in Problems 1 to 7:

1. $14 + 3 \times 15$

2. $17 - 12 \div 4$

3. $86 + 24 \div (14 - 2)$

4. $7(23 - 18) \div (12 - 5)$

5. $63 - 28(14 \div 2) + 26$

6. $\dfrac{112}{16} - 119 \div 17 + (3 \times 19)$

7. $\dfrac{(50 - 14)}{3} + 7(16 - 7) - 7$

2

Fractions, decimals and percentages

2.1 Fractions

When 2 is divided by 3, it may be written as $\frac{2}{3}$ or 2/3. $\frac{2}{3}$ is called a **fraction**. The number above the line, i.e. 2, is called the **numerator** and the number below the line, i.e. 3, is called the **denominator**.

When the value of the numerator is less than the value of the denominator, the fraction is called a **proper fraction**; thus $\frac{2}{3}$ is a proper fraction. When the value of the numerator is greater than the denominator, the fraction is called an **improper fraction**. Thus $\frac{7}{3}$ is an improper fraction and can also be expressed as a **mixed number**, that is, an integer and a proper fraction. Thus the improper fraction $\frac{7}{3}$ is equal to the mixed number $2\frac{1}{3}$.

When a fraction is simplified by dividing the numerator and denominator by the same number, the process is called **cancelling**. Cancelling by 0 is not permissible.

Problem 1. Simplify $\frac{1}{3} + \frac{2}{7}$

The LCM of the two denominators is 3×7, i.e. 21.

Expressing each fraction so that their denominators are 21, gives:

$$\frac{1}{3} + \frac{2}{7} = \frac{1}{3} \times \frac{7}{7} + \frac{2}{7} \times \frac{3}{3} = \frac{7}{21} + \frac{6}{21}$$

$$= \frac{7+6}{21} = \frac{\mathbf{13}}{\mathbf{21}}$$

Alternatively:

$$\begin{array}{cc} \text{Step (2)} & \text{Step (3)} \\ \downarrow & \downarrow \end{array}$$

$$\frac{1}{3} + \frac{2}{7} = \frac{(7 \times 1) + (3 \times 2)}{21}$$

$$\uparrow$$
$$\text{Step (1)}$$

Step 1: the LCM of the two denominators;

Step 2: for the fraction $\frac{1}{3}$, 3 into 21 goes 7 times, $7 \times$ the numerator is 7×1;

Step 3: for the fraction $\frac{2}{7}$, 7 into 21 goes 3 times, $3 \times$ the numerator is 3×2.

Thus $\frac{1}{3} + \frac{2}{7} = \frac{7+6}{21} = \frac{\mathbf{13}}{\mathbf{21}}$ as obtained previously.

Problem 2. Find the value of $3\frac{2}{3} - 2\frac{1}{6}$

One method is to split the mixed numbers into integers and their fractional parts. Then

$$3\frac{2}{3} - 2\frac{1}{6} = \left(3 + \frac{2}{3}\right) - \left(2 + \frac{1}{6}\right) = 3 + \frac{2}{3} - 2 - \frac{1}{6}$$

$$= 1 + \frac{4}{6} - \frac{1}{6} = 1\frac{3}{6} = 1\frac{\mathbf{1}}{\mathbf{2}}$$

Another method is to express the mixed numbers as improper fractions.

Since $3 = \frac{9}{3}$, then $3\frac{2}{3} = \frac{9}{3} + \frac{2}{3} = \frac{11}{3}$

Similarly, $2\frac{1}{6} = \frac{12}{6} + \frac{1}{6} = \frac{13}{6}$

Thus $3\frac{2}{3} - 2\frac{1}{6} = \frac{11}{3} - \frac{13}{6} = \frac{22}{6} - \frac{13}{6} = \frac{9}{6} = 1\frac{\mathbf{1}}{\mathbf{2}}$ as obtained previously.

Problem 3. Evaluate $7\frac{1}{8} - 5\frac{3}{7}$

$$7\frac{1}{8} - 5\frac{3}{7} = \left(7 + \frac{1}{8}\right) - \left(5 + \frac{3}{7}\right) = 7 + \frac{1}{8} - 5 - \frac{3}{7}$$

$$= 2 + \frac{1}{8} - \frac{3}{7} = 2 + \frac{7 \times 1 - 8 \times 3}{56}$$

$$= 2 + \frac{7 - 24}{56} = 2 + \frac{-17}{56}$$

$$= 2 - \frac{17}{56} = \frac{112}{56} - \frac{17}{56} = \frac{112 - 17}{56}$$

$$= \frac{95}{56} = \mathbf{1\frac{39}{56}}$$

Problem 4. Determine the value of $4\frac{5}{8} - 3\frac{1}{4} + 1\frac{2}{5}$

$$4\frac{5}{8} - 3\frac{1}{4} + 1\frac{2}{5} = (4 - 3 + 1) + \left(\frac{5}{8} - \frac{1}{4} + \frac{2}{5}\right)$$

$$= 2 + \frac{5 \times 5 - 10 \times 1 + 8 \times 2}{40}$$

$$= 2 + \frac{25 - 10 + 16}{40}$$

$$= 2 + \frac{31}{40} = \mathbf{2\frac{31}{40}}$$

Problem 5. Find the value of $\frac{3}{7} \times \frac{14}{15}$

Dividing numerator and denominator by 3 gives:

$$\frac{^1\cancel{3}}{7} \times \frac{14}{\cancel{15}_5} = \frac{1}{7} \times \frac{14}{5} = \frac{1 \times 14}{7 \times 5}$$

Dividing numerator and denominator by 7 gives:

$$\frac{1 \times \cancel{14}^{\,2}}{_1\cancel{7} \times 5} = \frac{1 \times 2}{1 \times 5} = \frac{2}{5}$$

This process of dividing both the numerator and denominator of a fraction by the same factor(s) is called **cancelling**.

Problem 6. Evaluate $1\frac{3}{5} \times 2\frac{1}{3} \times 3\frac{3}{7}$

Mixed numbers **must** be expressed as improper fractions before multiplication can be performed. Thus,

$$1\frac{3}{5} \times 2\frac{1}{3} \times 3\frac{3}{7} = \left(\frac{5}{5} + \frac{3}{5}\right) \times \left(\frac{6}{3} + \frac{1}{3}\right) \times \left(\frac{21}{7} + \frac{3}{7}\right)$$

$$= \frac{8}{5} \times \frac{^1\cancel{7}}{_1\cancel{3}} \times \frac{\cancel{24}^{\,8}}{\cancel{7}_1} = \frac{8 \times 1 \times 8}{5 \times 1 \times 1}$$

$$= \frac{64}{5} = \mathbf{12\frac{4}{5}}$$

Problem 7. Simplify $\frac{3}{7} \div \frac{12}{21}$

$$\frac{3}{7} \div \frac{12}{21} = \frac{\dfrac{3}{7}}{\dfrac{12}{21}}$$

Multiplying both numerator and denominator by the reciprocal of the denominator gives:

$$\frac{\dfrac{3}{7}}{\dfrac{12}{21}} = \frac{\dfrac{^1\cancel{3}}{_1\cancel{7}} \times \dfrac{\cancel{21}^{\,3}}{\cancel{12}_4}}{\dfrac{_1\cancel{12}}{_1\cancel{21}} \times \dfrac{\cancel{21}^{\,1}}{\cancel{12}_1}} = \frac{\dfrac{3}{4}}{1} = \frac{\mathbf{3}}{\mathbf{4}}$$

This method can be remembered by the rule: invert the second fraction and change the operation from division to multiplication. Thus:

$$\frac{3}{7} \div \frac{12}{21} = \frac{^1\cancel{3}}{_1\cancel{7}} \times \frac{\cancel{21}^{\,3}}{\cancel{12}_4} = \frac{\mathbf{3}}{\mathbf{4}} \text{ as obtained previously.}$$

Problem 8. Find the value of $5\frac{3}{5} \div 7\frac{1}{3}$

The mixed numbers must be expressed as improper fractions. Thus,

$$5\frac{3}{5} \div 7\frac{1}{3} = \frac{28}{5} \div \frac{22}{3} = \frac{^{14}\cancel{28}}{5} \times \frac{3}{\cancel{22}_{11}} = \frac{\mathbf{42}}{\mathbf{55}}$$

Problem 9. Simplify $\frac{1}{3} - \left(\frac{2}{5} + \frac{1}{4}\right) \div \left(\frac{3}{8} \times \frac{1}{3}\right)$

The order of precedence of operations for problems containing fractions is the same as that for integers, i.e. remembered by **BODMAS** (**B**rackets, **O**f, **D**ivision, **M**ultiplication, **A**ddition and **S**ubtraction). Thus,

$$\frac{1}{3} - \left(\frac{2}{5} + \frac{1}{4}\right) \div \left(\frac{3}{8} \times \frac{1}{3}\right)$$

$$= \frac{1}{3} - \frac{4 \times 2 + 5 \times 1}{20} \div \frac{\cancel{8}^{\,1}}{\cancel{24}_8} \qquad (B)$$

$$= \frac{1}{3} - \frac{13}{_5\cancel{20}} \times \frac{\cancel{8}^{\,2}}{1} \qquad (D)$$

$$= \frac{1}{3} - \frac{26}{5} \qquad (M)$$

$$= \frac{(5 \times 1) - (3 \times 26)}{15} \qquad (S)$$

$$= \frac{-73}{15} = \mathbf{-4\frac{13}{15}}$$

Problem 10. Determine the value of

$$\frac{7}{6} \text{ of } \left(3\frac{1}{2} - 2\frac{1}{4}\right) + 5\frac{1}{8} \div \frac{3}{16} - \frac{1}{2}$$

$$\frac{7}{6} \text{ of } \left(3\frac{1}{2} - 2\frac{1}{4}\right) + 5\frac{1}{8} \div \frac{3}{16} - \frac{1}{2}$$

$$= \frac{7}{6} \text{ of } 1\frac{1}{4} + \frac{41}{8} \div \frac{3}{16} - \frac{1}{2} \qquad \text{(B)}$$

$$= \frac{7}{6} \times \frac{5}{4} + \frac{41}{8} \div \frac{3}{16} - \frac{1}{2} \qquad \text{(O)}$$

$$= \frac{7}{6} \times \frac{5}{4} + \frac{41}{\cancel{8}_{1}} \times \frac{\cancel{16}^{2}}{3} - \frac{1}{2} \qquad \text{(D)}$$

$$= \frac{35}{24} + \frac{82}{3} - \frac{1}{2} \qquad \text{(M)}$$

$$= \frac{35 + 656}{24} - \frac{1}{2} \qquad \text{(A)}$$

$$= \frac{691}{24} - \frac{1}{2} \qquad \text{(A)}$$

$$= \frac{691 - 12}{24} \qquad \text{(S)}$$

$$= \frac{679}{24} = 28\frac{7}{24}$$

Now try the following exercise

Exercise 4   Further problems on fractions (Answers on page 252)

Evaluate the expressions given in Problems 1 to 13:

1. (a) $\frac{1}{2} + \frac{2}{5}$ (b) $\frac{7}{16} - \frac{1}{4}$

2. (a) $\frac{2}{7} + \frac{3}{11}$ (b) $\frac{2}{9} - \frac{1}{7} + \frac{2}{3}$

3. (a) $5\frac{3}{13} + 3\frac{3}{4}$ (b) $4\frac{5}{8} - 3\frac{2}{5}$

4. (a) $10\frac{3}{7} - 8\frac{2}{3}$ (b) $3\frac{1}{4} - 4\frac{4}{5} + 1\frac{5}{6}$

5. (a) $\frac{3}{4} \times \frac{5}{9}$ (b) $\frac{17}{35} \times \frac{15}{119}$

6. (a) $\frac{3}{5} \times \frac{7}{9} \times 1\frac{2}{7}$ (b) $\frac{13}{17} \times 4\frac{7}{11} \times 3\frac{4}{39}$

7. (a) $\frac{1}{4} \times \frac{3}{11} \times 1\frac{5}{39}$ (b) $\frac{3}{4} \div 1\frac{4}{5}$

8. (a) $\frac{3}{8} \div \frac{45}{64}$ (b) $1\frac{1}{3} \div 2\frac{5}{9}$

9. $\frac{1}{3} - \frac{3}{4} \times \frac{16}{27}$

10. $\frac{1}{2} + \frac{3}{5} \div \frac{9}{15} - \frac{1}{3}$

11. $\frac{7}{15} \text{ of } \left(15 \times \frac{5}{7}\right) + \left(\frac{3}{4} \div \frac{15}{16}\right)$

12. $\frac{1}{4} \times \frac{2}{3} - \frac{1}{3} \div \frac{3}{5} + \frac{2}{7}$

13. $\left(\frac{2}{3} \times 1\frac{1}{4}\right) \div \left(\frac{2}{3} + \frac{1}{4}\right) + 1\frac{3}{5}$

2.2 Ratio and proportion

The ratio of one quantity to another is a fraction, and is the number of times one quantity is contained in another quantity **of the same kind.** If one quantity is **directly proportional** to another, then as one quantity doubles, the other quantity also doubles. When a quantity is **inversely proportional** to another, then as one quantity doubles, the other quantity is halved.

Problem 11. Divide 126 in the ratio of 5 to 13

Because the ratio is to be 5 parts to 13 parts, then the total number of parts is $5 + 13$, that is 18. Then,

18 parts correspond to 126

Hence   1 part corresponds to $\dfrac{126}{18} = 7$,

5 parts correspond to $5 \times 7 = \mathbf{35}$ and 13 parts correspond to $13 \times 7 = \mathbf{91}$

(Check: the parts must add up to the total $35 + 91 = 126 =$ the total.)

Problem 12. A piece of timber 273 cm long is cut into three pieces in the ratio of 3 to 7 to 11. Determine the lengths of the three pieces.

The total number of parts is $3 + 7 + 11$, that is, 21. Hence 21 parts correspond to 273 cm

1 part corresponds to $\dfrac{273}{21} = 13$ cm

3 parts correspond to $3 \times 13 = 39$ cm

7 parts correspond to $7 \times 13 = 91$ cm

11 parts correspond to $11 \times 13 = 143$ cm

i.e. **the lengths of the three pieces are 39 cm, 91 cm and 143 cm.**

(Check: $39 + 91 + 143 = 273$)

Problem 13.   A gear wheel having 80 teeth is in mesh with a 25 tooth gear. What is the gear ratio?

Gear ratio $= 80{:}25 = \dfrac{80}{25} = \dfrac{16}{5} = 3.2$

i.e. gear ratio $= \mathbf{16{:}5}$ or $\mathbf{3.2{:}1}$

Problem 14.   Express 25p as a ratio of £4.25

Working in quantities **of the same kind**, the required ratio is

$\dfrac{25}{425}$ i.e. $\dfrac{1}{17}$

That is, 25p is $\dfrac{1}{17}$th of £4.25. This may be written either as:

25:425: :1:17 (stated as '25 is to 425 as 1 is to 17') or as

$$\dfrac{\mathbf{25}}{\mathbf{425}} = \dfrac{\mathbf{1}}{\mathbf{17}}$$

Problem 15.   An alloy is made up of metals A and B in the ratio 2.5:1 by mass. How much of A has to be added to 6 kg of B to make the alloy?

Ratio A:B: :2.5:1 i.e. $\dfrac{A}{B} = \dfrac{2.5}{1} = 2.5$

When B $= 6$ kg, $\dfrac{A}{6} = 2.5$ from which, $\mathbf{A} = 6 \times 2.5 = \mathbf{15\,kg}$

Problem 16.   If 3 people can complete a task in 4 hours, find how long it will take 5 people to complete the same task, assuming the rate of work remains constant.

The more the number of people, the more quickly the task is done, hence inverse proportion exists.

3 people complete the task in 4 hours,

1 person takes three times as long, i.e. $4 \times 3 = 12$ hours,

5 people can do it in one fifth of the time that one person takes, that is $\dfrac{12}{5}$ hours or **2 hours 24 minutes**.

Now try the following exercise

Exercise 5   Further problems on ratio and proportion (Answers on page 252)

1. Divide 312 mm in the ratio of 7 to 17.

2. Divide 621 cm in the ratio of 3 to 7 to 13.

3. £4.94 is to be divided between two people in the ratio of 9 to 17. Determine how much each person will receive.

4. When mixing a quantity of paints, dyes of four different colours are used in the ratio of 7:3:19:5. If the mass of the first dye used is $3\frac{1}{2}$ g, determine the total mass of the dyes used.

5. Determine how much copper and how much zinc is needed to make a 99 kg brass ingot if they have to be in the proportions copper:zinc: :8:3 by mass.

6. It takes 21 hours for 12 men to resurface a stretch of road. Find how many men it takes to resurface a similar stretch of road in 50 hours 24 minutes, assuming the work rate remains constant.

7. It takes 3 hours 15 minutes to fly from city A to city B at a constant speed. Find how long the journey takes if

   (a) the speed is $1\frac{1}{2}$ times that of the original speed and

   (b) if the speed is three-quarters of the original speed.

2.3   Decimals

The decimal system of numbers is based on the **digits** 0 to 9. A number such as 53.17 is called a **decimal fraction**, a decimal point separating the integer part, i.e. 53, from the fractional part, i.e. 0.17

A number which can be expressed exactly as a decimal fraction is called a **terminating decimal** and those which cannot be expressed exactly as a decimal fraction are called **non-terminating decimals**. Thus, $\frac{3}{2} = 1.5$ is a terminating decimal, but $\frac{4}{3} = 1.33333\ldots$ is a non-terminating decimal. $1.33333\ldots$ can be written as $1.\dot{3}$, called 'one point-three recurring'.

The answer to a non-terminating decimal may be expressed in two ways, depending on the accuracy required:

(i) correct to a number of **significant figures**, that is, figures which signify something, and

(ii) correct to a number of **decimal places**, that is, the number of figures after the decimal point.

The last digit in the answer is unaltered if the next digit on the right is in the group of numbers 0, 1, 2, 3 or 4, but is increased by 1 if the next digit on the right is in the group of numbers 5, 6, 7, 8 or 9. Thus the non-terminating decimal $7.6183\ldots$ becomes 7.62, correct to 3 significant figures, since the next digit on the right is 8, which is in the group of numbers 5, 6, 7, 8 or 9. Also $7.6183\ldots$ becomes 7.618, correct to 3 decimal places, since the next digit on the right is 3, which is in the group of numbers 0, 1, 2, 3 or 4.

Problem 17.   Evaluate $42.7 + 3.04 + 8.7 + 0.06$

The numbers are written so that the decimal points are under each other. Each column is added, starting from the right.

$$\begin{array}{r} 42.7 \\ 3.04 \\ 8.7 \\ 0.06 \\ \hline 54.50 \\ \hline \end{array}$$

Thus **42.7 + 3.04 + 8.7 + 0.06 = 54.50**

Problem 18. Take 81.70 from 87.23

The numbers are written with the decimal points under each other.

$$\begin{array}{r} 87.23 \\ -81.70 \\ \hline 5.53 \\ \hline \end{array}$$

Thus **87.23 − 81.70 = 5.53**

Problem 19. Find the value of

$$23.4 - 17.83 - 57.6 + 32.68$$

The sum of the positive decimal fractions is

$$23.4 + 32.68 = 56.08$$

The sum of the negative decimal fractions is

$$17.83 + 57.6 = 75.43$$

Taking the sum of the negative decimal fractions from the sum of the positive decimal fractions gives:

$$56.08 - 75.43$$

i.e. $-(75.43 - 56.08) = \mathbf{-19.35}$

Problem 20. Determine the value of 74.3×3.8

When multiplying decimal fractions: (i) the numbers are multiplied as if they are integers, and (ii) the position of the decimal point in the answer is such that there are as many digits to the right of it as the sum of the digits to the right of the decimal points of the two numbers being multiplied together. Thus

$$\text{(i)} \quad \begin{array}{r} 743 \\ 38 \\ \hline 5944 \\ 22\,290 \\ \hline 28\,234 \\ \hline \end{array}$$

(ii) As there are $(1+1) = 2$ digits to the right of the decimal points of the two numbers being multiplied together, $(74.\underline{3} \times 3.\underline{8})$, then

$$\mathbf{74.3 \times 3.8 = 282.34}$$

Problem 21. Evaluate $37.81 \div 1.7$, correct to (i) 4 significant figures and (ii) 4 decimal places.

$$37.81 \div 1.7 = \frac{37.81}{1.7}$$

The denominator is changed into an integer by multiplying by 10. The numerator is also multiplied by 10 to keep the fraction the same. Thus

$$37.81 \div 1.7 = \frac{37.81 \times 10}{1.7 \times 10} = \frac{378.1}{17}$$

The long division is similar to the long division of integers and the first four steps are as shown:

$$\begin{array}{r} 22.24117.. \\ 17 \,\overline{)\, 378.100000} \\ \underline{34} \\ 38 \\ \underline{34} \\ 41 \\ \underline{34} \\ 70 \\ \underline{68} \\ 20 \end{array}$$

(i) $\mathbf{37.81 \div 1.7 = 22.24}$, **correct to 4 significant figures,** and

(ii) $\mathbf{37.81 \div 1.7 = 22.2412}$, **correct to 4 decimal places.**

Problem 22. Convert (a) 0.4375 to a proper fraction and (b) 4.285 to a mixed number.

(a) 0.4375 can be written as $\dfrac{0.4375 \times 10\,000}{10\,000}$ without changing its value,

i.e. $0.4375 = \dfrac{4375}{10\,000}$

By cancelling $\dfrac{4375}{10\,000} = \dfrac{875}{2000} = \dfrac{175}{400} = \dfrac{35}{80} = \dfrac{7}{16}$

i.e. $\mathbf{0.4375 = \dfrac{7}{16}}$

(b) Similarly, $\mathbf{4.285 = 4\dfrac{285}{1000} = 4\dfrac{57}{200}}$

Problem 23. Express as decimal fractions:

(a) $\dfrac{9}{16}$ and (b) $5\dfrac{7}{8}$

(a) To convert a proper fraction to a decimal fraction, the numerator is divided by the denominator. Division by 16 can be done by the long division method, or, more simply, by dividing by 2 and then 8:

$$
\begin{array}{r} 4.50 \\ 2\,)\,\overline{9.00} \end{array}
\qquad
\begin{array}{r} 0.5\ 6\ 2\ 5 \\ 8\,)\,\overline{4.5^50^20^40} \end{array}
$$

Thus, $\dfrac{9}{16} = \mathbf{0.5625}$

(b) For mixed numbers, it is only necessary to convert the proper fraction part of the mixed number to a decimal fraction. Thus, dealing with the $\frac{7}{8}$ gives:

$$
\begin{array}{r} 0.875 \\ 8\,)\,\overline{7.000} \end{array}
\qquad \text{i.e.} \quad \frac{7}{8} = 0.875
$$

Thus $5\dfrac{7}{8} = \mathbf{5.875}$

Now try the following exercise

Exercise 6   Further problems on decimals (Answers on page 252)

In Problems 1 to 7, determine the values of the expressions given:

1. $23.6 + 14.71 - 18.9 - 7.421$
2. $73.84 - 113.247 + 8.21 - 0.068$
3. 5.73×4.2
4. $3.8 \times 4.1 \times 0.7$
5. 374.1×0.006
6. $421.8 \div 17$, (a) correct to 4 significant figures and (b) correct to 3 decimal places.
7. $\dfrac{0.0147}{2.3}$, (a) correct to 5 decimal places and

   (b) correct to 2 significant figures.
8. Convert to proper fractions:

   (a) 0.65 (b) 0.84 (c) 0.0125 (d) 0.282 and (e) 0.024
9. Convert to mixed numbers:

   (a) 1.82 (b) 4.275 (c) 14.125 (d) 15.35 and

   (e) 16.2125

In Problems 10 to 15, express as decimal fractions to the accuracy stated:

10. $\dfrac{4}{9}$, correct to 5 significant figures.

11. $\dfrac{17}{27}$, correct to 5 decimal place.

12. $1\dfrac{9}{16}$, correct to 4 significant figures.

13. $53\dfrac{5}{11}$, correct to 3 decimal places.

14. $13\dfrac{31}{37}$, correct to 2 decimal places.

15. $8\dfrac{9}{13}$, correct to 3 significant figures.

2.4 Percentages

Percentages are used to give a common standard and are fractions having the number 100 as their denominators. For example, 25 per cent means $\dfrac{25}{100}$ i.e. $\dfrac{1}{4}$ and is written 25%.

Problem 24. Express as percentages: (a) 1.875 and (b) 0.0125

A decimal fraction is converted to a percentage by multiplying by 100. Thus,

(a) 1.875 corresponds to $1.875 \times 100\%$, i.e. **187.5%**

(b) 0.0125 corresponds to $0.0125 \times 100\%$, i.e. **1.25%**

Problem 25. Express as percentages:

(a) $\dfrac{5}{16}$ and (b) $1\dfrac{2}{5}$

To convert fractions to percentages, they are (i) converted to decimal fractions and (ii) multiplied by 100

(a) By division, $\dfrac{5}{16} = 0.3125$, hence $\dfrac{5}{16}$ corresponds to $0.3125 \times 100\%$, i.e. **31.25%**

(b) Similarly, $1\dfrac{2}{5} = 1.4$ when expressed as a decimal fraction.

   Hence $1\dfrac{2}{5} = 1.4 \times 100\% = \mathbf{140\%}$

Problem 26. It takes 50 minutes to machine a certain part. Using a new type of tool, the time can be reduced by 15%. Calculate the new time taken.

15% of 50 minutes $= \dfrac{15}{100} \times 50 = \dfrac{750}{100} = 7.5$ minutes.

Hence the **new time taken** is $50 - 7.5 = $ **42.5 minutes**.

Alternatively, if the time is reduced by 15%, then it now takes 85% of the original time, i.e. 85% of $50 = \frac{85}{100} \times 50 = \frac{4250}{100} = $ **42.5 minutes**, as above.

Problem 27. Find 12.5% of £378

12.5% of £378 means $\frac{12.5}{100} \times 378$, since per cent means 'per hundred'.

Hence 12.5% of £378 $= \frac{12.5^1}{100_8} \times 378 = \frac{378}{8} = $ **£47.25**

Problem 28. Express 25 minutes as a percentage of 2 hours, correct to the nearest 1%.

Working in minute units, 2 hours $= 120$ minutes. Hence 25 minutes is $\frac{25}{120}$ ths of 2 hours.

By cancelling, $\frac{25}{120} = \frac{5}{24}$

Expressing $\frac{5}{24}$ as a decimal fraction gives $0.208\dot{3}$

Multiplying by 100 to convert the decimal fraction to a percentage gives:

$$0.208\dot{3} \times 100 = 20.8\dot{3}\%$$

Thus **25 minutes is 21% of 2 hours**, correct to the nearest 1%.

Problem 29. A German silver alloy consists of 60% copper, 25% zinc and 15% nickel. Determine the masses of the copper, zinc and nickel in a 3.74 kilogram block of the alloy.

By direct proportion:

100% corresponds to $3.74 \, \text{kg}$

1% corresponds to $\frac{3.74}{100} = 0.0374 \, \text{kg}$

60% corresponds to $60 \times 0.0374 = 2.244 \, \text{kg}$

25% corresponds to $25 \times 0.0374 = 0.935 \, \text{kg}$

15% corresponds to $15 \times 0.0374 = 0.561 \, \text{kg}$

Thus, the masses of the copper, zinc and nickel are **2.244 kg, 0.935 kg and 0.561 kg**, respectively.

(Check: $2.244 + 0.935 + 0.561 = 3.74$)

Now try the following exercise

Exercise 7   Further problems percentages (Answers on page 253)

1. Convert to percentages:

    (a) 0.057   (b) 0.374   (c) 1.285

2. Express as percentages, correct to 3 significant figures:

    (a) $\frac{7}{33}$ (b) $\frac{19}{24}$ (c) $1\frac{11}{16}$

3. Calculate correct to 4 significant figures:

    (a) 18% of 2758 tonnes   (b) 47% of 18.42 grams

    (c) 147% of 14.1 seconds

4. When 1600 bolts are manufactured, 36 are unsatisfactory. Determine the percentage unsatisfactory.

5. Express:

    (a) 140 kg as a percentage of 1 t

    (b) 47 s as a percentage of 5 min

    (c) 13.4 cm as a percentage of 2.5 m

6. A block of monel alloy consists of 70% nickel and 30% copper. If it contains 88.2 g of nickel, determine the mass of copper in the block.

7. A drilling machine should be set to 250 rev/min. The nearest speed available on the machine is 268 rev/min. Calculate the percentage overspeed.

8. Two kilograms of a compound contains 30% of element A, 45% of element B and 25% of element C. Determine the masses of the three elements present.

9. A concrete mixture contains seven parts by volume of ballast, four parts by volume of sand and two parts by volume of cement. Determine the percentage of each of these three constituents correct to the nearest 1% and the mass of cement in a two tonne dry mix, correct to 1 significant figure.

Assignment 1

This assignment covers the material contained in Chapters 1 and 2. The marks for each question are shown in brackets at the end of each question.

1. Evaluate the following:

    (a) $2016 - (-593)$

    (b) $73 + 35 \div (11 - 4)$

(c) $\dfrac{120}{15} - 133 \div 19 + (2 \times 17)$

(d) $\dfrac{2}{5} - \dfrac{1}{15} + \dfrac{5}{6}$

(e) $49.31 - 97.763 + 9.44 - 0.079$ (14)

2. Determine, by long multiplication 37×42 (3)

3. Evaluate, by long division $\dfrac{4675}{11}$ (3)

4. Find (a) the highest common factor, and (b) the lowest common multiple of the following numbers:

   15   40   75   120   (6)

5. Simplify (a) $2\dfrac{2}{3} \div 3\dfrac{1}{3}$

(b) $\dfrac{1}{\left(\dfrac{4}{7} \times 2\dfrac{1}{4}\right)} \div \left(\dfrac{1}{3} + \dfrac{1}{5}\right) + 2\dfrac{7}{24}$ (9)

6. A piece of steel, 1.69 m long, is cut into three pieces in the ratio 2 to 5 to 6. Determine, in centimeters, the lengths of the three pieces.   (4)

7. Evaluate $\dfrac{576.29}{19.3}$

   (a) correct to 4 significant figures

   (b) correct to 1 decimal place   (4)

8. Express $7\dfrac{9}{46}$ correct to 2 decimal places   (2)

9. Determine, correct to 1 decimal places, 57% of 17.64 g   (2)

10. Express 54.7 mm as a percentage of 1.15 m, correct to 3 significant figures.   (3)

3

Indices and standard form

3.1 Indices

The lowest factors of 2000 are $2 \times 2 \times 2 \times 2 \times 5 \times 5 \times 5$. These factors are written as $2^4 \times 5^3$, where 2 and 5 are called **bases** and the numbers 4 and 3 are called **indices**.

When an index is an integer it is called a **power**. Thus, 2^4 is called 'two to the power of four', and has a base of 2 and an index of 4. Similarly, 5^3 is called 'five to the power of 3' and has a base of 5 and an index of 3.

Special names may be used when the indices are 2 and 3, these being called 'squared' and 'cubed', respectively. Thus 7^2 is called 'seven squared' and 9^3 is called 'nine cubed'. When no index is shown, the power is 1, i.e. 2 means 2^1.

Reciprocal

The **reciprocal** of a number is when the index is -1 and its value is given by 1 divided by the base. Thus the reciprocal of 2 is 2^{-1} and its value is $\frac{1}{2}$ or 0.5. Similarly, the reciprocal of 5 is 5^{-1} which means $\frac{1}{5}$ or 0.2

Square root

The **square root** of a number is when the index is $\frac{1}{2}$, and the square root of 2 is written as $2^{1/2}$ or $\sqrt{2}$. The value of a square root is the value of the base which when multiplied by itself gives the number. Since $3 \times 3 = 9$, then $\sqrt{9} = 3$. However, $(-3) \times (-3) = 9$, so $\sqrt{9} = -3$. There are always two answers when finding the square root of a number and this is shown by putting both a $+$ and a $-$ sign in front of the answer to a square root problem. Thus $\sqrt{9} = \pm 3$ and $4^{1/2} = \sqrt{4} = \pm 2$, and so on.

Laws of indices

When simplifying calculations involving indices, certain basic rules or laws can be applied, called the **laws of indices**. These are given below.

(i) When multiplying two or more numbers having the same base, the indices are added. Thus
$$3^2 \times 3^4 = 3^{2+4} = 3^6$$

(ii) When a number is divided by a number having the same base, the indices are subtracted. Thus
$$\frac{3^5}{3^2} = 3^{5-2} = 3^3$$

(iii) When a number which is raised to a power is raised to a further power, the indices are multiplied. Thus
$$(3^5)^2 = 3^{5 \times 2} = 3^{10}$$

(iv) When a number has an index of 0, its value is 1. Thus
$3^0 = 1$

(v) A number raised to a negative power is the reciprocal of that number raised to a positive power. Thus $3^{-4} = \dfrac{1}{3^4}$

Similarly, $\dfrac{1}{2^{-3}} = 2^3$

(vi) When a number is raised to a fractional power the denominator of the fraction is the root of the number and the numerator is the power.

Thus $\quad 8^{2/3} = \sqrt[3]{8^2} = (2)^2 = 4$

and $\quad 25^{1/2} = \sqrt[2]{25^1} = \sqrt{25^1} = \pm 5$

(Note that $\sqrt{} \equiv \sqrt[2]{}$)

3.2 Worked problems on indices

Problem 1. Evaluate: (a) $5^2 \times 5^3$, (b) $3^2 \times 3^4 \times 3$ and (c) $2 \times 2^2 \times 2^5$

From law (i):

(a) $5^2 \times 5^3 = 5^{(2+3)} = 5^5 = 5 \times 5 \times 5 \times 5 \times 5 = \mathbf{3125}$

(b) $3^2 \times 3^4 \times 3 = 3^{(2+4+1)} = 3^7$

$$= 3 \times 3 \times \ldots \text{ to 7 terms}$$
$$= \mathbf{2187}$$

(c) $2 \times 2^2 \times 2^5 = 2^{(1+2+5)} = 2^8 = \mathbf{256}$

Problem 2. Find the value of: (a) $\dfrac{7^5}{7^3}$ and (b) $\dfrac{5^7}{5^4}$

From law (ii):

(a) $\dfrac{7^5}{7^3} = 7^{(5-3)} = 7^2 = \mathbf{49}$

(b) $\dfrac{5^7}{5^4} = 5^{(7-4)} = 5^3 = \mathbf{125}$

Problem 3. Evaluate: (a) $5^2 \times 5^3 \div 5^4$ and
(b) $(3 \times 3^5) \div (3^2 \times 3^3)$

From laws (i) and (ii):

(a) $5^2 \times 5^3 \div 5^4 = \dfrac{5^2 \times 5^3}{5^4} = \dfrac{5^{(2+3)}}{5^4}$

$$= \dfrac{5^5}{5^4} = 5^{(5-4)} = 5^1 = \mathbf{5}$$

(b) $(3 \times 3^5) \div (3^2 \times 3^3) = \dfrac{3 \times 3^5}{3^2 \times 3^3} = \dfrac{3^{(1+5)}}{3^{(2+3)}}$

$$= \dfrac{3^6}{3^5} = 3^{6-5} = 3^1 = \mathbf{3}$$

Problem 4. Simplify: (a) $(2^3)^4$ (b) $(3^2)^5$, expressing
the answers in index form.

From law (iii):

(a) $(2^3)^4 = 2^{3\times4} = \mathbf{2^{12}}$

(b) $(3^2)^5 = 3^{2\times5} = \mathbf{3^{10}}$

Problem 5. Evaluate: $\dfrac{(10^2)^3}{10^4 \times 10^2}$

From the laws of indices:

$$\dfrac{(10^2)^3}{10^4 \times 10^2} = \dfrac{10^{(2\times3)}}{10^{(4+2)}} = \dfrac{10^6}{10^6} = 10^{6-6} = 10^0 = 1$$

Problem 6. Find the value of

(a) $\dfrac{2^3 \times 2^4}{2^7 \times 2^5}$ and (b) $\dfrac{(3^2)^3}{3 \times 3^9}$

From the laws of indices:

(a) $\dfrac{2^3 \times 2^4}{2^7 \times 2^5} = \dfrac{2^{(3+4)}}{2^{(7+5)}} = \dfrac{2^7}{2^{12}} = 2^{7-12} = 2^{-5} = \dfrac{1}{2^5} = \dfrac{1}{32}$

(b) $\dfrac{(3^2)^3}{3 \times 3^9} = \dfrac{3^{2\times3}}{3^{1+9}} = \dfrac{3^6}{3^{10}} = 3^{6-10} = 3^{-4} = \dfrac{1}{3^4} = \dfrac{1}{81}$

Problem 7. Evaluate (a) $4^{1/2}$ (b) $16^{3/4}$ (c) $27^{2/3}$
(d) $9^{-1/2}$

(a) $4^{1/2} = \sqrt{4} = \pm 2$

(b) $16^{3/4} = \sqrt[4]{16^3} = (2)^3 = \mathbf{8}$

(Note that it does not matter whether the 4th root of 16
is found first or whether 16 cubed is found first–the same
answer will result.)

(c) $27^{2/3} = \sqrt[3]{27^2} = (3)^2 = \mathbf{9}$

(d) $9^{-1/2} = \dfrac{1}{9^{1/2}} = \dfrac{1}{\sqrt{9}} = \dfrac{1}{\pm3} = \pm\dfrac{1}{3}$

Now try the following exercise

**Exercise 8   Further problems on indices (Answers on
page 253)**

In Problems 1 to 12, simplify the expressions given,
expressing the answers in index form and with positive
indices:

1. (a) $3^3 \times 3^4$ (b) $4^2 \times 4^3 \times 4^4$

2. (a) $2^3 \times 2 \times 2^2$ (b) $7^2 \times 7^4 \times 7 \times 7^3$

3. (a) $\dfrac{2^4}{2^3}$ (b) $\dfrac{3^7}{3^2}$

4. (a) $5^6 \div 5^3$ (b) $7^{13}/7^{10}$

5. (a) $(7^2)^3$ (b) $(3^3)^2$

6. (a) $(15^3)^5$ (b) $(17^2)^4$

7. (a) $\dfrac{2^2 \times 2^3}{2^4}$ (b) $\dfrac{3^7 \times 3^4}{3^5}$

8. (a) $\dfrac{5^7}{5^2 \times 5^3}$ (b) $\dfrac{13^5}{13 \times 13^2}$

9. (a) $\dfrac{(9 \times 3^2)^3}{(3 \times 27)^2}$ (b) $\dfrac{(16 \times 4)^2}{(2 \times 8)^3}$

10. (a) $\dfrac{5^{-2}}{5^{-4}}$ (b) $\dfrac{3^2 \times 3^{-4}}{3^3}$

11. (a) $\dfrac{7^2 \times 7^{-3}}{7 \times 7^{-4}}$ (b) $\dfrac{2^3 \times 2^{-4} \times 2^5}{2 \times 2^{-2} \times 2^6}$

12. (a) $13 \times 13^{-2} \times 13^4 \times 13^{-3}$ (b) $\dfrac{5^{-7} \times 5^2}{5^{-8} \times 5^3}$

3.3 Further worked problems on indices

> **Problem 8.**   Evaluate $\dfrac{3^3 \times 5^7}{5^3 \times 3^4}$

The laws of indices only apply to terms **having the same base**. Grouping terms having the same base, and then applying the laws of indices to each of the groups independently gives:

$$\frac{3^3 \times 5^7}{5^3 \times 3^4} = \frac{3^3}{3^4} \times \frac{5^7}{5^3} = 3^{(3-4)} \times 5^{(7-3)}$$

$$= 3^{-1} \times 5^4 = \frac{5^4}{3^1} = \frac{625}{3} = 208\frac{1}{3}$$

> **Problem 9.**   Find the value of $\dfrac{2^3 \times 3^5 \times (7^2)^2}{7^4 \times 2^4 \times 3^3}$

$$\frac{2^3 \times 3^5 \times (7^2)^2}{7^4 \times 2^4 \times 3^3} = 2^{3-4} \times 3^{5-3} \times 7^{2 \times 2 - 4}$$

$$= 2^{-1} \times 3^2 \times 7^0 = \frac{1}{2} \times 3^2 \times 1$$

$$= \frac{9}{2} = 4\frac{1}{2}$$

> **Problem 10.**   Evaluate: $\dfrac{4^{1.5} \times 8^{1/3}}{2^2 \times 32^{-2/5}}$

$4^{1.5} = 4^{3/2} = \sqrt{4^3} = 2^3 = 8$, $8^{1/3} = \sqrt[3]{8} = 2$, $2^2 = 4$

$$32^{-2/5} = \frac{1}{32^{2/5}} = \frac{1}{\sqrt[5]{32^2}} = \frac{1}{2^2} = \frac{1}{4}$$

Hence $\dfrac{4^{1.5} \times 8^{1/3}}{2^2 \times 32^{-2/5}} = \dfrac{8 \times 2}{4 \times \frac{1}{4}} = \dfrac{16}{1} = \mathbf{16}$

Alternatively,

$$\frac{4^{1.5} \times 8^{1/3}}{2^2 \times 32^{-2/5}} = \frac{[(2)^2]^{3/2} \times (2^3)^{1/3}}{2^2 \times (2^5)^{-2/5}} = \frac{2^3 \times 2^1}{2^2 \times 2^{-2}}$$

$$= 2^{3+1-2-(-2)} = 2^4 = \mathbf{16}$$

> **Problem 11.**   Evaluate: $\dfrac{3^2 \times 5^5 + 3^3 \times 5^3}{3^4 \times 5^4}$

Dividing each term by the HCF (i.e. highest common factor) of the three terms, i.e. $3^2 \times 5^3$, gives:

$$\frac{3^2 \times 5^5 + 3^3 \times 5^3}{3^4 \times 5^4} = \frac{\dfrac{3^2 \times 5^5}{3^2 \times 5^3} + \dfrac{3^3 \times 5^3}{3^2 \times 5^3}}{\dfrac{3^4 \times 5^4}{3^2 \times 5^3}}$$

$$= \frac{3^{(2-2)} \times 5^{(5-3)} + 3^{(3-2)} \times 5^0}{3^{(4-2)} \times 5^{(4-3)}}$$

$$= \frac{3^0 \times 5^2 + 3^1 \times 5^0}{3^2 \times 5^1}$$

$$= \frac{1 \times 25 + 3 \times 1}{9 \times 5} = \frac{\mathbf{28}}{\mathbf{45}}$$

> **Problem 12.**   Find the value of $\dfrac{3^2 \times 5^5}{3^4 \times 5^4 + 3^3 \times 5^3}$

To simplify the arithmetic, each term is divided by the HCF of all the terms, i.e. $3^2 \times 5^3$. Thus

$$\frac{3^2 \times 5^5}{3^4 \times 5^4 + 3^3 \times 5^3} = \frac{\dfrac{3^2 \times 5^5}{3^2 \times 5^3}}{\dfrac{3^4 \times 5^4}{3^2 \times 5^3} + \dfrac{3^3 \times 5^3}{3^2 \times 5^3}}$$

$$= \frac{3^{(2-2)} \times 5^{(5-3)}}{3^{(4-2)} \times 5^{(4-3)} + 3^{(3-2)} \times 5^{(3-3)}}$$

$$= \frac{3^0 \times 5^2}{3^2 \times 5^1 + 3^1 \times 5^0}$$

$$= \frac{25}{45 + 3} = \frac{\mathbf{25}}{\mathbf{48}}$$

> **Problem 13.**   Simplify $\dfrac{7^{-3} \times 3^4}{3^{-2} \times 7^5 \times 5^{-2}}$, expressing the answer in index form with positive indices.

Since $7^{-3} = \dfrac{1}{7^3}$, $\dfrac{1}{3^{-2}} = 3^2$ and $\dfrac{1}{5^{-2}} = 5^2$ then

$$\frac{7^{-3} \times 3^4}{3^{-2} \times 7^5 \times 5^{-2}} = \frac{3^4 \times 3^2 \times 5^2}{7^3 \times 7^5}$$

$$= \frac{3^{(4+2)} \times 5^2}{7^{(3+5)}} = \frac{\mathbf{3^6} \times \mathbf{5^2}}{\mathbf{7^8}}$$

Problem 14. Simplify $\dfrac{16^2 \times 9^{-2}}{4 \times 3^3 - 2^{-3} \times 8^2}$ expressing the answer in index form with positive indices.

Expressing the numbers in terms of their lowest prime numbers gives:

$$\frac{16^2 \times 9^{-2}}{4 \times 3^3 - 2^{-3} \times 8^2} = \frac{(2^4)^2 \times (3^2)^{-2}}{2^2 \times 3^3 - 2^{-3} \times (2^3)^2}$$

$$= \frac{2^8 \times 3^{-4}}{2^2 \times 3^3 - 2^{-3} \times 2^6}$$

$$= \frac{2^8 \times 3^{-4}}{2^2 \times 3^3 - 2^3}$$

Dividing each term by the HCF (i.e. 2^2) gives:

$$\frac{2^8 \times 3^{-4}}{2^2 \times 3^3 - 2^3} = \frac{2^6 \times 3^{-4}}{3^3 - 2} = \frac{2^6}{3^4(3^3 - 2)}$$

Problem 15. Simplify

$$\frac{\left(\dfrac{4}{3}\right)^3 \times \left(\dfrac{3}{5}\right)^{-2}}{\left(\dfrac{2}{5}\right)^{-3}}$$

giving the answer with positive indices.

A fraction raised to a power means that both the numerator and the denominator of the fraction are raised to that power,

i.e. $\left(\dfrac{4}{3}\right)^3 = \dfrac{4^3}{3^3}$

A fraction raised to a negative power has the same value as the inverse of the fraction raised to a positive power.

Thus, $\left(\dfrac{3}{5}\right)^{-2} = \dfrac{1}{\left(\dfrac{3}{5}\right)^2} = \dfrac{1}{\dfrac{3^2}{5^2}} = 1 \times \dfrac{5^2}{3^2} = \dfrac{5^2}{3^2}$

Similarly, $\left(\dfrac{2}{5}\right)^{-3} = \left(\dfrac{5}{2}\right)^3 = \dfrac{5^3}{2^3}$

Thus, $\dfrac{\left(\dfrac{4}{3}\right)^3 \times \left(\dfrac{3}{5}\right)^{-2}}{\left(\dfrac{2}{5}\right)^{-3}} = \dfrac{\dfrac{4^3}{3^3} \times \dfrac{5^2}{3^2}}{\dfrac{5^3}{2^3}}$

$$= \frac{4^3}{3^3} \times \frac{5^2}{3^2} \times \frac{2^3}{5^3}$$

$$= \frac{(2^2)^3 \times 2^3}{3^{(3+2)} \times 5^{(3-2)}} = \frac{2^9}{3^5 \times 5}$$

Now try the following exercise

Exercise 9 Further problems on indices (Answers on page 253)

In Problems 1 and 2, simplify the expressions given, expressing the answers in index form and with positive indices:

1. (a) $\dfrac{3^3 \times 5^2}{5^4 \times 3^4}$ (b) $\dfrac{7^{-2} \times 3^{-2}}{3^5 \times 7^4 \times 7^{-3}}$

2. (a) $\dfrac{4^2 \times 9^3}{8^3 \times 3^4}$ (b) $\dfrac{8^{-2} \times 5^2 \times 3^{-4}}{25^2 \times 2^4 \times 9^{-2}}$

3. Evaluate (a) $\left(\dfrac{1}{3^2}\right)^{-1}$ (b) $81^{0.25}$

    (c) $16^{(-1/4)}$ (d) $\left(\dfrac{4}{9}\right)^{1/2}$

In problems 4 to 10, evaluate the expressions given.

4. $\dfrac{9^2 \times 7^4}{3^4 \times 7^4 + 3^3 \times 7^2}$

5. $\dfrac{3^3 \times 5^2}{2^3 \times 3^2 - 8^2 \times 9}$

6. $\dfrac{3^3 \times 7^2 - 5^2 \times 7^3}{3^2 \times 5 \times 7^2}$

7. $\dfrac{(2^4)^2 - 3^{-2} \times 4^4}{2^3 \times 16^2}$

8. $\dfrac{\left(\dfrac{1}{2}\right)^3 - \left(\dfrac{2}{3}\right)^{-2}}{\left(\dfrac{3}{5}\right)^2}$

9. $\dfrac{\left(\dfrac{4}{3}\right)^4}{\left(\dfrac{2}{9}\right)^2}$

10. $\dfrac{(3^2)^{3/2} \times (8^{1/3})^2}{(3)^2 \times (4^3)^{1/2} \times (9)^{-1/2}}$

3.4 Standard form

A number written with one digit to the left of the decimal point and multiplied by 10 raised to some power is said to be written in **standard form**. Thus: 5837 is written as 5.837×10^3 in standard form, and 0.0415 is written as 4.15×10^{-2} in standard form.

When a number is written in standard form, the first factor is called the **mantissa** and the second factor is called the **exponent**. Thus the number 5.8×10^3 has a mantissa of 5.8 and an exponent of 10^3

(i) Numbers having the same exponent can be added or subtracted in standard form by adding or subtracting the mantissae and keeping the exponent the same. Thus:

$$2.3 \times 10^4 + 3.7 \times 10^4 = (2.3 + 3.7) \times 10^4$$

$$= 6.0 \times 10^4$$

and $5.9 \times 10^{-2} - 4.6 \times 10^{-2} = (5.9 - 4.6) \times 10^{-2}$

$$= 1.3 \times 10^{-2}$$

When the numbers have different exponents, one way of adding or subtracting the numbers is to express one of the numbers in non-standard form, so that both numbers have the same exponent. Thus:

$$2.3 \times 10^4 + 3.7 \times 10^3 = 2.3 \times 10^4 + 0.37 \times 10^4$$

$$= (2.3 + 0.37) \times 10^4$$

$$= 2.67 \times 10^4$$

Alternatively,

$$2.3 \times 10^4 + 3.7 \times 10^3 = 23\,000 + 3700 = 26\,700$$

$$= 2.67 \times 10^4$$

(ii) The laws of indices are used when multiplying or dividing numbers given in standard form. For example,

$$(2.5 \times 10^3) \times (5 \times 10^2) = (2.5 \times 5) \times (10^{3+2})$$

$$= 12.5 \times 10^5 \text{ or } 1.25 \times 10^6$$

Similarly, $\dfrac{6 \times 10^4}{1.5 \times 10^2} = \dfrac{6}{1.5} \times (10^{4-2}) = 4 \times 10^2$

3.5   Worked problems on standard form

Problem 16.   Express in standard form:

(a) 38.71   (b) 3746   (c) 0.0124

For a number to be in standard form, it is expressed with only one digit to the left of the decimal point. Thus:

(a) 38.71 must be divided by 10 to achieve one digit to the left of the decimal point and it must also be multiplied by 10 to maintain the equality, i.e.

$$38.71 = \frac{38.71}{10} \times 10 = \mathbf{3.871 \times 10} \text{ in standard form}$$

(b) $3746 = \dfrac{3746}{1000} \times 1000 = \mathbf{3.746 \times 10^3}$ in standard form.

(c) $0.0124 = 0.0124 \times \dfrac{100}{100} = \dfrac{1.24}{100} = \mathbf{1.24 \times 10^{-2}}$ in standard form

---

Problem 17.   Express the following numbers, which are in standard form, as decimal numbers:

(a) 1.725×10^{-2} (b) 5.491×10^4 (c) 9.84×10^0

(a) $1.725 \times 10^{-2} = \dfrac{1.725}{100} = \mathbf{0.01725}$

(b) $5.491 \times 10^4 = 5.491 \times 10\,000 = \mathbf{54\,910}$

(c) $9.84 \times 10^0 = 9.84 \times 1 = \mathbf{9.84}$ (since $10^0 = 1$)

Problem 18.   Express in standard form, correct to 3 significant figures:

(a) $\dfrac{3}{8}$ (b) $19\dfrac{2}{3}$ (c) $741\dfrac{9}{16}$

(a) $\dfrac{3}{8} = 0.375$, and expressing it in standard form gives:

$$0.375 = \mathbf{3.75 \times 10^{-1}}$$

(b) $19\dfrac{2}{3} = 19.\dot{6} = \mathbf{1.97 \times 10}$ in standard form, correct to 3 significant figures.

(c) $741\dfrac{9}{16} = 741.5625 = \mathbf{7.42 \times 10^2}$ in standard form, correct to 3 significant figures.

Problem 19.   Express the following numbers, given in standard form, as fractions or mixed numbers:

(a) 2.5×10^{-1} (b) 6.25×10^{-2} (c) 1.354×10^2

(a) $2.5 \times 10^{-1} = \dfrac{2.5}{10} = \dfrac{25}{100} = \dfrac{\mathbf{1}}{\mathbf{4}}$

(b) $6.25 \times 10^{-2} = \dfrac{6.25}{100} = \dfrac{625}{10\,000} = \dfrac{\mathbf{1}}{\mathbf{16}}$

(c) $1.354 \times 10^2 = 135.4 = 135\dfrac{4}{10} = \mathbf{135\dfrac{2}{5}}$

Now try the following exercise

Exercise 10   Further problems on standard form
(Answers on page 253)

In Problems 1 to 5, express in standard form:

1. (a) 73.9   (b) 28.4   (c) 197.72

2. (a) 2748   (b) 33170   (c) 274218

3. (a) 0.2401   (b) 0.0174   (c) 0.00923

4. (a) 1702.3   (b) 10.04   (c) 0.0109

5. (a) $\dfrac{1}{2}$ (b) $11\dfrac{7}{8}$ (c) $130\dfrac{3}{5}$ (d) $\dfrac{1}{32}$

In Problems 6 and 7, express the numbers given as integers or decimal fractions:

6. (a) 1.01×10^3 (b) 9.327×10^2 (c) 5.41×10^4
   (d) 7×10^0

7. (a) 3.89×10^{-2} (b) 6.741×10^{-1} (c) 8×10^{-3}

3.6 Further worked problems on standard form

Problem 20. Find the value of

(a) $7.9 \times 10^{-2} - 5.4 \times 10^{-2}$

(b) $8.3 \times 10^3 + 5.415 \times 10^3$ and

(c) $9.293 \times 10^2 + 1.3 \times 10^3$ expressing the answers in standard form.

Numbers having the same exponent can be added or subtracted by adding or subtracting the mantissae and keeping the exponent the same. Thus:

(a) $7.9 \times 10^{-2} - 5.4 \times 10^{-2} = (7.9 - 5.4) \times 10^{-2}$

$$= \mathbf{2.5 \times 10^{-2}}$$

(b) $8.3 \times 10^3 + 5.415 \times 10^3 = (8.3 + 5.415) \times 10^3$

$$= 13.715 \times 10^3$$

$$= \mathbf{1.3715 \times 10^4}$$

in standard form.

(c) Since only numbers having the same exponents can be added by straight addition of the mantissae, the numbers are converted to this form before adding. Thus:

$$9.293 \times 10^2 + 1.3 \times 10^3 = 9.293 \times 10^2 + 13 \times 10^2$$

$$= (9.293 + 13) \times 10^2$$

$$= 22.293 \times 10^2$$

$$= \mathbf{2.2293 \times 10^3}$$

in standard form.
Alternatively, the numbers can be expressed as decimal fractions, giving:

$$9.293 \times 10^2 + 1.3 \times 10^3$$

$$= 929.3 + 1300$$

$$= 2229.3$$

$$= \mathbf{2.2293 \times 10^3}$$

in standard form as obtained previously. This method is often the 'safest' way of doing this type of problem.

Problem 21. Evaluate (a) $(3.75 \times 10^3)(6 \times 10^4)$ and (b) $\dfrac{3.5 \times 10^5}{7 \times 10^2}$ expressing answers in standard form.

(a) $(3.75 \times 10^3)(6 \times 10^4) = (3.75 \times 6)(10^{3+4})$

$$= 22.50 \times 10^7$$

$$= \mathbf{2.25 \times 10^8}$$

(b) $\dfrac{3.5 \times 10^5}{7 \times 10^2} = \dfrac{3.5}{7} \times 10^{5-2} = 0.5 \times 10^3 = \mathbf{5 \times 10^2}$

Now try the following exercise

Exercise 11 Further problems on standard form (Answers on page 253)

In Problems 1 to 4, find values of the expressions given, stating the answers in standard form:

1. (a) $3.7 \times 10^2 + 9.81 \times 10^2$

   (b) $1.431 \times 10^{-1} + 7.3 \times 10^{-1}$

   (c) $8.414 \times 10^{-2} - 2.68 \times 10^{-2}$

2. (a) $4.831 \times 10^2 + 1.24 \times 10^3$

   (b) $3.24 \times 10^{-3} - 1.11 \times 10^{-4}$

   (c) $1.81 \times 10^2 + 3.417 \times 10^2 + 5.972 \times 10^2$

3. (a) $(4.5 \times 10^{-2})(3 \times 10^3)$ (b) $2 \times (5.5 \times 10^4)$

4. (a) $\dfrac{6 \times 10^{-3}}{3 \times 10^{-5}}$ (b) $\dfrac{(2.4 \times 10^3)(3 \times 10^{-2})}{(4.8 \times 10^4)}$

5. Write the following statements in standard form.

   (a) The density of aluminium is $2710 \, \text{kg m}^{-3}$

   (b) Poisson's ratio for gold is 0.44

   (c) The impedance of free space is $376.73 \, \Omega$

   (d) The electron rest energy is $0.511 \, \text{MeV}$

   (e) Proton charge-mass ratio is $95\,789\,700 \, \text{C kg}^{-1}$

   (f) The normal volume of a perfect gas is $0.02241 \, \text{m}^3 \, \text{mol}^{-1}$

4

Calculations and evaluation of formulae

4.1 Errors and approximations

(i) In all problems in which the measurement of distance, time, mass or other quantities occurs, an exact answer cannot be given; only an answer which is correct to a stated degree of accuracy can be given. To take account of this an **error due to measurement** is said to exist.

(ii) To take account of measurement errors it is usual to limit answers so that the result given is **not more than one significant figure greater than the least accurate number given in the data**.

(iii) **Rounding-off errors** can exist with decimal fractions. For example, to state that $\pi = 3.142$ is not strictly correct, but '$\pi = 3.142$ correct to 4 significant figures' is a true statement.
(Actually, $\pi = 3.14159265\ldots$)

(iv) It is possible, through an incorrect procedure, to obtain the wrong answer to a calculation. This type of error is known as **a blunder**.

(v) An **order of magnitude error** is said to exist if incorrect positioning of the decimal point occurs after a calculation has been completed.

(vi) Blunders and order of magnitude errors can be reduced by determining **approximate values of calculations**. Answers which do not seem feasible must be checked and the calculation must be repeated as necessary. An engineer will often need to make a quick mental approximation for a calculation. For example, $\dfrac{49.1 \times 18.4 \times 122.1}{61.2 \times 38.1}$ may be approximated to $\dfrac{50 \times 20 \times 120}{60 \times 40}$ and then, by cancelling, $\dfrac{50 \times \overset{1}{\cancel{20}} \times \overset{1}{\cancel{120}}}{\underset{1}{\cancel{60}} \times \underset{1}{\cancel{40}}} = 50$. An accurate answer somewhere between 45 and 55 could therefore be expected. Certainly an answer around 500 or 5 would not be expected. Actually, by calculator $\dfrac{49.1 \times 18.4 \times 122.1}{61.2 \times 38.1} = 47.31$, correct to 4 significant figures.

Problem 1. The area A of a triangle is given by $A = \frac{1}{2}bh$. The base b when measured is found to be $3.26\,\text{cm}$, and the perpendicular height h is $7.5\,\text{cm}$. Determine the area of the triangle.

Area of triangle $= \frac{1}{2}bh = \frac{1}{2} \times 3.26 \times 7.5 = 12.225\,\text{cm}^2$ (by calculator).

The approximate value is $\frac{1}{2} \times 3 \times 8 = 12\,\text{cm}^2$, so there are no obvious blunder or magnitude errors. However, it is not usual in a measurement type problem to state the answer to an accuracy greater than 1 significant figure more than the least accurate number in the data: this is 7.5 cm, so the result should not have more than 3 significant figures

Thus **area of triangle $= 12.2\,\text{cm}^2$**

Problem 2. State which type of error has been made in the following statements:

(a) $72 \times 31.429 = 2262.9$

(b) $16 \times 0.08 \times 7 = 89.6$

(c) $11.714 \times 0.0088 = 0.3247$, correct to 4 decimal places.

(d) $\dfrac{29.74 \times 0.0512}{11.89} = 0.12$, correct to 2 significant figures.

(a) $72 \times 31.429 = 2262.888$ (by calculator), hence a **rounding-off error** has occurred. The answer should have stated:

$72 \times 31.429 = 2262.9$, correct to 5 significant figures or 2262.9, correct to 1 decimal place.

(b) $16 \times 0.08 \times 7 = \cancel{16}^{4} \times \dfrac{8}{\cancel{100}_{25}} \times 7$

$$= \frac{32 \times 7}{25} = \frac{224}{25} = 8\frac{24}{25}$$

$$= 8.96$$

Hence an **order of magnitude** error has occurred.

(c) 11.714×0.0088 is approximately equal to $12 \times 9 \times 10^{-3}$, i.e. about 108×10^{-3} or 0.108. Thus a **blunder** has been made.

(d) $\dfrac{29.74 \times 0.0512}{11.89} \approx \dfrac{30 \times 5 \times 10^{-2}}{12} = \dfrac{150}{12 \times 10^2}$

$$= \frac{15}{120} = \frac{1}{8} \quad \text{or} \quad 0.125$$

hence no order of magnitude error has occurred. However, $\dfrac{29.74 \times 0.0512}{11.89} = 0.128$ correct to 3 significant figures, which equals 0.13 correct to 2 significant figures.

Hence a **rounding-off error** has occurred.

Problem 3. Without using a calculator, determine an approximate value of

(a) $\dfrac{11.7 \times 19.1}{9.3 \times 5.7}$ (b) $\dfrac{2.19 \times 203.6 \times 17.91}{12.1 \times 8.76}$

(a) $\dfrac{11.7 \times 19.1}{9.3 \times 5.7}$ is approximately equal to $\dfrac{10 \times 20}{10 \times 5}$, i.e. about **4**

(By calculator, $\dfrac{11.7 \times 19.1}{9.3 \times 5.7} = \mathbf{4.22}$, correct to 3 significant figures.)

(b) $\dfrac{2.19 \times 203.6 \times 17.91}{12.1 \times 8.76} \approx \dfrac{2 \times 200 \times 20}{10 \times 10} = 2 \times 2 \times 20$ after cancelling

i.e. $\dfrac{2.19 \times 203.6 \times 17.91}{12.1 \times 8.76} \approx \mathbf{80}$

(By calculator, $\dfrac{2.19 \times 203.6 \times 17.91}{12.1 \times 8.76} = \mathbf{75.3}$, correct to 3 significant figures.)

Now try the following exercise

Exercise 12   Further problems on errors (Answers on page 253)

In Problems 1 to 5 state which type of error, or errors, have been made:

1. $25 \times 0.06 \times 1.4 = 0.21$

2. $137 \times 6.842 = 937.4$

3. $\dfrac{24 \times 0.008}{12.6} = 10.42$

4. For a gas $pV = c$. When pressure $p = 103\,400\,\text{Pa}$ and $V = 0.54\,\text{m}^3$ then $c = 55\,836\,\text{Pa}\,\text{m}^3$.

5. $\dfrac{4.6 \times 0.07}{52.3 \times 0.274} = 0.225$

In Problems 6 to 8, evaluate the expressions approximately, without using a calculator.

6. 4.7×6.3

7. $\dfrac{2.87 \times 4.07}{6.12 \times 0.96}$

8. $\dfrac{72.1 \times 1.96 \times 48.6}{139.3 \times 5.2}$

4.2   Use of calculator

The most modern aid to calculations is the pocket-sized electronic calculator. With one of these, calculations can be quickly and accurately performed, correct to about 9 significant figures. The scientific type of calculator has made the use of tables and logarithms largely redundant.

To help you to become competent at using your calculator check that you agree with the answers to the following problems:

Problem 4. Evaluate the following, correct to 4 significant figures:

(a) $4.7826 + 0.02713$ (b) $17.6941 - 11.8762$

(c) 21.93×0.012981

(a) $4.7826 + 0.02713 = 4.80973 = \mathbf{4.810}$, correct to 4 significant figures.

(b) $17.6941 - 11.8762 = 5.8179 = \mathbf{5.818}$, correct to 4 significant figures.

(c) $21.93 \times 0.012981 = 0.2846733\ldots = \mathbf{0.2847}$, correct to 4 significant figures.

Problem 5. Evaluate the following, correct to 4 decimal places:

(a) $46.32 \times 97.17 \times 0.01258$ (b) $\dfrac{4.621}{23.76}$

(c) $\dfrac{1}{2}(62.49 \times 0.0172)$

(a) $46.32 \times 97.17 \times 0.01258 = 56.6215031\ldots = \mathbf{56.6215}$, correct to 4 decimal places.

(b) $\dfrac{4.621}{23.76} = 0.19448653\ldots = \mathbf{0.1945}$, correct to 4 decimal places.

(c) $\dfrac{1}{2}(62.49 \times 0.0172) = 0.537414 = \mathbf{0.5374}$, correct to 4 decimal places.

Problem 6. Evaluate the following, correct to 3 decimal places:

(a) $\dfrac{1}{52.73}$ (b) $\dfrac{1}{0.0275}$ (c) $\dfrac{1}{4.92} + \dfrac{1}{1.97}$

(a) $\dfrac{1}{52.73} = 0.01896453\ldots = \mathbf{0.019}$, correct to 3 decimal places.

(b) $\dfrac{1}{0.0275} = 36.3636363\ldots = \mathbf{36.364}$, correct to 3 decimal places.

(c) $\dfrac{1}{4.92} + \dfrac{1}{1.97} = 0.71086624\ldots = \mathbf{0.711}$, correct to 3 decimal places.

Problem 7. Evaluate the following, expressing the answers in standard form, correct to 4 significant figures.

(a) $(0.00451)^2$ (b) $631.7 - (6.21 + 2.95)^2$

(c) $46.27^2 - 31.79^2$

(a) $(0.00451)^2 = 2.03401 \times 10^{-5} = \mathbf{2.034 \times 10^{-5}}$, correct to 4 significant figures.

(b) $631.7 - (6.21 + 2.95)^2 = 547.7944 = 5.477944 \times 10^2 = \mathbf{5.478 \times 10^2}$, correct to 4 significant figures.

(c) $46.27^2 - 31.79^2 = 1130.3088 = \mathbf{1.130 \times 10^3}$, correct to 4 significant figures.

Problem 8. Evaluate the following, correct to 3 decimal places:

(a) $\dfrac{(2.37)^2}{0.0526}$ (b) $\left(\dfrac{3.60}{1.92}\right)^2 + \left(\dfrac{5.40}{2.45}\right)^2$

(c) $\dfrac{15}{7.6^2 - 4.8^2}$

(a) $\dfrac{(2.37)^2}{0.0526} = 106.785171\ldots = \mathbf{106.785}$, correct to 3 decimal places.

(b) $\left(\dfrac{3.60}{1.92}\right)^2 + \left(\dfrac{5.40}{2.45}\right)^2 = 8.37360084\ldots = \mathbf{8.374}$, correct to 3 decimal places.

(c) $\dfrac{15}{7.6^2 - 4.8^2} = 0.43202764\ldots = \mathbf{0.432}$, correct to 3 decimal places.

Problem 9. Evaluate the following, correct to 4 significant figures:

(a) $\sqrt{5.462}$ (b) $\sqrt{54.62}$ (c) $\sqrt{546.2}$

(a) $\sqrt{5.462} = 2.3370922\ldots = \mathbf{2.337}$, correct to 4 significant figures.

(b) $\sqrt{54.62} = 7.39053448\ldots = \mathbf{7.391}$, correct to 4 significant figures.

(c) $\sqrt{546.2} = 23.370922\ldots = \mathbf{23.37}$, correct to 4 significant figures.

Problem 10. Evaluate the following, correct to 3 decimal places:

(a) $\sqrt{0.007328}$ (b) $\sqrt{52.91} - \sqrt{31.76}$

(c) $\sqrt{(1.6291 \times 10^4)}$

(a) $\sqrt{0.007328} = 0.08560373 = \mathbf{0.086}$, correct to 3 decimal places.

(b) $\sqrt{52.91} - \sqrt{31.76} = 1.63832491\ldots = \mathbf{1.638}$, correct to 3 decimal places.

(c) $\sqrt{(1.6291 \times 10^4)} = \sqrt{(16291)} = 127.636201\ldots = \mathbf{127.636}$, correct to 3 decimal places.

Problem 11. Evaluate the following, correct to 4 significant figures:

(a) 4.72^3 (b) $(0.8316)^4$ (c) $\sqrt{(76.21^2 - 29.10^2)}$

(a) $4.72^3 = 105.15404\ldots = \mathbf{105.2}$, correct to 4 significant figures.

(b) $(0.8316)^4 = 0.47825324\ldots = \mathbf{0.4783}$, correct to 4 significant figures.

(c) $\sqrt{(76.21^2 - 29.10^2)} = 70.4354605\ldots = \mathbf{70.44}$, correct to 4 significant figures.

Problem 12. Evaluate the following, correct to 3 significant figures:

(a) $\sqrt{\left[\dfrac{6.09^2}{25.2 \times \sqrt{7}}\right]}$

(b) $\sqrt[3]{47.291}$

(c) $\sqrt{(7.213^2 + 6.418^3 + 3.291^4)}$

(a) $\sqrt{\left[\dfrac{6.09^2}{25.2 \times \sqrt{7}}\right]} = 0.74583457\ldots = \mathbf{0.746}$, correct to 3 significant figures.

(b) $\sqrt[3]{47.291} = 3.61625876\ldots = \mathbf{3.62}$, correct to 3 significant figures.

(c) $\sqrt{(7.213^2 + 6.418^3 + 3.291^4)} = 20.8252991\ldots = \mathbf{20.8}$, correct to 3 significant figures.

Problem 13. Evaluate the following, expressing the answers in standard form, correct to 4 decimal places:

(a) $(5.176 \times 10^{-3})^2$

(b) $\left(\dfrac{1.974 \times 10^1 \times 8.61 \times 10^{-2}}{3.462}\right)^4$

(c) $\sqrt{(1.792 \times 10^{-4})}$

(a) $(5.176 \times 10^{-3})^2 = 2.679097\ldots \times 10^{-5} = \mathbf{2.6791 \times 10^{-5}}$, correct to 4 decimal places.

(b) $\left(\dfrac{1.974 \times 10^1 \times 8.61 \times 10^{-2}}{3.462}\right)^4 = 0.05808887\ldots$

$= \mathbf{5.8089 \times 10^{-2}}$, correct to 4 decimal places

(c) $\sqrt{(1.792 \times 10^{-4})} = 0.0133865\ldots = \mathbf{1.3387 \times 10^{-2}}$, correct to 4 decimal places.

Now try the following exercise

Exercise 13 Further problems on use of calculator
(Answers on page 253)

In Problems 1 to 3, use a calculator to evaluate the quantities shown correct to 4 significant figures:

1. (a) 3.249^2 (b) 73.78^2 (c) 311.4^2 (d) 0.0639^2

2. (a) $\sqrt{4.735}$ (b) $\sqrt{35.46}$ (c) $\sqrt{73280}$

   (d) $\sqrt{0.0256}$

3. (a) $\dfrac{1}{7.768}$ (b) $\dfrac{1}{48.46}$ (c) $\dfrac{1}{0.0816}$ (d) $\dfrac{1}{1.118}$

In Problems 4 to 11, use a calculator to evaluate correct to 4 significant figures:

4. (a) 43.27×12.91 (b) 54.31×0.5724

5. (a) $127.8 \times 0.0431 \times 19.8$ (b) $15.76 \div 4.329$

6. (a) $\dfrac{137.6}{552.9}$ (b) $\dfrac{11.82 \times 1.736}{0.041}$

7. (a) $\dfrac{1}{17.31}$ (b) $\dfrac{1}{0.0346}$ (c) $\dfrac{1}{147.9}$

8. (a) 13.6^3 (b) 3.476^4 (c) 0.124^5

9. (a) $\sqrt{347.1}$ (b) $\sqrt{7632}$ (c) $\sqrt{0.027}$

   (d) $\sqrt{0.004168}$

10. (a) $\left(\dfrac{24.68 \times 0.0532}{7.412}\right)^3$ (b) $\left(\dfrac{0.2681 \times 41.2^2}{32.6 \times 11.89}\right)^4$

11. (a) $\dfrac{14.32^3}{21.68^2}$ (b) $\dfrac{4.821^3}{17.33^2 - 15.86 \times 11.6}$

12. Evaluate correct to 3 decimal places:

    (a) $\dfrac{29.12}{(5.81)^2 - (2.96)^2}$ (b) $\sqrt{53.98} - \sqrt{21.78}$

13. Evaluate correct to 4 significant figures:

    (a) $\sqrt{\left[\dfrac{(15.62)^2}{29.21 \times \sqrt{10.52}}\right]}$

    (b) $\sqrt{(6.921^2 + 4.816^3 - 2.161^4)}$

14. Evaluate the following, expressing the answers in standard form, correct to 3 decimal places:

    (a) $(8.291 \times 10^{-2})^2$ (b) $\sqrt{(7.623 \times 10^{-3})}$

4.3   Conversion tables and charts

It is often necessary to make calculations from various conversion tables and charts. Examples include currency exchange rates, imperial to metric unit conversions, train or bus timetables, production schedules and so on.

Problem 14. Currency exchange rates for five countries are shown in Table 4.1

Table 4.1

France	£1 = 1.50 euros
Japan	£1 = 175 yen
Norway	£1 = 11.25 kronor
Switzerland	£1 = 2.20 francs
U.S.A.	£1 = 1.52 dollars ($)

Calculate:

(a) how many French euros £27.80 will buy,

(b) the number of Japanese yen which can be bought for £23,

(c) the pounds sterling which can be exchanged for 6412.5 Norwegian kronor,

(d) the number of American dollars which can be purchased for £90, and

(e) the pounds sterling which can be exchanged for 2794 Swiss francs.

(a) £1 = 1.50 euros, hence
  £27.80 = 27.80 × 1.50 euros = **41.70 euros**

(b) £1 = 175 yen, hence
  £23 = 23 × 175 yen = **4025 yen**

(c) £1 = 11.25 kronor, hence
  6412.5 kronor = £$\dfrac{6412.5}{11.25}$ = **£570**

(d) £1 = 1.52 dollars, hence
  £90 = 90 × 1.52 dollars = **$136.80**

(e) £1 = 2.20 Swiss francs, hence
  2794 franc = £$\dfrac{2794}{2.20}$ = **£1270**

Problem 15. Some approximate imperial to metric conversions are shown in Table 4.2

Table 4.2

length	1 inch = 2.54 cm	
	1 mile = 1.61 km	
weight	2.2 lb = 1 kg	
	(1 lb = 16 oz)	
capacity	1.76 pints = 1 litre	
	(8 pints = 1 gallon)	

Use the table to determine:

(a) the number of millimetres in 9.5 inches,

(b) a speed of 50 miles per hour in kilometres per hour,

(c) the number of miles in 300 km,

(d) the number of kilograms in 30 pounds weight,

(e) the number of pounds and ounces in 42 kilograms (correct to the nearest ounce),

(f) the number of litres in 15 gallons, and

(g) the number of gallons in 40 litres.

(a) 9.5 inches = 9.5 × 2.54 cm = 24.13 cm
  24.13 cm = 24.13 × 10 mm = **241.3 mm**

(b) 50 m.p.h. = 50 × 1.61 km/h = **80.5 km/h**

(c) 300 km = $\dfrac{300}{1.61}$ miles = **186.3 miles**

(d) 30 lb = $\dfrac{30}{2.2}$ kg = **13.64 kg**

(e) 42 kg = 42 × 2.2 lb = 92.4 lb
  0.4 lb = 0.4 × 16 oz = 6.4 oz = 6 oz, correct to the nearest ounce. Thus 42 kg = **92 lb 6 oz**, correct to the nearest ounce.

(f) 15 gallons = 15 × 8 pints = 120 pints
  120 pints = $\dfrac{120}{1.76}$ litres = **68.18 litres**

(g) 40 litres = 40 × 1.76 pints = 70.4 pints
  70.4 pints = $\dfrac{70.4}{8}$ gallons = **8.8 gallons**

Now try the following exercise

Exercise 14   Further problems conversion tables and charts (Answers on page 254)

1. Currency exchange rates listed in a newspaper included the following:

Italy	£1 = 1.52 euro
Japan	£1 = 180 yen
Australia	£1 = 2.80 dollars
Canada	£1 = $2.35
Sweden	£1 = 13.90 kronor

Calculate (a) how many Italian euros £32.50 will buy, (b) the number of Canadian dollars that can be purchased for £74.80, (c) the pounds sterling which can be exchanged for 14 040 yen, (d) the pounds sterling which can be exchanged for 1751.4 Swedish kronor, and (e) the Australian dollars which can be bought for £55

2. Below is a list of some metric to imperial conversions.

Length   2.54 cm = 1 inch

            1.61 km = 1 mile

Weight   1 kg = 2.2 lb (1 lb = 16 ounces)

Capacity 1 litre = 1.76 pints

           (8 pints = 1 gallon)

Use the list to determine (a) the number of millimetres in 15 inches, (b) a speed of 35 mph in km/h, (c) the number of kilometres in 235 miles, (d) the number of pounds and ounces in 24 kg (correct to the nearest ounce), (e) the number of kilograms in 15 lb, (f) the number of litres in 12 gallons and (g) the number of gallons in 25 litres.

3. Deduce the following information from the BR train timetable shown in Table 4.3:

(a) At what time should a man catch a train at Mossley Hill to enable him to be in Manchester Piccadilly by 8.15 a.m.?

Table 4.3 Liverpool, Hunt's Cross and Warrington → Manchester

Reproduced with permission of British Rail

(b) A girl leaves Hunts Cross at 8.17 a.m. and travels to Manchester Oxford Road. How long does the journey take. What is the average speed of the journey?

(c) A man living at Edge Hill has to be at work at Trafford Park by 8.45 a.m. It takes him 10 minutes to walk to his work from Trafford Park station. What time train should he catch from Edge Hill ?

4.4 Evaluation of formulae

The statement $v = u + at$ is said to be a **formula** for v in terms of u, a and t.

v, u, a and t are called **symbols**.

The single term on the left-hand side of the equation, v, is called the **subject of the formulae**.

Provided values are given for all the symbols in a formula except one, the remaining symbol can be made the subject of the formula and may be evaluated by using a calculator.

Problem 16. In an electrical circuit the voltage V is given by Ohm's law, i.e. $V = IR$. Find, correct to 4 significant figures, the voltage when $I = 5.36$ A and $R = 14.76$ Ω.

$$V = IR = (5.36)(14.76)$$

Hence voltage $V = 79.11$ V, correct to 4 significant figures.

Problem 17. The surface area A of a hollow cone is given by $A = \pi r l$. Determine, correct to 1 decimal place, the surface area when $r = 3.0$ cm and $l = 8.5$ cm.

$$A = \pi r l = \pi(3.0)(8.5)\,\text{cm}^2$$

Hence surface area $A = 80.1$ cm^2, correct to 1 decimal place.

Problem 18. Velocity v is given by $v = u + at$. If $u = 9.86$ m/s, $a = 4.25$ m/s^2 and $t = 6.84$ s, find v, correct to 3 significant figures.

$$v = u + at = 9.86 + (4.25)(6.84)$$
$$= 9.86 + 29.07$$
$$= 38.93$$

Hence velocity $v = 38.9$ m/s, correct to 3 significant figures.

Problem 19. The area, A, of a circle is given by $A = \pi r^2$. Determine the area correct to 2 decimal places, given radius $r = 5.23$ m.

$$A = \pi r^2 = \pi(5.23)^2 = \pi(27.3529)$$

Hence area, $A = 85.93$ m^2, correct to 2 decimal places.

Problem 20. The power P watts dissipated in an electrical circuit may be expressed by the formula $P = \dfrac{V^2}{R}$. Evaluate the power, correct to 3 significant figures, given that $V = 17.48$ V and $R = 36.12$ Ω.

$$P = \frac{V^2}{R} = \frac{(17.48)^2}{36.12} = \frac{305.5504}{36.12}$$

Hence power, $P = 8.46$ W, correct to 3 significant figures.

Problem 21. The volume V cm^3 of a right circular cone is given by $V = \frac{1}{3}\pi r^2 h$. Given that $r = 4.321$ cm and $h = 18.35$ cm, find the volume, correct to 4 significant figures.

$$V = \tfrac{1}{3}\pi r^2 h = \tfrac{1}{3}\pi(4.321)^2(18.35)$$
$$= \tfrac{1}{3}\pi(18.671041)(18.35)$$

Hence volume, $V = 358.8$ cm^3, correct to 4 significant figures.

Problem 22. Force F newtons is given by the formula $F = \dfrac{Gm_1 m_2}{d^2}$, where m_1 and m_2 are masses, d their distance apart and G is a constant. Find the value of the force given that $G = 6.67 \times 10^{-11}$, $m_1 = 7.36$, $m_2 = 15.5$ and $d = 22.6$. Express the answer in standard form, correct to 3 significant figures.

$$F = \frac{Gm_1 m_2}{d^2} = \frac{(6.67 \times 10^{-11})(7.36)(15.5)}{(22.6)^2}$$
$$= \frac{(6.67)(7.36)(15.5)}{(10^{11})(510.76)} = \frac{1.490}{10^{11}}$$

Hence force $F = 1.49 \times 10^{-11}$ newtons, correct to 3 significant figures.

Problem 23. The time of swing t seconds, of a simple pendulum is given by $t = 2\pi\sqrt{\dfrac{l}{g}}$. Determine the time, correct to 3 decimal places, given that $l = 12.0$ and $g = 9.81$

$$t = 2\pi\sqrt{\frac{l}{g}} = (2)\pi\sqrt{\frac{12.0}{9.81}}$$

$$= (2)\pi\sqrt{1.22324159}$$

$$= (2)\pi(1.106002527)$$

Hence time $t = 6.950$ seconds, correct to 3 decimal places.

Problem 24. Resistance, $R\,\Omega$, varies with temperature according to the formula $R = R_0(1 + \alpha t)$. Evaluate R, correct to 3 significant figures, given $R_0 = 14.59$, $\alpha = 0.0043$ and $t = 80$.

$$R = R_0(1 + \alpha t) = 14.59[1 + (0.0043)(80)]$$

$$= 14.59(1 + 0.344) = 14.59(1.344)$$

Hence resistance, $R = 19.6\,\Omega$, correct to 3 significant figures.

Now try the following exercise

Exercise 15   Further problems on evaluation of formulae (Answers on page 254)

1. The area A of a rectangle is given by the formula $A = lb$. Evaluate the area when $l = 12.4$ cm and $b = 5.37$ cm.

2. The circumference C of a circle is given by the formula $C = 2\pi r$. Determine the circumference given $\pi = 3.14$ and $r = 8.40$ mm.

3. A formula used in connection with gases is $R = (PV)/T$. Evaluate R when $P = 1500$, $V = 5$ and $T = 200$.

4. The velocity of a body is given by $v = u + at$. The initial velocity u is measured when time t is 15 seconds and found to be 12 m/s. If the acceleration a is 9.81 m/s^2 calculate the final velocity v.

5. Calculate the current I in an electrical circuit, when $I = V/R$ amperes when the voltage V is measured and found to be 7.2 V and the resistance R is 17.7 Ω.

6. Find the distance s, given that $s = \frac{1}{2}gt^2$. Time $t = 0.032$ seconds and acceleration due to gravity $g = 9.81$ m/s^2.

7. The energy stored in a capacitor is given by $E = \frac{1}{2}CV^2$ joules. Determine the energy when capacitance $C = 5 \times 10^{-6}$ farads and voltage $V = 240$ V.

8. Find the area A of a triangle, given $A = \frac{1}{2}bh$, when the base length b is 23.42 m and the height h is 53.7 m.

9. Resistance R_2 is given by $R_2 = R_1(1 + \alpha t)$. Find R_2, correct to 4 significant figures, when $R_1 = 220$, $\alpha = 0.00027$ and $t = 75.6$

10. Density $= \dfrac{\text{mass}}{\text{volume}}$. Find the density when the mass is 2.462 kg and the volume is 173 cm^3. Give the answer in units of kg/m^3.

11. Velocity $=$ frequency $\times$ wavelength. Find the velocity when the frequency is 1825 Hz and the wavelength is 0.154 m.

12. Evaluate resistance R_T, given

$$\frac{1}{R_T} = \frac{1}{R_1} + \frac{1}{R_2} + \frac{1}{R_3}$$

when $R_1 = 5.5\ \Omega$, $R_2 = 7.42\ \Omega$ and $R_3 = 12.6\ \Omega$.

13. Find the total cost of 37 calculators costing £12.65 each and 19 drawing sets costing £6.38 each.

14. Power $= \dfrac{\text{force} \times \text{distance}}{\text{time}}$. Find the power when a force of 3760 N raises an object a distance of 4.73 m in 35 s.

15. The potential difference, V volts, available at battery terminals is given by $V = E - Ir$. Evaluate V when $E = 5.62$, $I = 0.70$ and $R = 4.30$

16. Given force $F = \frac{1}{2}m(v^2 - u^2)$, find F when $m = 18.3$, $v = 12.7$ and $u = 8.24$

17. The current I amperes flowing in a number of cells is given by $I = \dfrac{nE}{R + nr}$. Evaluate the current when $n = 36$, $E = 2.20$, $R = 2.80$ and $r = 0.50$

18. The time, t seconds, of oscillation for a simple pendulum is given by $t = 2\pi\sqrt{\dfrac{l}{g}}$. Determine the time when $\pi = 3.142$, $l = 54.32$ and $g = 9.81$

19. Energy, E joules, is given by the formula $E = \frac{1}{2}LI^2$. Evaluate the energy when $L = 5.5$ and $I = 1.2$

20. The current I amperes in an a.c. circuit is given by $I = \dfrac{V}{\sqrt{(R^2 + X^2)}}$. Evaluate the current when $V = 250$, $R = 11.0$ and $X = 16.2$

21. Distance s metres is given by the formula $s = ut + \frac{1}{2}at^2$. If $u = 9.50$, $t = 4.60$ and $a = -2.50$, evaluate the distance.

22. The area, A, of any triangle is given by $A = \sqrt{[s(s-a)(s-b)(s-c)]}$ where $s = \dfrac{a+b+c}{2}$. Evaluate the area given $a = 3.60\,\text{cm}$, $b = 4.00\,\text{cm}$ and $c = 5.20\,\text{cm}$.

23. Given that $a = 0.290$, $b = 14.86$, $c = 0.042$, $d = 31.8$ and $e = 0.650$, evaluate v given that

$$v = \sqrt{\left(\frac{ab}{c} - \frac{d}{e}\right)}$$

Assignment 2

This assignment covers the material contained in Chapters 3 and 4. The marks for each question are shown in brackets at the end of each question.

1. Evaluate the following:

   (a) $\dfrac{2^3 \times 2 \times 2^2}{2^4}$ (b) $\dfrac{(2^3 \times 16)^2}{(8 \times 2)^3}$ (c) $\left(\dfrac{1}{4^2}\right)^{-1}$

   (d) $(27)^{-1/3}$ (e) $\dfrac{\left(\dfrac{3}{2}\right)^{-2} - \dfrac{2}{9}}{\left(\dfrac{2}{3}\right)^2}$ (14)

2. Express the following in standard form:

   (a) 1623 (b) 0.076 (c) $145\frac{2}{5}$ (6)

3. Determine the value of the following, giving the answer in standard form:

   (a) $5.9 \times 10^2 + 7.31 \times 10^2$

   (b) $2.75 \times 10^{-2} - 2.65 \times 10^{-3}$ (6)

4. Is the statement $\dfrac{73 \times 0.009}{15.7} = 4.18 \times 10^{-2}$ correct? If not, what should it be? (2)

5. Evaluate the following, each correct to 4 significant figures:

   (a) 61.22^2 (b) $\dfrac{1}{0.0419}$ (c) $\sqrt{0.0527}$ (6)

6. Evaluate the following, each correct to 2 decimal places:

   (a) $\left(\dfrac{36.2^2 \times 0.561}{27.8 \times 12.83}\right)^3$

   (b) $\sqrt{\left(\dfrac{14.69^2}{\sqrt{17.42} \times 37.98}\right)}$ (7)

7. If $1.6\,\text{km} = 1$ mile, determine the speed of 45 miles/hour in kilometres per hour. (3)

8. The area A of a circle is given by $A = \pi r^2$. Find the area of a circle of radius $r = 3.73\,\text{cm}$, correct to 2 decimal places. (3)

9. Evaluate B, correct to 3 significant figures, when $W = 7.20$, $v = 10.0$ and $g = 9.81$, given that

$$B = \frac{Wv^2}{2g}$$ (3)

5

Computer numbering systems

5.1 Binary numbers

The system of numbers in everyday use is the **denary** or **decimal** system of numbers, using the digits 0 to 9. It has ten different digits (0, 1, 2, 3, 4, 5, 6, 7, 8 and 9) and is said to have a **radix** or **base** of 10.

The **binary** system of numbers has a radix of 2 and uses only the digits 0 and 1.

5.2 Conversion of binary to denary

The denary number 234.5 is equivalent to

$$2 \times 10^2 + 3 \times 10^1 + 4 \times 10^0 + 5 \times 10^{-1}$$

i.e. is the sum of terms comprising: (a digit) multiplied by (the base raised to some power).

In the binary system of numbers, the base is 2, so 1101.1 is equivalent to:

$$1 \times 2^3 + 1 \times 2^2 + 0 \times 2^1 + 1 \times 2^0 + 1 \times 2^{-1}$$

Thus the denary number equivalent to the binary number 1101.1 is

$8 + 4 + 0 + 1 + \frac{1}{2}$, that is 13.5

i.e. **$1101.1_2 = 13.5_{10}$**, the suffixes 2 and 10 denoting binary and denary systems of numbers respectively.

Problem 1. Convert 11011_2 to a denary number.

From above: $11011_2 = 1 \times 2^4 + 1 \times 2^3 + 0 \times 2^2$
$$+ 1 \times 2^1 + 1 \times 2^0$$
$$= 16 + 8 + 0 + 2 + 1$$
$$= 27_{10}$$

Problem 2. Convert 0.1011_2 to a decimal fraction.

$0.1011_2 = 1 \times 2^{-1} + 0 \times 2^{-2} + 1 \times 2^{-3} + 1 \times 2^{-4}$
$$= 1 \times \frac{1}{2} + 0 \times \frac{1}{2^2} + 1 \times \frac{1}{2^3} + 1 \times \frac{1}{2^4}$$
$$= \frac{1}{2} + \frac{1}{8} + \frac{1}{16}$$
$$= 0.5 + 0.125 + 0.0625$$
$$= \mathbf{0.6875_{10}}$$

Problem 3. Convert 101.0101_2 to a denary number.

$101.0101_2 = 1 \times 2^2 + 0 \times 2^1 + 1 \times 2^0 + 0 \times 2^{-1}$
$$+ 1 \times 2^{-2} + 0 \times 2^{-3} + 1 \times 2^{-4}$$
$$= 4 + 0 + 1 + 0 + 0.25 + 0 + 0.0625$$
$$= \mathbf{5.3125_{10}}$$

Now try the following exercise

Exercise 16 Further problems on conversion of binary to denary numbers (Answers on page 254)

In Problems 1 to 4, convert the binary numbers given to denary numbers.

1. (a) 110   (b) 1011   (c) 1110   (d) 1001
2. (a) 10101   (b) 11001   (c) 101101   (d) 110011

3. (a) 0.1101   (b) 0.11001   (c) 0.00111   (d) 0.01011

4. (a) 11010.11   (b) 10111.011

   (c) 110101.0111   (d) 11010101.10111

5.3   Conversion of denary to binary

An integer denary number can be converted to a corresponding binary number by repeatedly dividing by 2 and noting the remainder at each stage, as shown below for 39_{10}

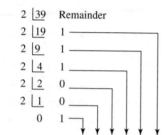

(most significant bit) $\rightarrow$ 1   0   0   1   1   1 $\leftarrow$ (least significant bit)

The result is obtained by writing the top digit of the remainder as the least significant bit, (a bit is a **b**inary dig**it** and the least significant bit is the one on the right). The bottom bit of the remainder is the most significant bit, i.e. the bit on the left.

Thus $39_{10} = 100111_2$

The fractional part of a denary number can be converted to a binary number by repeatedly multiplying by 2, as shown below for the fraction 0.625

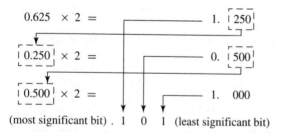

(most significant bit) . 1   0   1   (least significant bit)

For fractions, the most significant bit of the result is the top bit obtained from the integer part of multiplication by 2. The least significant bit of the result is the bottom bit obtained from the integer part of multiplication by 2.

Thus $0.625_{10} = 0.101_2$

Problem 4.   Convert 47_{10} to a binary number.

From above, repeatedly dividing by 2 and noting the remainder gives:

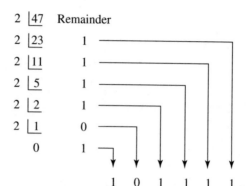

Thus $47_{10} = 101111_2$

Problem 5.   Convert 0.40625_{10} to a binary number.

From above, repeatedly multiplying by 2 gives:

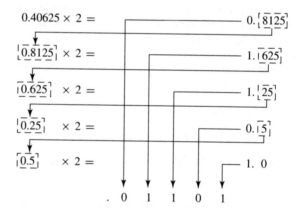

i.e. **$0.40625_{10} = 0.01101_2$**

Problem 6.   Convert 58.3125_{10} to a binary number.

The integer part is repeatedly divided by 2, giving:

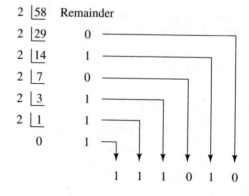

The fractional part is repeatedly multiplied by 2 giving:

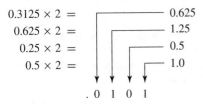

$0.3125 \times 2 = 0.625$

$0.625 \times 2 = 1.25$

$0.25 \times 2 = 0.5$

$0.5 \times 2 = 1.0$

$. \quad 0 \quad 1 \quad 0 \quad 1$

Thus $58.3125_{10} = 111010.0101_2$

Now try the following exercise

Exercise 17   Further problems on conversion of denary to binary numbers (Answers on page 254)

In Problems 1 to 4, convert the denary numbers given to binary numbers.

1. (a) 5   (b) 15   (c) 19   (d) 29

2. (a) 31   (b) 42   (c) 57   (d) 63

3. (a) 0.25   (b) 0.21875   (c) 0.28125   (d) 0.59375

4. (a) 47.40625   (b) 30.8125

   (c) 53.90625   (d) 61.65625

5.4   Conversion of denary to binary via octal

For denary integers containing several digits, repeatedly dividing by 2 can be a lengthy process. In this case, it is usually easier to convert a denary number to a binary number via the octal system of numbers. This system has a radix of 8, using the digits 0, 1, 2, 3, 4, 5, 6 and 7. The denary number equivalent to the octal number 4317_8 is

$$4 \times 8^3 + 3 \times 8^2 + 1 \times 8^1 + 7 \times 8^0$$

i.e. $4 \times 512 + 3 \times 64 + 1 \times 8 + 7 \times 1$ or 2255_{10}

An integer denary number can be converted to a corresponding octal number by repeatedly dividing by 8 and noting the remainder at each stage, as shown below for 493_{10}

```
8 |493   Remainder
8 | 61      5
8 | 7       5
    0       7

         7  5  5
```

Thus $493_{10} = 755_8$

The fractional part of a denary number can be converted to an octal number by repeatedly multiplying by 8, as shown below for the fraction 0.4375_{10}

$0.4375 \times 8 = 3 . 5$

$0.5 \times 8 = 4 . 0$

$. 3 \quad 4$

For fractions, the most significant bit is the top integer obtained by multiplication of the denary fraction by 8, thus

$$0.4375_{10} = 0.34_8$$

The natural binary code for digits 0 to 7 is shown in Table 5.1, and an octal number can be converted to a binary number by writing down the three bits corresponding to the octal digit.

Table 5.1

Octal digit	Natural binary number
0	000
1	001
2	010
3	011
4	100
5	101
6	110
7	111

Thus $437_8 = 100\ 011\ 111_2$

and $26.35_8 = 010\ 110.011\ 101_2$

The '0' on the extreme left does not signify anything, thus $26.35_8 = 10\ 110.011\ 101_2$

Conversion of denary to binary via octal is demonstrated in the following worked problems.

Problem 7.   Convert 3714_{10} to a binary number, via octal.

Dividing repeatedly by 8, and noting the remainder gives:

```
8 |3714   Remainder
8 | 464     2
8 | 58      0
8 | 7       2
    0       7

         7  2  0  2
```

From Table 5.1, $7202_8 = 111\ 010\ 000\ 010_2$

i.e. $\mathbf{3714_{10} = 111\ 010\ 000\ 010_2}$

Problem 8.   Convert 0.59375_{10} to a binary number, via octal.

Multiplying repeatedly by 8, and noting the integer values, gives:

$$0.59375 \times 8 = \quad 4.75$$
$$0.75 \times 8 = \quad 6.00$$

. 4 6

Thus $0.59375_{10} = 0.46_8$

From Table 5.1, $0.46_8 = 0.100\ 110_2$

i.e. $\mathbf{0.59375_{10} = 0.100\ 11_2}$

Problem 9.   Convert 5613.90625_{10} to a binary number, via octal.

The integer part is repeatedly divided by 8, noting the remainder, giving:

```
8 | 5613    Remainder
8 |  701        5  ————————————————
8 |   87        5  ——————————————
8 |   10        7  ————————————
8 |    1        2  ——————————
        0        1  ————————
                  1   2   7   5   5
```

This octal number is converted to a binary number, (see Table 5.1)

$$12755_8 = 001\ 010\ 111\ 101\ 101_2$$

i.e. $5613_{10} = 1\ 010\ 111\ 101\ 101_2$

The fractional part is repeatedly multiplied by 8, and noting the integer part, giving:

$$0.90625 \times 8 = \quad 7.25$$
$$0.25 \times 8 = \quad 2.00$$

.7 2

This octal fraction is converted to a binary number, (see Table 5.1)

$$0.72_8 = 0.111\ 010_2$$

i.e. $0.90625_{10} = 0.111\ 01_2$

Thus, $5613.90625_{10} = 1\ 010\ 111\ 101\ 101.111\ 01_2$

Problem 10.   Convert $11\ 110\ 011.100\ 01_2$ to a denary number via octal.

Grouping the binary number in three's from the binary point gives: $011\ 110\ 011.100\ 010_2$

Using Table 5.1 to convert this binary number to an octal number gives: 363.42_8 and

$$363.42_8 = 3 \times 8^2 + 6 \times 8^1 + 3 \times 8^0 + 4 \times 8^{-1} + 2 \times 8^{-2}$$
$$= 192 + 48 + 3 + 0.5 + 0.03125$$
$$= \mathbf{243.53125_{10}}$$

Now try the following exercise

Exercise 18   Further problems on conversion between denary and binary numbers via octal (Answers on page 254)

In Problems 1 to 3, convert the denary numbers given to binary numbers, via octal.

1. (a) 343   (b) 572   (c) 1265

2. (a) 0.46875   (b) 0.6875   (c) 0.71875

3. (a) 247.09375   (b) 514.4375   (c) 1716.78125

4. Convert the following binary numbers to denary numbers via octal:

   (a) 111.011 1   (b) 101 001.01

   (c) 1 110 011 011 010.001 1

5.5   Hexadecimal numbers

The complexity of computers requires higher order numbering systems such as octal (base 8) and hexadecimal (base 16) which are merely extensions of the binary system. A **hexadecimal numbering system** has a radix of 16 and uses the following 16 distinct digits:

0, 1, 2, 3, 4, 5, 6, 7, 8, 9, A, B, C, D, E and F

'A' corresponds to 10 in the denary system, B to 11, C to 12, and so on.

To convert from hexadecimal to decimal:

For example $1A_{16} = 1 \times 16^1 + A \times 16^0$

$$= 1 \times 16^1 + 10 \times 1 = 16 + 10 = 26$$

i.e. $\mathbf{1A_{16} = 26_{10}}$

Similarly, $\mathbf{2E_{16}} = 2 \times 16^1 + E \times 16^0$

$$= 2 \times 16^1 + 14 \times 16^0 = 32 + 14 = \mathbf{46_{10}}$$

and
$$1BF_{16} = 1 \times 16^2 + B \times 16^1 + F \times 16^0$$
$$= 1 \times 16^2 + 11 \times 16^1 + 15 \times 16^0$$
$$= 256 + 176 + 15 = 447_{10}$$

Table 5.2 compares decimal, binary, octal and hexadecimal numbers and shows, for example, that

$$23_{10} = 10111_2 = 27_8 = 17_{16}$$

Table 5.2

Decimal	Binary	Octal	Hexadecimal
0	0000	0	0
1	0001	1	1
2	0010	2	2
3	0011	3	3
4	0100	4	4
5	0101	5	5
6	0110	6	6
7	0111	7	7
8	1000	10	8
9	1001	11	9
10	1010	12	A
11	1011	13	B
12	1100	14	C
13	1101	15	D
14	1110	16	E
15	1111	17	F
16	10000	20	10
17	10001	21	11
18	10010	22	12
19	10011	23	13
20	10100	24	14
21	10101	25	15
22	10110	26	16
23	10111	27	17
24	11000	30	18
25	11001	31	19
26	11010	32	1A
27	11011	33	1B
28	11100	34	1C
29	11101	35	1D
30	11110	36	1E
31	11111	37	1F
32	100000	40	20

Problem 11. Convert the following hexadecimal numbers into their decimal equivalents: (a) $7A_{16}$ (b) $3F_{16}$

(a) $7A_{16} = 7 \times 16^1 + A \times 16^0 = 7 \times 16 + 10 \times 1$
$$= 112 + 10 = 122$$

Thus $\mathbf{7A_{16} = 122_{10}}$

(b) $3F_{16} = 3 \times 16^1 + F \times 16^0 = 3 \times 16 + 15 \times 1$
$$= 48 + 15 = 63$$

Thus $\mathbf{3F_{16} = 63_{10}}$

Problem 12. Convert the following hexadecimal numbers into their decimal equivalents: (a) $C9_{16}$ (b) BD_{16}

(a) $C9_{16} = C \times 16^1 + 9 \times 16^0 = 12 \times 16 + 9 \times 1$
$$= 192 + 9 = 201$$

Thus $\mathbf{C9_{16} = 201_{10}}$

(b) $BD_{16} = B \times 16^1 + D \times 16^0 = 11 \times 16 + 13 \times 1$
$$= 176 + 13 = 189$$

Thus $\mathbf{BD_{16} = 189_{10}}$

Problem 13. Convert $1A4E_{16}$ into a denary number.

$$1A4E_{16} = 1 \times 16^3 + A \times 16^2 + 4 \times 16^1 + E \times 16^0$$
$$= 1 \times 16^3 + 10 \times 16^2 + 4 \times 16^1 + 14 \times 16^0$$
$$= 1 \times 4096 + 10 \times 256 + 4 \times 16 + 14 \times 1$$
$$= 4096 + 2560 + 64 + 14 = 6734$$

Thus $\mathbf{1A4E_{16} = 6734_{10}}$

To convert from decimal to hexadecimal:

This is achieved by repeatedly dividing by 16 and noting the remainder at each stage, as shown below for 26_{10}

```
16 | 26    Remainder
16 |  1    10 ≡ A₁₆
        0    1 ≡ 1₁₆
```

most significant bit → 1 A ← least significant bit

Hence $\mathbf{26_{10} = 1A_{16}}$

Similarly, for 447_{10}

```
16 | 447    Remainder
16 |  27    15 ≡ F₁₆
16 |   1    11 ≡ B₁₆
        0    1 ≡ 1₁₆
```

1 B F

Thus $\mathbf{447_{10} = 1BF_{16}}$

Problem 14. Convert the following decimal numbers into their hexadecimal equivalents: (a) 37_{10} (b) 108_{10}

(a)
$$16 \underline{\lfloor 37 } \quad \text{Remainder}$$
$$16 \underline{\lfloor 2 } \quad 5 = 5_{16}$$
$$0 \quad\quad 2 = 2_{16}$$

most significant bit $\rightarrow$ 2   5 $\leftarrow$ least significant bit

Hence $37_{10} = 25_{16}$

(b)
$$16 \underline{\lfloor 108 } \quad \text{Remainder}$$
$$16 \underline{\lfloor 6 } \quad 12 = C_{16}$$
$$0 \quad\quad 6 = 6_{16}$$

6 C

Hence $108_{10} = 6C_{16}$

Problem 15. Convert the following decimal numbers into their hexadecimal equivalents: (a) 162_{10} (b) 239_{10}

(a)
$$16 \underline{\lfloor 162 } \quad \text{Remainder}$$
$$16 \underline{\lfloor 10 } \quad 2 = 2_{16}$$
$$0 \quad\quad 10 = A_{16}$$

A 2

Hence $162_{10} = A2_{16}$

(b)
$$16 \underline{\lfloor 239 } \quad \text{Remainder}$$
$$16 \underline{\lfloor 14 } \quad 15 = F_{16}$$
$$0 \quad\quad 14 = E_{16}$$

E F

Hence $239_{10} = EF_{16}$

To convert from binary to hexadecimal:

The binary bits are arranged in groups of four, starting from right to left, and a hexadecimal symbol is assigned to each group. For example, the binary number 1110011110101001 is initially grouped in fours as: 1110 0111 1010 1001

and a hexadecimal symbol assigned   E   7   A   9

to each group   from Table 5.2

Hence $1110011110101001_2 = E7A9_{16}$

To convert from hexadecimal to binary:

The above procedure is reversed, thus, for example,

$6CF3_{16} = 0110\ 1100\ 1111\ 0011$ from Table 5.2

i.e. $6CF3_{16} = 1101100111100011_2$

Problem 16. Convert the following binary numbers into their hexadecimal equivalents:

(a) 11010110_2 (b) 1100111_2

(a) Grouping bits in fours from the

right gives: 1101 0110

and assigning hexadecimal symbols

to each group gives: D   6

from Table 5.2

Thus, $11010110_2 = D6_{16}$

(b) Grouping bits in fours from the

right gives: 0110 0111

and assigning hexadecimal symbols

to each group gives: 6   7

from Table 5.2

Thus, $1100111_2 = 67_{16}$

Problem 17. Convert the following binary numbers into their hexadecimal equivalents:

(a) 11001111_2 (b) 110011110_2

(a) Grouping bits in fours from the

right gives: 1100   1111

and assigning hexadecimal symbols

to each group gives: C   F

from Table 5.2

Thus, $11001111_2 = CF_{16}$

(b) Grouping bits in fours from the

right gives: 0001 1001 1110

and assigning hexadecimal symbols

to each group gives: 1   9   E

from Table 5.2

Thus, $110011110_2 = 19E_{16}$

Problem 18. Convert the following hexadecimal numbers into their binary equivalents: (a) $3F_{16}$ (b) $A6_{16}$

(a) Spacing out hexadecimal digits gives: 3   F

and converting each into binary gives: 0011 1111

from Table 5.2

Thus, $3F_{16} = 111111_2$

(b) Spacing out hexadecimal digits gives: A   6

and converting each into binary gives: 1010 0110

from Table 5.2

Thus, $A6_{16} = 10100110_2$

Problem 19. Convert the following hexadecimal numbers into their binary equivalents: (a) $7B_{16}$ (b) $17D_{16}$

(a) Spacing out hexadecimal digits gives: 7   B

and converting each into binary gives: 0111 1011

from Table 5.2

Thus, $7B_{16} = 1111011_2$

(b) Spacing out hexadecimal digits gives: 1   7   D

and converting each into binary

gives: 0001 0111 1101

from Table 5.2

Thus, $17D_{16} = 101111101_2$

Now try the following exercise

Exercise 19   Further problems on hexadecimal numbers (Answers on page 254)

In Problems 1 to 4, convert the given hexadecimal numbers into their decimal equivalents.

1. $E7_{16}$ 2. $2C_{16}$ 3. 98_{16} 4. $2F1_{16}$

In Problems 5 to 8, convert the given decimal numbers into their hexadecimal equivalents.

5. 54_{10} 6. 200_{10} 7. 91_{10} 8. 238_{10}

In Problems 9 to 12, convert the given binary numbers into their hexadecimal equivalents.

9. 11010111_2 10. 11101010_2

11. 10001011_2 12. 10100101_2

In Problems 13 to 16, convert the given hexadecimal numbers into their binary equivalents.

13. 37_{16} 14. ED_{16} 15. $9F_{16}$ 16. $A21_{16}$

6

Algebra

6.1 Basic operations

Algebra is that part of mathematics in which the relations and properties of numbers are investigated by means of general symbols. For example, the area of a rectangle is found by multiplying the length by the breadth; this is expressed algebraically as $A = l \times b$, where A represents the area, l the length and b the breadth.

The basic laws introduced in arithmetic are generalized in algebra.

Let a, b, c and d represent any four numbers. Then:

(i) $a + (b + c) = (a + b) + c$

(ii) $a(bc) = (ab)c$

(iii) $a + b = b + a$

(iv) $ab = ba$

(v) $a(b + c) = ab + ac$

(vi) $\dfrac{a + b}{c} = \dfrac{a}{c} + \dfrac{b}{c}$

(vii) $(a + b)(c + d) = ac + ad + bc + bd$

Problem 1. Evaluate $3ab - 2bc + abc$ when $a = 1$, $b = 3$ and $c = 5$

Replacing a, b and c with their numerical values gives:

$$3ab - 2bc + abc = 3 \times 1 \times 3 - 2 \times 3 \times 5 + 1 \times 3 \times 5$$

$$= 9 - 30 + 15 = -6$$

Problem 2. Find the value of $4p^2qr^3$, given that $p = 2$, $q = \frac{1}{2}$ and $r = 1\frac{1}{2}$

Replacing p, q and r with their numerical values gives:

$$4p^2qr^3 = 4(2)^2 \left(\tfrac{1}{2}\right) \left(\tfrac{3}{2}\right)^3$$

$$= 4 \times 2 \times 2 \times \tfrac{1}{2} \times \tfrac{3}{2} \times \tfrac{3}{2} \times \tfrac{3}{2} = 27$$

Problem 3. Find the sum of $3x$, $2x$, $-x$ and $-7x$

The sum of the positive terms is: $3x + 2x = 5x$

The sum of the negative terms is: $x + 7x = 8x$

Taking the sum of the negative terms from the sum of the positive terms gives:

$$5x - 8x = -3x$$

Alternatively

$$3x + 2x + (-x) + (-7x) = 3x + 2x - x - 7x = -3x$$

Problem 4. Find the sum of $4a$, $3b$, c, $-2a$, $-5b$ and $6c$

Each symbol must be dealt with individually.

For the 'a' terms: $+4a - 2a = 2a$

For the 'b' terms: $+3b - 5b = -2b$

For the 'c' terms: $+c + 6c = 7c$

Thus $4a + 3b + c + (-2a) + (-5b) + 6c$

$$= 4a + 3b + c - 2a - 5b + 6c$$

$$= 2a - 2b + 7c$$

Problem 5. Find the sum of $5a - 2b$, $2a + c$, $4b - 5d$ and $b - a + 3d - 4c$

The algebraic expressions may be tabulated as shown below, forming columns for the a's, b's, c's and d's. Thus:

$$
\begin{array}{l}
+5a - 2b \\
+2a \qquad\quad + c \\
\qquad\ + 4b \qquad\quad - 5d \\
\underline{- a + \ \ b - 4c + 3d} \\
\end{array}
$$

Adding gives: $\quad \mathbf{6a + 3b - 3c - 2d}$

Problem 6. Subtract $2x + 3y - 4z$ from $x - 2y + 5z$

$$
\begin{array}{l}
x - 2y + 5z \\
\underline{2x + 3y - 4z} \\
\end{array}
$$

Subtracting gives: $\quad \mathbf{-x - 5y + 9z}$

(Note that $+5z - -4z = +5z + 4z = 9z$)

An alternative method of subtracting algebraic expressions is to 'change the signs of the bottom line and add'. Hence:

$$
\begin{array}{l}
x - 2y + 5z \\
\underline{-2x - 3y + 4z} \\
\end{array}
$$

Adding gives: $\quad \mathbf{-x \ - 5y + 9z}$

Problem 7. Multiply $2a + 3b$ by $a + b$

Each term in the first expression is multiplied by a, then each term in the first expression is multiplied by b, and the two results are added. The usual layout is shown below.

$$
\begin{array}{l}
2a \ + \ 3b \\
\underline{a \ + \ \ b} \\
\end{array}
$$

Multiplying by $a \rightarrow \quad 2a^2 + 3ab$

Multiplying by $b \rightarrow \quad \underline{\quad\ + 2ab + 3b^2}$

Adding gives: $\quad \mathbf{2a^2 + 5ab + 3b^2}$

Problem 8. Multiply $3x - 2y^2 + 4xy$ by $2x - 5y$

$$
\begin{array}{l}
3x \ - \quad 2y^2 + 4xy \\
\underline{2x \ - \quad 5y} \\
\end{array}
$$

Multiplying by $2x \rightarrow \quad 6x^2 - 4xy^2 + 8x^2y$

Multiplying by $-5y \rightarrow \quad\quad - 20xy^2 \qquad\qquad - 15xy + 10y^3$

Adding gives: $\quad \mathbf{6x^2 - 24xy^2 + 8x^2y - 15xy + 10y^3}$

Problem 9. Simplify $2p \div 8pq$

$2p \div 8pq$ means $\dfrac{2p}{8pq}$. This can be reduced by cancelling as in arithmetic.

Thus: $\dfrac{2p}{8pq} = \dfrac{2 \times p}{8 \times p \times q} = \dfrac{1}{4q}$

Problem 10. Divide $2x^2 + x - 3$ by $x - 1$

$2x^2 + x - 3$ is called the **dividend** and $x - 1$ the **divisor**. The usual layout is shown below with the dividend and divisor both arranged in descending powers of the symbols.

$$
\begin{array}{r}
2x + 3 \\
x - 1 \ \overline{\smash{)}\ 2x^2 + \ x - 3} \\
\underline{2x^2 - 2x} \\
3x - 3 \\
\underline{3x - 3} \\
0 \quad 0 \\
\end{array}
$$

Dividing the first term of the dividend by the first term of the divisor, i.e. $\dfrac{2x^2}{x}$ gives $2x$, which is put above the first term of the dividend as shown. The divisor is then multiplied by $2x$, i.e. $2x(x - 1) = 2x^2 - 2x$, which is placed under the dividend as shown. Subtracting gives $3x - 3$.

The process is then repeated, i.e. the first term of the divisor, x, is divided into $3x$, giving 3, which is placed above the dividend as shown.

Then $3(x - 1) = 3x - 3$, which is placed under the $3x - 3$. The remainder, on subtraction, is zero, which completes the process.

Thus $(2x^2 + x - 3) \div (x - 1) = \mathbf{(2x + 3)}$

(A check can be made on this answer by multiplying $(2x + 3)$ by $(x - 1)$, which equals $2x^2 + x - 3$)

Problem 11. Simplify $\dfrac{x^3 + y^3}{x + y}$

$$
\begin{array}{r}
\text{(1)} \quad\ \text{(4)} \quad\ \text{(7)} \\
x^2 - \ xy + y^2 \\
x + y \ \overline{\smash{)}\ x^3 + \ 0 \ + \ 0 + y^3} \\
\underline{x^3 + x^2y} \\
- x^2y \qquad\quad + y^3 \\
\underline{- x^2y - xy^2} \\
xy^2 + y^3 \\
\underline{xy^2 + y^3} \\
0 \quad\ 0 \\
\end{array}
$$

(1) x into x^3 goes x^2. Put x^2 above x^3

(2) $x^2(x + y) = x^3 + x^2y$

(3) Subtract

(4) x into $-x^2y$ goes $-xy$. Put $-xy$ above dividend

(5) $-xy(x + y) = -x^2y - xy^2$

(6) Subtract

(7) x into xy^2 goes y^2. Put y^2 above dividend

(8) $y^2(x + y) = xy^2 + y^3$

(9) Subtract

Thus $\dfrac{x^3+y^3}{x+y} = x^2 - xy + y^2$

The zeros shown in the dividend are not normally shown, but are included to clarify the subtraction process and to keep similar terms in their respective columns.

Problem 12. Divide $4a^3 - 6a^2b + 5b^3$ by $2a - b$

$$
\begin{array}{r}
2a^2 - 2ab - b^2 \\
2a - b \overline{)\,4a^3 - 6a^2b \qquad\quad + 5b^3} \\
4a^3 - 2a^2b \\
\hline
-4a^2b \qquad\quad + 5b^3 \\
-4a^2b + 2ab^2 \\
\hline
-2ab^2 + 5b^3 \\
-2ab^2 + b^3 \\
\hline
4b^3 \\
\hline
\end{array}
$$

Thus $\dfrac{4a^3 - 6a^2b + 5b^3}{2a - b} = \mathbf{2a^2 - 2ab - b^2}$**, remainder** $\mathbf{4b^3}$**.**

Alternatively, the answer may be expressed as

$$2a^2 - 2ab - b^2 + \frac{4b^3}{2a - b}$$

Now try the following exercise

Exercise 20   Further problems on basic operations (Answers on page 254)

1. Find the value of $2xy + 3yz - xyz$, when $x = 2$, $y = -2$ and $z = 4$

2. Evaluate $3pq^2r^3$ when $p = \dfrac{2}{3}$, $q = -2$ and $r = -1$

3. Find the sum of $3a$, $-2a$, $-6a$, $5a$ and $4a$

4. Simplify $\dfrac{4}{3}c + 2c - \dfrac{1}{6}c - c$

5. Find the sum of $3x$, $2y$, $-5x$, $2z$, $-\dfrac{1}{2}y$, $-\dfrac{1}{4}x$

6. Add together $2a + 3b + 4c$, $-5a - 2b + c$, $4a - 5b - 6c$

7. Add together $3d + 4e$, $-2e + f$, $2d - 3f$, $4d - e + 2f - 3e$

8. From $4x - 3y + 2z$ subtract $x + 2y - 3z$

9. Subtract $\dfrac{3}{2}a - \dfrac{b}{3} + c$ from $\dfrac{b}{2} - 4a - 3c$

10. Multiply $3x + 2y$ by $x - y$

11. Multiply $2a - 5b + c$ by $3a + b$

12. Simplify (i) $3a \div 9ab$ (ii) $4a^2b \div 2a$

---

13. Divide $2x^2 + xy - y^2$ by $x + y$

14. Divide $3p^2 - pq - 2q^2$ by $p - q$

6.2 Laws of Indices

The laws of indices are:

(i) $a^m \times a^n = a^{m+n}$

(ii) $\dfrac{a^m}{a^n} = a^{m-n}$

(iii) $(a^m)^n = a^{mn}$

(iv) $a^{m/n} = \sqrt[n]{a^m}$

(v) $a^{-n} = \dfrac{1}{a^n}$

(vi) $a^0 = 1$

Problem 13. Simplify $a^3b^2c \times ab^3c^5$

Grouping like terms gives: $a^3 \times a \times b^2 \times b^3 \times c \times c^5$

Using the first law of indices gives: $a^{3+1} \times b^{2+3} \times c^{1+5}$

i.e. $a^4 \times b^5 \times c^6 = \mathbf{a^4 b^5 c^6}$

Problem 14. Simplify $a^{1/2}b^2c^{-2} \times a^{1/6}b^{1/2}c$

Using the first law of indices,

$$a^{1/2}b^2c^{-2} \times a^{(1/6)}b^{(1/2)}c = a^{(1/2)+(1/6)} \times b^{2+(1/2)} \times c^{-2+1}$$

$$= \mathbf{a^{2/3} b^{5/2} c^{-1}}$$

Problem 15. Simplify $\dfrac{a^3b^2c^4}{abc^{-2}}$ and evaluate when $a = 3$, $b = \dfrac{1}{8}$ and $c = 2$

Using the second law of indices,

$$\frac{a^3}{a} = a^{3-1} = a^2, \qquad \frac{b^2}{b} = b^{2-1} = b$$

and

$$\frac{c^4}{c^{-2}} = c^{4--2} = c^6$$

Thus

$$\frac{a^3b^2c^4}{abc^{-2}} = a^2bc^6$$

When $a = 3$, $b = \tfrac{1}{8}$ and $c = 2$,

$$a^2bc^6 = (3)^2 \left(\tfrac{1}{8}\right)(2)^6 = (9)\left(\tfrac{1}{8}\right)(64) = \mathbf{72}$$

Problem 16. Simplify $\dfrac{p^{1/2}q^2r^{2/3}}{p^{1/4}q^{1/2}r^{1/6}}$ and evaluate when $p = 16$, $q = 9$ and $r = 4$, taking positive roots only.

Using the second law of indices gives:

$$p^{(1/2)-(1/4)}q^{2-(1/2)}r^{(2/3)-(1/6)} = p^{1/4}q^{3/2}r^{1/2}$$

When $p = 16$, $q = 9$ and $r = 4$,

$$p^{1/4}q^{3/2}r^{1/2} = (16)^{1/4}(9)^{3/2}(4)^{1/2}$$

$$= (\sqrt[4]{16})(\sqrt{9^3})(\sqrt{4}) = (2)(3^3)(2) = \mathbf{108}$$

Problem 17. Simplify $\dfrac{x^2y^3 + xy^2}{xy}$

Algebraic expressions of the form $\dfrac{a+b}{c}$ can be split into $\dfrac{a}{c} + \dfrac{b}{c}$. Thus

$$\frac{x^2y^3 + xy^2}{xy} = \frac{x^2y^3}{xy} + \frac{xy^2}{xy} = x^{2-1}y^{3-1} + x^{1-1}y^{2-1}$$

$$= xy^2 + y$$

(since $x^0 = 1$, from the sixth law of indices.)

Problem 18. Simplify $\dfrac{x^2y}{xy^2 - xy}$

The highest common factor (HCF) of each of the three terms comprising the numerator and denominator is xy. Dividing each term by xy gives:

$$\frac{x^2y}{xy^2 - xy} = \frac{\dfrac{x^2y}{xy}}{\dfrac{xy^2}{xy} - \dfrac{xy}{xy}} = \frac{x}{y-1}$$

Problem 19. Simplify $\dfrac{a^2b}{ab^2 - a^{1/2}b^3}$

The HCF of each of the three terms is $a^{1/2}b$. Dividing each term by $a^{1/2}b$ gives:

$$\frac{a^2b}{ab^2 - a^{1/2}b^3} = \frac{\dfrac{a^2b}{a^{1/2}b}}{\dfrac{ab^2}{a^{1/2}b} - \dfrac{a^{1/2}b^3}{a^{1/2}b}} = \frac{a^{3/2}}{a^{1/2}b - b^2}$$

Problem 20. Simplify $(p^3)^{1/2}(q^2)^4$

Using the third law of indices gives:

$$p^{3\times(1/2)}q^{2\times4} = p^{(3/2)}q^8$$

Problem 21. Simplify $\dfrac{(mn^2)^3}{(m^{1/2}n^{1/4})^4}$

The brackets indicate that each letter in the bracket must be raised to the power outside.

Using the third law of indices gives:

$$\frac{(mn^2)^3}{(m^{1/2}n^{1/4})^4} = \frac{m^{1\times3}n^{2\times3}}{m^{(1/2)\times4}n^{(1/4)\times4}} = \frac{m^3n^6}{m^2n^1}$$

Using the second law of indices gives:

$$\frac{m^3n^6}{m^2n^1} = m^{3-2}n^{6-1} = mn^5$$

Problem 22. Simplify $(a^3\sqrt{b}\sqrt{c^5})(\sqrt{a}\sqrt[3]{b^2}c^3)$ and evaluate when $a = \frac{1}{4}$, $b = 6$ and $c = 1$

Using the fourth law of indices, the expression can be written as:

$$(a^3b^{1/2}c^{5/2})(a^{1/2}b^{2/3}c^3)$$

Using the first law of indices gives:

$$a^{3+(1/2)}b^{(1/2)+(2/3)}c^{(5/2)+3} = a^{7/2}b^{7/6}c^{11/2}$$

It is usual to express the answer in the same form as the question. Hence

$$a^{7/2}b^{7/6}c^{11/2} = \sqrt{a^7}\sqrt[6]{b^7}\sqrt{c^{11}}$$

When $a = \frac{1}{4}$, $b = 64$ and $c = 1$,

$$\sqrt{a^7}\sqrt[6]{b^7}\sqrt{c^{11}} = \sqrt{\left(\tfrac{1}{4}\right)^7}(\sqrt[6]{64^7})(\sqrt{1^{11}})$$

$$= \left(\tfrac{1}{2}\right)^7 (2)^7(1) = 1$$

Problem 23. Simplify $(a^3b)(a^{-4}b^{-2})$, expressing the answer with positive indices only.

Using the first law of indices gives: $a^{3+-4}b^{1+-2} = a^{-1}b^{-1}$

Using the fifth law of indices gives: $a^{-1}b^{-1} = \dfrac{1}{a^{+1}b^{+1}} = \dfrac{1}{ab}$

Problem 24. Simplify $\dfrac{d^2e^2f^{1/2}}{(d^{3/2}ef^{5/2})^2}$ expressing the answer with positive indices only.

Using the third law of indices gives:

$$\frac{d^2 e^2 f^{1/2}}{(d^{3/2} e f^{5/2})^2} = \frac{d^2 e^2 f^{1/2}}{d^3 e^2 f^5}$$

Using the second law of indices gives:

$$d^{2-3} e^{2-2} f^{(1/2)-5} = d^{-1} e^0 f^{-9/2}$$

$$= d^{-1} f^{(-9/2)} \quad \text{since } e^0 = 1 \text{ from}$$
$$\text{the sixth law of indices}$$

$$= \frac{1}{df^{9/2}} \quad \text{from the fifth law of indices}$$

Problem 25. Simplify $\dfrac{(x^2 y^{1/2})(\sqrt{x}\sqrt[3]{y^2})}{(x^5 y^3)^{1/2}}$

Using the third and fourth laws of indices gives:

$$\frac{(x^2 y^{1/2})(\sqrt{x}\sqrt[3]{y^2})}{(x^5 y^3)^{1/2}} = \frac{(x^2 y^{1/2})(x^{1/2} y^{2/3})}{x^{5/2} y^{3/2}}$$

Using the first and second laws of indices gives:

$$x^{2+(1/2)-(5/2)} y^{(1/2)+(2/3)-(3/2)} = x^0 y^{-1/3}$$

$$= y^{-1/3} \quad \text{or} \quad \frac{1}{y^{1/3}} \quad \text{or} \quad \frac{1}{\sqrt[3]{y}}$$

from the fifth and sixth laws of indices.

Now try the following exercise

**Exercise 21   Further problems on laws of indices
(Answers on page 255)**

1. Simplify $(x^2 y^3 z)(x^3 yz^2)$ and evaluate when $x = \dfrac{1}{2}$, $y = 2$ and $z = 3$

2. Simplify $(a^{3/2} bc^{-3})(a^{1/2} b^{-1/2} c)$ and evaluate when $a = 3$, $b = 4$ and $c = 2$

3. Simplify $\dfrac{a^5 bc^3}{a^2 b^3 c^2}$ and evaluate when $a = \dfrac{3}{2}$, $b = \dfrac{1}{2}$ and $c = \dfrac{2}{3}$

In Problems 4 to 11, simplify the given expressions:

4. $\dfrac{x^{1/5} y^{1/2} z^{1/3}}{x^{-1/2} y^{1/3} z^{-1/6}}$

5. $\dfrac{a^2 b + a^3 b}{a^2 b^2}$

6. $\dfrac{p^3 q^2}{pq^2 - p^2 q}$

7. $(a^2)^{1/2}(b^2)^3(c^{1/2})^3$

8. $\dfrac{(abc)^2}{(a^2 b^{-1} c^{-3})^3}$

9. $(\sqrt{x}\sqrt{y^3}\sqrt[3]{z^2})(\sqrt{x}\sqrt{y^3}\sqrt{z^3})$

10. $(e^2 f^3)(e^{-3} f^{-5})$, expressing the answer with positive indices only

11. $\dfrac{(a^3 b^{1/2} c^{-1/2})(ab)^{1/3}}{(\sqrt{a^3}\sqrt{b}\,c)}$

6.3 Brackets and factorization

When two or more terms in an algebraic expression contain a common factor, then this factor can be shown outside of a **bracket**. For example

$$ab + ac = a(b + c)$$

which is simply the reverse of law (v) of algebra on page 36, and

$$6px + 2py - 4pz = 2p(3x + y - 2z)$$

This process is called **factorization**.

Problem 26. Remove the brackets and simplify the expression $(3a + b) + 2(b + c) - 4(c + d)$

Both b and c in the second bracket have to be multiplied by 2, and c and d in the third bracket by -4 when the brackets are removed. Thus:

$$(3a + b) + 2(b + c) - 4(c + d)$$
$$= 3a + b + 2b + 2c - 4c - 4d$$

Collecting similar terms together gives: $\mathbf{3a + 3b - 2c - 4d}$

Problem 27. Simplify $a^2 - (2a - ab) - a(3b + a)$

When the brackets are removed, both $2a$ and $-ab$ in the first bracket must be multiplied by -1 and both $3b$ and a in the second bracket by $-a$. Thus

$$a^2 - (2a - ab) - a(3b + a)$$
$$= a^2 - 2a + ab - 3ab - a^2$$

Collecting similar terms together gives: $-2a - 2ab$.

Since $-2a$ is a common factor, the answer can be expressed as $-2a(1 + b)$

Problem 28. Simplify $(a + b)(a - b)$

Each term in the second bracket has to be multiplied by each term in the first bracket. Thus:

$$(a + b)(a - b) = a(a - b) + b(a - b)$$
$$= a^2 - ab + ab - b^2 = a^2 - b^2$$

Alternatively

$$
\begin{array}{r}
a + b \\
a - b \\
\hline
\end{array}
$$

Multiplying by $a \rightarrow$ $\quad a^2 + ab$
Multiplying by $-b \rightarrow$ $\quad -ab - b^2$

$$
\begin{array}{r}
\hline
\end{array}
$$

Adding gives: $\quad a^2 \qquad - b^2$

$$
\begin{array}{r}
\hline
\end{array}
$$

Problem 29. Remove the brackets from the expression $(x - 2y)(3x + y^2)$

$$(x - 2y)(3x + y^2) = x(3x + y^2) - 2y(3x + y^2)$$
$$= 3x^2 + xy^2 - 6xy - 2y^3$$

Problem 30. Simplify $(2x - 3y)^2$

$$(2x - 3y)^2 = (2x - 3y)(2x - 3y)$$
$$= 2x(2x - 3y) - 3y(2x - 3y)$$
$$= 4x^2 - 6xy - 6xy + 9y^2$$
$$= 4x^2 - 12xy + 9y^2$$

Alternatively,

$$
\begin{array}{r}
2x - 3y \\
2x - 3y \\
\hline
\end{array}
$$

Multiplying by $2x \rightarrow$ $\quad 4x^2 - 6xy$
Multiplying by $-3y \rightarrow$ $\qquad - 6xy + 9y^2$

$$
\begin{array}{r}
\hline
\end{array}
$$

Adding gives: $\quad 4x^2 - 12xy + 9y^2$

$$
\begin{array}{r}
\hline
\end{array}
$$

Problem 31. Remove the brackets from the expression

$$2[p^2 - 3(q + r) + q^2]$$

In this problem there are two brackets and the 'inner' one is removed first.

Hence $2[p^2 - 3(q + r) + q^2] = 2[p^2 - 3q - 3r + q^2]$
$$= 2p^2 - 6q - 6r + 2q^2$$

Problem 32. Remove the brackets and simplify the expression: $2a - [3\{2(4a - b) - 5(a + 2b)\} + 4a]$

Removing the innermost brackets gives:

$$2a - [3\{8a - 2b - 5a - 10b\} + 4a]$$

Collecting together similar terms gives:

$$2a - [3\{3a - 12b\} + 4a]$$

Removing the 'curly' brackets gives:

$$2a - [9a - 36b + 4a]$$

Collecting together similar terms gives:

$$2a - [13a - 36b]$$

Removing the outer brackets gives:

$$2a - 13a + 36b$$

i.e. $-11a + 36b$ or $36b - 11a$ (see law (iii), page 36)

Problem 33. Simplify $x(2x - 4y) - 2x(4x + y)$

Removing brackets gives:

$$2x^2 - 4xy - 8x^2 - 2xy$$

Collecting together similar terms gives:

$$- 6x^2 - 6xy$$

Factorizing gives:

$$-6x(x + y)$$

since $-6x$ is common to both terms.

Problem 34. Factorize (a) $xy - 3xz$ (b) $4a^2 + 16ab^3$ (c) $3a^2b - 6ab^2 + 15ab$

For each part of this problem, the HCF of the terms will become one of the factors. Thus:

(a) $xy - 3xz = x(y - 3z)$

(b) $4a^2 + 16ab^3 = 4a(a + 4b^3)$

(c) $3a^2b - 6ab^2 + 15ab = 3ab(a - 2b + 5)$

Problem 35. Factorize $ax - ay + bx - by$

The first two terms have a common factor of a and the last two terms a common factor of b. Thus:

$$ax - ay + bx - by = a(x - y) + b(x - y)$$

The two newly formed terms have a common factor of $(x - y)$. Thus:

$$a(x - y) + b(x - y) = (x - y)(a + b)$$

Problem 36. Factorize $2ax - 3ay + 2bx - 3by$

---

a is a common factor of the first two terms and *b* a common factor of the last two terms. Thus:

$$2ax - 3ay + 2bx - 3by = a(2x - 3y) + b(2x - 3y)$$

$(2x - 3y)$ is now a common factor thus:

$$a(2x - 3y) + b(2x - 3y) = (2x - 3y)(a + b)$$

Alternatively, $2x$ is a common factor of the original first and third terms and $-3y$ is a common factor of the second and fourth terms. Thus:

$$2ax - 3ay + 2bx - 3by = 2x(a + b) - 3y(a + b)$$

$(a + b)$ is now a common factor thus:

$$2x(a + b) - 3y(a + b) = (a + b)(2x - 3y)$$

as before

Problem 37. Factorize $x^3 + 3x^2 - x - 3$

---

x^2 is a common factor of the first two terms, thus:

$$x^3 + 3x^2 - x - 3 = x^2(x + 3) - x - 3$$

-1 is a common factor of the last two terms, thus:

$$x^2(x + 3) - x - 3 = x^2(x + 3) - 1(x + 3)$$

$(x + 3)$ is now a common factor, thus:

$$x^2(x + 3) - 1(x + 3) = (x + 3)(x^2 - 1)$$

Now try the following exercise

Exercise 22   Further problems on brackets and factorization (Answers on page 255)

In Problems 1 to 13, remove the brackets and simplify where possible:

1. $(x + 2y) + (2x - y)$

2. $(4a + 3y) - (a - 2y)$

3. $2(x - y) - 3(y - x)$

4. $2x^2 - 3(x - xy) - x(2y - x)$

5. $2(p + 3q - r) - 4(r - q + 2p) + p$

6. $(a + b)(a + 2b)$

7. $(p + q)(3p - 2q)$

8. (i) $(x - 2y)^2$ (ii) $(3a - b)^2$

9. $3a(b + c) + 4c(a - b)$

10. $2x + [y - (2x + y)]$

11. $3a + 2[a - (3a - 2)]$

12. $2 - 5[a(a - 2b) - (a - b)^2]$

13. $24p - [2(3(5p - q) - 2(p + 2q)) + 3q]$

In Problems 14 to 17, factorize:

14. (i) $pb + 2pc$ (ii) $2q^2 + 8qn$

15. (i) $21a^2b^2 - 28ab$ (ii) $2xy^2 + 6x^2y + 8x^3y$

16. (i) $ay + by + a + b$ (ii) $px + qx + py + qy$

17. (i) $ax - ay + bx - by$ (ii) $2ax + 3ay - 4bx - 6by$

6.4   Fundamental laws and precedence

The **laws of precedence** which apply to arithmetic also apply to algebraic expressions. The order is **B**rackets, **O**f, **D**ivision, **M**ultiplication, **A**ddition and **S**ubtraction (i.e. **BODMAS**)

Problem 38. Simplify $2a + 5a \times 3a - a$

Multiplication is performed before addition and subtraction thus:

$$2a + 5a \times 3a - a = 2a + 15a^2 - a$$
$$= a + 15a^2 = a(1 + 15a)$$

Problem 39. Simplify $(a + 5a) \times 2a - 3a$

The order of precedence is brackets, multiplication, then subtraction. Hence

$$(a + 5a) \times 2a - 3a = 6a \times 2a - 3a = 12a^2 - 3a$$
$$= 3a(4a - 1)$$

Problem 40. Simplify $a + 5a \times (2a - 3a)$

The order of precedence is brackets, multiplication, then subtraction. Hence

$$a + 5a \times (2a - 3a) = a + 5a \times -a = a + -5a^2$$
$$= a - 5a^2 = a(1 - 5a)$$

Problem 41. Simplify $a \div 5a + 2a - 3a$

The order of precedence is division, then addition and subtraction. Hence

$$a \div 5a + 2a - 3a = \frac{a}{5a} + 2a - 3a$$
$$= \frac{1}{5} + 2a - 3a = \frac{1}{5} - a$$

Problem 42.   Simplify $a \div (5a + 2a) - 3a$

The order of precedence is brackets, division and subtraction. Hence

$$a \div (5a + 2a) - 3a = a \div 7a - 3a$$

$$= \frac{a}{7a} - 3a = \frac{1}{7} - 3a$$

Problem 43.   Simplify $a \div (5a + 2a - 3a)$

The order of precedence is brackets, then division. Hence:

$$a \div (5a + 2a - 3a) = a \div 4a = \frac{a}{4a} = \frac{1}{4}$$

Problem 44.   Simplify $3c + 2c \times 4c + c \div 5c - 8c$

The order of precedence is division, multiplication, addition and subtraction. Hence:

$$3c + 2c \times 4c + c \div 5c - 8c = 3c + 2c \times 4c + \frac{c}{5c} - 8c$$

$$= 3c + 8c^2 + \frac{1}{5} - 8c$$

$$= 8c^2 - 5c + \frac{1}{5} \quad \text{or}$$

$$c(8c - 5) + \frac{1}{5}$$

Problem 45.   Simplify $(3c + 2c)4c + c \div 5c - 8c$

The order of precedence is brackets, division, multiplication, addition and subtraction. Hence

$$(3c + 2c)4c + c \div 5c - 8c = 5c \times 4c + c \div 5c - 8c$$

$$= 5c \times 4c + \frac{c}{5c} - 8c$$

$$= 20c^2 + \frac{1}{5} - 8c \quad \text{or}$$

$$4c(5c - 2) + \frac{1}{5}$$

Problem 46.   Simplify $3c + 2c \times 4c + c \div (5c - 8c)$

The order of precedence is brackets, division, multiplication and addition. Hence:

$$3c + 2c \times 4c + c \div (5c - 8c) = 3c + 2c \times 4c + c \div -3c$$

$$= 3c + 2c \times 4c + \frac{c}{-3c}$$

Now $\dfrac{c}{-3c} = \dfrac{1}{-3}$

Multiplying numerator and denominator by -1 gives

$$\frac{1 \times -1}{-3 \times -1} \quad \text{i.e.} \quad -\frac{1}{3}$$

Hence:

$$3c + 2c \times 4c + \frac{c}{-3c} = 3c + 2c \times 4c - \frac{1}{3}$$

$$= 3c + 8c^2 - \frac{1}{3} \quad \text{or}$$

$$c(3 + 8c) - \frac{1}{3}$$

Problem 47.   Simplify $(3c + 2c)(4c + c) \div (5c - 8c)$

The order of precedence is brackets, division and multiplication. Hence

$$(3c + 2c)(4c + c) \div (5c - 8c) = 5c \times 5c \div -3c$$

$$= 5c \times \frac{5c}{-3c}$$

$$= 5c \times -\frac{5}{3} = -\frac{25}{3}c$$

Problem 48.   Simplify $(2a - 3) \div 4a + 5 \times 6 - 3a$

The bracket around the $(2a - 3)$ shows that both $2a$ and -3 have to be divided by $4a$, and to remove the bracket the expression is written in fraction form. Hence:

$$(2a - 3) \div 4a + 5 \times 6 - 3a = \frac{2a - 3}{4a} + 5 \times 6 - 3a$$

$$= \frac{2a - 3}{4a} + 30 - 3a$$

$$= \frac{2a}{4a} - \frac{3}{4a} + 30 - 3a$$

$$= \frac{1}{2} - \frac{3}{4a} + 30 - 3a$$

$$= 30\frac{1}{2} - \frac{3}{4a} - 3a$$

Problem 49.   Simplify $\frac{1}{3}$ of $3p + 4p(3p - p)$

Applying BODMAS, the expression becomes

$$\tfrac{1}{3} \text{ of } 3p + 4p \times 2p$$

and changing 'of' to '×' gives:

$$\tfrac{1}{3} \times 3p + 4p \times 2p$$

i.e. $p + 8p^2$ or $p(1 + 8p)$

Now try the following exercise

Exercise 23   Further problems on fundamental laws and precedence (Answers on page 255)

In Problems 1 to 12, simplify:

1. $2x \div 4x + 6x$ 2. $2x \div (4x + 6x)$

3. $3a - 2a \times 4a + a$ 4. $(3a - 2a)4a + a$

5. $3a - 2a(4a + a)$ 6. $2y + 4 \div 6y + 3 \times 4 - 5y$

7. $(2y + 4) \div 6y + 3 \times 4 - 5y$

8. $2y + 4 \div 6y + 3(4 - 5y)$

9. $3 \div y + 2 \div y + 1$

10. $p^2 - 3pq \times 2p \div 6q + pq$

11. $(x + 1)(x - 4) \div (2x + 2)$

12. $\frac{1}{4}$ of $2y + 3y(2y - y)$

6.5   Direct and inverse proportionality

An expression such as $y = 3x$ contains two variables. For every value of x there is a corresponding value of y. The variable x is called the **independent variable** and y is called the **dependent variable**.

When an increase or decrease in an independent variable leads to an increase or decrease of the same proportion in the dependent variable this is termed **direct proportion**. If $y = 3x$ then y is directly proportional to x, which may be written as $y \propto x$ or $y = kx$, where k is called the **coefficient of proportionality** (in this case, k being equal to 3).

When an increase in an independent variable leads to a decrease of the same proportion in the dependent variable (or vice versa) this is termed **inverse proportion**. If y is inversely proportional to x then $y \propto \frac{1}{x}$ or $y = k/x$. Alternatively, $k = xy$, that is, for inverse proportionality the product of the variables is constant.

Examples of laws involving direct and inverse proportional in science include:

(i) **Hooke's law**, which states that within the elastic limit of a material, the strain ε produced is directly proportional to the stress, σ, producing it, i.e. $\varepsilon \propto \sigma$ or $\varepsilon = k\sigma$.

(ii) **Charles's law**, which states that for a given mass of gas at constant pressure the volume V is directly proportional to its thermodynamic temperature T, i.e. $V \propto T$ or $V = kT$.

(iii) **Ohm's law**, which states that the current I flowing through a fixed resistor is directly proportional to the applied voltage V, i.e. $I \propto V$ or $I = kV$.

(iv) **Boyle's law**, which states that for a gas at constant temperature, the volume V of a fixed mass of gas is inversely proportional to its absolute pressure p, i.e. $p \propto (1/V)$ or $p = k/V$, i.e. $pV = k$

Problem 50. If y is directly proportional to x and $y = 2.48$ when $x = 0.4$, determine (a) the coefficient of proportionality and (b) the value of y when $x = 0.65$

(a) $y \propto x$, i.e. $y = kx$. If $y = 2.48$ when $x = 0.4$, $2.48 = k(0.4)$

Hence the coefficient of proportionality,

$$k = \frac{2.48}{0.4} = \textbf{6.2}$$

(b) $y = kx$, hence, when $x = 0.65$, $y = (6.2)(0.65) = \textbf{4.03}$

Problem 51. Hooke's law states that stress σ is directly proportional to strain ε within the elastic limit of a material. When, for mild steel, the stress is 25×10^6 pascals, the strain is 0.000125. Determine (a) the coefficient of proportionality and (b) the value of strain when the stress is 18×10^6 pascals.

(a) $\sigma \propto \varepsilon$, i.e. $\sigma = k\varepsilon$, from which $k = \sigma/\varepsilon$. Hence the coefficient of proportionality,

$$k = \frac{25 \times 10^6}{0.000125} = \textbf{200} \times \textbf{10}^9 \textbf{ pascals}$$

(The coefficient of proportionality k in this case is called Young's Modulus of Elasticity)

(b) Since $\sigma = k\varepsilon$, $\varepsilon = \sigma/k$

Hence when $\sigma = 18 \times 10^6$, strain

$$\varepsilon = \frac{18 \times 10^6}{200 \times 10^9} = \textbf{0.00009}$$

Problem 52. The electrical resistance R of a piece of wire is inversely proportional to the cross-sectional area A. When $A = 5$ mm^2, $R = 7.02$ ohms. Determine (a) the coefficient of proportionality and (b) the cross-sectional area when the resistance is 4 ohms.

(a) $R \propto \frac{1}{A}$, i.e. $R = k/A$ or $k = RA$. Hence, when $R = 7.2$ and $A = 5$, the coefficient of proportionality, $k = (7.2)(5) = \textbf{36}$

(b) Since $k = RA$ then $A = k/R$

When $R = 4$, the cross sectional area, $A = \dfrac{36}{4} = \textbf{9 mm}^2$

Problem 53. Boyle's law states that at constant temperature, the volume V of a fixed mass of gas is inversely proportional to its absolute pressure p. If a gas occupies a volume of $0.08\,\text{m}^3$ at a pressure of 1.5×10^6 pascals determine (a) the coefficient of proportionality and (b) the volume if the pressure is changed to 4×10^6 pascals..

(a) $V \propto \dfrac{1}{p}$, i.e. $V = k/p$ or $k = pV$

Hence the coefficient of proportionality,

$$k = (1.5 \times 10^6)(0.08) = \mathbf{0.12 \times 10^6}$$

(b) Volume $V = \dfrac{k}{p} = \dfrac{0.12 \times 10^6}{4 \times 10^6} = \mathbf{0.03\,\text{m}^3}$

Now try the following exercise

Exercise 24   Further problems on direct and inverse proportionality (Answers on page 255)

1. If p is directly proportional to q and $p = 37.5$ when $q = 2.5$, determine (a) the constant of proportionality and (b) the value of p when q is 5.2

2. Charles's law states that for a given mass of gas at constant pressure the volume is directly proportional to its thermodynamic temperature. A gas occupies a volume of 2.25 litres at 300 K. Determine (a) the constant of proportionality, (b) the volume at 420 K and (c) the temperature when the volume is 2.625 litres.

3. Ohm's law states that the current flowing in a fixed resistor is directly proportional to the applied voltage. When 30 volts is applied across a resistor the current flowing through the resistor is 2.4×10^{-3} amperes. Determine (a) the constant of proportionality, (b) the current when the voltage is 52 volts and (c) the voltage when the current is 3.6×10^{-3} amperes.

4. If y is inversely proportional to x and $y = 15.3$ when $x = 0.6$, determine (a) the coefficient of proportionality, (b) the value of y when x is 1.5, and (c) the value of x when y is 27.2

5. Boyle's law states that for a gas at constant temperature, the volume of a fixed mass of gas is inversely proportional to its absolute pressure. If a gas occupies a volume of $1.5\,\text{m}^3$ at a pressure of 200×10^3 pascals, determine (a) the constant of proportionality, (b) the

volume when the pressure is 800×10^3 pascals and (c) the pressure when the volume is $1.25\,\text{m}^3$.

Assignment 3

This assignment covers the material contained in Chapters 5 and 6. The marks for each question are shown in brackets at the end of each question

1. Convert the following binary numbers to decimal form:
   (a) 1101   (b) 101101.0101   (5)

2. Convert the following decimal number to binary form:
   (a) 27   (b) 44.1875   (6)

3. Convert the following denary numbers to binary, via octal:
   (a) 479   (b) 185.2890625   (7)

4. Convert (a) $5F_{16}$ into its decimal equivalent

   (b) 132_{10} into its hexadecimal equivalent

   (c) 110101011_2 into its hexadecimal equivalent   (6)

5. Evaluate $3xy^2z^3 - 2yz$ when $x = \frac{4}{3}$, $y = 2$ and $z = \frac{1}{2}$ (3)

6. Simplify the following:

   (a) $(2a + 3b)(x - y)$ (b) $\dfrac{8a^2b\sqrt{c^3}}{(2a^2)\sqrt{b}\sqrt{c}}$

   (c) $\dfrac{xy^2 + x^2y}{xy}$ (d) $3x + 4 \div 2x + 5 \times 2 - 4x$ (10)

7. Remove the brackets in the following expressions and simplify:

   (a) $3b^2 - 2(a - a^2b) - b(2a^2 - b)$ (b) $(2x - y)^2$

   (c) $4ab - [3\{2(4a - b) + b(2 - a)\}]$ (7)

8. Factorize $3x^2y + 9xy^2 + 6xy^3$ (2)

9. If x is inversely proportional to y and $x = 12$ when $y = 0.4$, determine (a) the value of x when y is 3, and (b) the value of y when $x = 2$ (4)

7

Simple equations

7.1 Expressions, equations and identities

$(3x - 5)$ is an example of an **algebraic expression**, whereas $3x - 5 = 1$ is an example of an **equation** (i.e. it contains an 'equals' sign).

An equation is simply a statement that two quantities are equal. For example, $1\,\text{m} = 1000\,\text{mm}$ or $F = \frac{9}{5}C + 32$ or $y = mx + c$.

An **identity** is a relationship which is true for all values of the unknown, whereas an equation is only true for particular values of the unknown. For example, $3x - 5 = 1$ is an equation, since it is only true when $x = 2$, whereas $3x \equiv 8x - 5x$ is an identity since it is true for all values of x.

(Note '$\equiv$' means 'is identical to').

Simple linear equations (or equations of the first degree) are those in which an unknown quantity is raised only to the power 1.

To **'solve an equation'** means 'to find the value of the unknown'.

Any arithmetic operation may be applied to an equation **as long as the equality of the equation is maintained**.

7.2 Worked problems on simple equations

Problem 1. Solve the equation $4x = 20$

Dividing each side of the equation by 4 gives: $\dfrac{4x}{4} = \dfrac{20}{4}$

(Note that the same operation has been applied to both the left-hand side (LHS) and the right-hand side (RHS) of the equation so the equality has been maintained) Cancelling gives: $x = 5$, which is the solution to the equation.

Solutions to simple equations should always be checked and this is accomplished by substituting the solution into the original equation. In this case, LHS $= 4(5) = 20 =$ RHS.

Problem 2. Solve $\dfrac{2x}{5} = 6$

The LHS is a fraction and this can be removed by multiplying both sides of the equation by 5. Hence $5\left(\dfrac{2x}{5}\right) = 5(6)$

Cancelling gives: $2x = 30$

Dividing both sides of the equation by 2 gives:

$$\frac{2x}{2} = \frac{30}{2} \quad \text{i.e.} \quad x = 15$$

Problem 3. Solve $a - 5 = 8$

Adding 5 to both sides of the equation gives:

$$a - 5 + 5 = 8 + 5$$

i.e. $\qquad a = 13$

The result of the above procedure is to move the '-5' from the LHS of the original equation, across the equals sign, to the RHS, but the sign is changed to $+$

Problem 4. Solve $x + 3 = 7$

Subtracting 3 from both sides of the equation gives:

$$x + 3 - 3 = 7 - 3$$

i.e. $\qquad x = 4$

The result of the above procedure is to move the '$+3$' from the LHS of the original equation, across the equals sign, to the RHS, but the sign is changed to $-$. Thus a term can be moved from one side of an equation to the other as long as a change in sign is made.

Problem 5. Solve $6x + 1 = 2x + 9$

In such equations the terms containing x are grouped on one side of the equation and the remaining terms grouped on the other side of the equation. As in Problems 3 and 4, changing from one side of an equation to the other must be accompanied by a change of sign.

Thus since $\qquad 6x + 1 = 2x + 9$

then $\qquad 6x - 2x = 9 - 1$

$$4x = 8$$

$$\frac{4x}{4} = \frac{8}{4}$$

i.e. $\qquad x = 2$

Check: LHS of original equation $= 6(2) + 1 = 13$
$\qquad$ RHS of original equation $= 2(2) + 9 = 13$

Hence the solution $x = 2$ is correct.

Problem 6. Solve $4 - 3p = 2p - 11$

In order to keep the p term positive the terms in p are moved to the RHS and the constant terms to the LHS.

Hence $\qquad 4 + 11 = 2p + 3p$

$$15 = 5p$$

$$\frac{15}{5} = \frac{5p}{5}$$

Hence $\qquad 3 = p \quad$ or $\quad p = 3$

Check: LHS $= 4 - 3(3) = 4 - 9 = -5$
$\qquad$ RHS $= 2(3) - 11 = 6 - 11 = -5$

Hence the solution $p = 3$ is correct.

If, in this example, the unknown quantities had been grouped initially on the LHS instead of the RHS then:

$$-3p - 2p = -11 - 4$$

i.e. $\qquad -5p = -15$

$$\frac{-5p}{-5} = \frac{-15}{-5}$$

and $\qquad p = 3, \text{ as before}$

It is often easier, however, to work with positive values where possible.

Problem 7. Solve $3(x - 2) = 9$

Removing the bracket gives: $\quad 3x - 6 = 9$

Rearranging gives: $\qquad\qquad\qquad 3x = 9 + 6$

$$3x = 15$$

$$\frac{3x}{3} = \frac{15}{3}$$

i.e. $\qquad\qquad\qquad\qquad\qquad x = 5$

Check: LHS $= 3(5 - 2) = 3(3) = 9 = $ RHS

Hence the solution $x = 5$ is correct.

Problem 8. Solve $4(2r - 3) - 2(r - 4) = 3(r - 3) - 1$

Removing brackets gives:

$$8r - 12 - 2r + 8 = 3r - 9 - 1$$

Rearranging gives: $\qquad 8r - 2r - 3r = -9 - 1 + 12 - 8$

i.e. $\qquad\qquad\qquad\qquad 3r = -6$

$$r = \frac{-6}{3} = -2$$

Check: LHS $= 4(-4 - 3) - 2(-2 - 4) = -28 + 12 = -16$
$\qquad$ RHS $= 3(-2 - 3) - 1 = -15 - 1 = -16$

Hence the solution $r = -2$ is correct.

Now try the following exercise

**Exercise 25 Further problems on simple equations
(Answers on page 255)**

Solve the following equations:

1. $2x + 5 = 7$
2. $8 - 3t = 2$
3. $\frac{2}{3}c - 1 = 3$
4. $2x - 1 = 5x + 11$
5. $7 - 4p = 2p - 3$
6. $2.6x - 1.3 = 0.9x + 0.4$
7. $2a + 6 - 5a = 0$
8. $3x - 2 - 5x = 2x - 4$
9. $20d - 3 + 3d = 11d + 5 - 8$
10. $2(x - 1) = 4$
11. $16 = 4(t + 2)$
12. $5(f - 2) - 3(2f + 5) + 15 = 0$
13. $2x = 4(x - 3)$
14. $6(2 - 3y) - 42 = -2(y - 1)$
15. $2(3g - 5) - 5 = 0$
16. $4(3x + 1) = 7(x + 4) - 2(x + 5)$

17. $10 + 3(r - 7) = 16 - (r + 2)$

18. $8 + 4(x - 1) - 5(x - 3) = 2(5 - 2x)$

7.3 Further worked problems on simple equations

Problem 9. Solve $\dfrac{3}{x} = \dfrac{4}{5}$

The lowest common multiple (LCM) of the denominators, i.e. the lowest algebraic expression that both x and 5 will divide into, is $5x$.

Multiplying both sides by $5x$ gives:

$$5x\left(\frac{3}{x}\right) = 5x\left(\frac{4}{5}\right)$$

Cancelling gives:

$$15 = 4x \qquad\qquad (1)$$

$$\frac{15}{4} = \frac{4x}{4}$$

i.e. $\qquad x = \dfrac{15}{4} \quad$ or $\quad 3\dfrac{3}{4}$

Check: LHS $= \dfrac{3}{3\frac{3}{4}} = \dfrac{3}{\frac{15}{4}} = 3\left(\dfrac{4}{15}\right) = \dfrac{12}{15} = \dfrac{4}{5} =$ RHS

(Note that when there is only one fraction on each side of an equation, 'cross-multiplication' can be applied. In this example, if $\dfrac{3}{x} = \dfrac{4}{5}$ then $(3)(5) = 4x$, which is a quicker way of arriving at equation (1) above.)

Problem 10. Solve $\dfrac{2y}{5} + \dfrac{3}{4} + 5 = \dfrac{1}{20} - \dfrac{3y}{2}$

The LCM of the denominators is 20. Multiplying each term by 20 gives:

$$20\left(\frac{2y}{5}\right) + 20\left(\frac{3}{4}\right) + 20(5) = 20\left(\frac{1}{20}\right) - 20\left(\frac{3y}{2}\right)$$

Cancelling gives:

$$4(2y) + 5(3) + 100 = 1 - 10(3y)$$

i.e. $\qquad 8y + 15 + 100 = 1 - 30y$

Rearranging gives:

$$8y + 30y = 1 - 15 - 100$$

$$38y = -114$$

$$y = \frac{-114}{38} = -3$$

Check: LHS $= \dfrac{2(-3)}{5} + \dfrac{3}{4} + 5 = \dfrac{-6}{5} + \dfrac{3}{4} + 5$

$$= \frac{-9}{20} + 5 = 4\frac{11}{20}$$

RHS $= \dfrac{1}{20} - \dfrac{3(-3)}{2} = \dfrac{1}{20} + \dfrac{9}{2} = 4\dfrac{11}{20}$

Hence the solution $y = -3$ is correct.

Problem 11. Solve $\dfrac{3}{t - 2} = \dfrac{4}{3t + 4}$

By 'cross-multiplication': $\quad 3(3t + 4) = 4(t - 2)$

Removing brackets gives: $\quad 9t + 12 = 4t - 8$

Rearranging gives: $\quad 9t - 4t = -8 - 12$

i.e. $\qquad\qquad 5t = -20$

$$t = \frac{-20}{5} = -4$$

Check: LHS $= \dfrac{3}{-4 - 2} = \dfrac{3}{-6} = -\dfrac{1}{2}$

RHS $= \dfrac{4}{3(-4) + 4} = \dfrac{4}{-12 + 4} = \dfrac{4}{-8} = -\dfrac{1}{2}$

Hence the solution $t = -4$ is correct.

Problem 12. Solve $\sqrt{x} = 2$

[$\sqrt{x} = 2$ is not a 'simple equation' since the power of x is $\frac{1}{2}$ i.e. $\sqrt{x} = x^{(1/2)}$; however, it is included here since it occurs often in practise].

Wherever square root signs are involved with the unknown quantity, both sides of the equation must be squared. Hence

$$(\sqrt{x})^2 = (2)^2$$

i.e. $\qquad x = 4$

Problem 13. Solve $2\sqrt{2} = 8$

To avoid possible errors it is usually best to arrange the term containing the square root on its own. Thus

$$\frac{2\sqrt{d}}{2} = \frac{8}{2}$$

i.e. $\qquad \sqrt{d} = 4$

Squaring both sides gives: $d = 16$, which may be checked in the original equation.

Problem 14. Solve $\left(\dfrac{\sqrt{b}+3}{\sqrt{b}}\right)=2$

To remove the fraction each term is multiplied by $\sqrt{b}$. Hence:

$$\sqrt{b}\left(\dfrac{\sqrt{b}+3}{\sqrt{b}}\right)=\sqrt{b}(2)$$

Cancelling gives: $\qquad\sqrt{b}+3=2\sqrt{b}$

Rearranging gives: $\qquad 3=2\sqrt{b}-\sqrt{b}=\sqrt{b}$

Squaring both sides gives: $\quad\mathbf{9=b}$

Check: LHS $=\dfrac{\sqrt{9}+3}{\sqrt{9}}=\dfrac{3+3}{3}=\dfrac{6}{3}=2=$ RHS

Problem 15. Solve $x^2=25$

This problem involves a square term and thus is not a simple equation (it is, in fact, a quadratic equation). However the solution of such an equation is often required and is therefore included here for completeness. Whenever a square of the unknown is involved, the square root of both sides of the equation is taken. Hence

$$\sqrt{x^2}=\sqrt{25}$$

i.e. $\qquad x=5$

However, $x=-5$ is also a solution of the equation because $(-5)\times(-5)=+25$ Therefore, whenever the square root of a number is required there are always two answers, one positive, the other negative.

The solution of $x^2=25$ is thus written as $\mathbf{x=\pm5}$

Problem 16. Solve $\dfrac{15}{4t^2}=\dfrac{2}{3}$

'Cross-multiplying' gives: $\quad 15(3)=2(4t^2)$

i.e. $\qquad\qquad\qquad 45=8t^2$

$$\dfrac{45}{8}=t^2$$

i.e. $\qquad\qquad\qquad t^2=5.625$

Hence $t=\sqrt{5.625}=\pm\mathbf{2.372}$, correct to 4 significant figures.

Now try the following exercise

Exercise 26   Further problems on simple equations (Answers on page 255)

Solve the following equations:

1. $\dfrac{1}{5}d+3=4$

2. $2+\dfrac{3}{4}y=1+\dfrac{2}{3}y+\dfrac{5}{6}$

3. $\dfrac{1}{4}(2x-1)+3=\dfrac{1}{2}$

4. $\dfrac{1}{5}(2f-3)+\dfrac{1}{6}(f-4)+\dfrac{2}{15}=0$

5. $\dfrac{1}{3}(3m-6)-\dfrac{1}{4}(5m+4)+\dfrac{t}{5}(2m-9)=-3$

6. $\dfrac{x}{3}-\dfrac{x}{5}=2$

7. $1-\dfrac{y}{3}=3+\dfrac{y}{3}-\dfrac{y}{6}$

8. $\dfrac{2}{a}=\dfrac{3}{8}$

9. $\dfrac{1}{3n}+\dfrac{1}{4n}=\dfrac{7}{24}$

10. $\dfrac{x+3}{4}=\dfrac{x-3}{5}+2$

11. $\dfrac{3t}{20}=\dfrac{6-t}{12}+\dfrac{2t}{15}-\dfrac{3}{2}$

12. $\dfrac{y}{5}+\dfrac{7}{20}=\dfrac{5-y}{4}$

13. $\dfrac{v-2}{2v-3}=\dfrac{1}{3}$

14. $\dfrac{2}{a-3}=\dfrac{3}{2a+1}$

15. $\dfrac{x}{4}-\dfrac{x+6}{5}=\dfrac{x+3}{2}$

16. $3\sqrt{t}=9$

17. $2\sqrt{y}=5$

18. $4=\sqrt{\left(\dfrac{3}{a}\right)}+3$

19. $\dfrac{3\sqrt{x}}{1-\sqrt{x}}=-6$

20. $10=5\sqrt{\left(\dfrac{x}{2}-1\right)}$

21. $16=\dfrac{t^2}{9}$

22. $\sqrt{\left(\dfrac{y+2}{y-2}\right)}=\dfrac{1}{2}$

23. $\dfrac{6}{a}=\dfrac{2a}{3}$

24. $\dfrac{11}{2}=5+\dfrac{8}{x^2}$

7.4   Practical problems involving simple equations

Problem 17.   A copper wire has a length l of 1.5 km, a resistance R of $5\,\Omega$ and a resistivity of $17.2\times10^{-6}\,\Omega\,\text{mm}$. Find the cross-sectional area, a, of the wire, given that $R=\rho l/a$

Since $\quad R=\rho l/a\quad$ then

$$5\,\Omega=\dfrac{(1.72\times10^{-6}\,\Omega\,\text{mm})\,(1500\times10^3\,\text{mm})}{a}$$

From the units given, a is measured in mm^2.

Thus $\quad 5a=17.2\times10^{-6}\times1500\times10^3$

and $\quad a = \dfrac{17.2 \times 10^{-6} \times 1500 \times 10^3}{5}$

$\qquad = \dfrac{17.2 \times 1500 \times 10^3}{10^6 \times 5} = \dfrac{17.2 \times 15}{10 \times 5} = 5.16$

Hence the cross-sectional area of the wire is 5.16 mm²

Problem 18. A rectangular box with square ends has its length 15 cm greater than its breadth and the total length of its edges is 2.04 m. Find the width of the box and its volume.

Let x cm = width = height of box. Then the length of the box is $(x + 15)$ cm. The length of the edges of the box is $2(4x) + 4(x + 15)$ cm.

Hence $\quad 204 = 2(4x) + 4(x + 15)$

$\qquad 204 = 8x + 4x + 60$

$\qquad 204 - 60 = 12x$

i.e. $\qquad 144 = 12x$

and $\qquad x = 12$ cm

Hence the width of the box is 12 cm.

$\qquad$ **Volume of box** = length × width × height

$\qquad\qquad = (x + 15)(x)(x) = (27)(12)(12)$

$\qquad\qquad = \mathbf{3888\ cm^3}$

Problem 19. The temperature coefficient of resistance α may be calculated from the formula $R_t = R_0(1 + \alpha t)$. Find α given $R_t = 0.928$, $R_0 = 0.8$ and $t = 40$

Since $R_t = R_0(1 + \alpha t)$ then $0.928 = 0.8[1 + \alpha(40)]$

$\qquad\qquad 0.928 = 0.8 + (0.8)(\alpha)(40)$

$\qquad\qquad 0.928 - 0.8 = 32\alpha$

$\qquad\qquad 0.128 = 32\alpha$

Hence $\qquad\qquad \alpha = \dfrac{0.128}{32} = \mathbf{0.004}$

Problem 20. The distance s metres travelled in time t seconds is given by the formula $s = ut + \frac{1}{2}at^2$, where u is the initial velocity in m/s and a is the acceleration in m/s². Find the acceleration of the body if it travels 168 m in 6 s, with an initial velocity of 10 m/s.

$s = ut + \frac{1}{2}at^2$, and $s = 168$, $u = 10$ and $t = 6$

Hence $\qquad 168 = (10)(6) + \frac{1}{2}a(6)^2$

$\qquad\qquad 168 = 60 + 18a$

$\qquad 168 - 60 = 18a$

$\qquad 108 = 18a$

$\qquad a = \dfrac{108}{18} = 6$

Hence the acceleration of the body is 6 m/s²

Problem 21. When three resistors in an electrical circuit are connected in parallel the total resistance R_T is given by:

$$\frac{1}{R_T} = \frac{1}{R_1} + \frac{1}{R_2} + \frac{1}{R_3}$$

Find the total resistance when $R_1 = 5\,\Omega$, $R_2 = 10\,\Omega$ and $R_3 = 30\,\Omega$.

$\dfrac{1}{R_T} = \dfrac{1}{5} + \dfrac{1}{10} + \dfrac{1}{30}$

$\qquad = \dfrac{6 + 3 + 1}{30} = \dfrac{10}{30} = \dfrac{1}{3}$

Taking the reciprocal of both sides gives: $\quad \mathbf{R_T = 3\,\Omega}$

Alternatively, if $\dfrac{1}{R_T} = \dfrac{1}{5} + \dfrac{1}{10} + \dfrac{1}{30}$ the LCM of the denominators is $30\,R_T$

Hence

$$30R_T\left(\frac{1}{R_T}\right) = 30R_T\left(\frac{1}{5}\right) + 30R_T\left(\frac{1}{10}\right)$$
$$+ 30R_T\left(\frac{1}{30}\right)$$

Cancelling gives:

$\qquad 30 = 6R_T + 3R_T + R_T$

$\qquad 30 = 10R_T$

$\qquad R_T = \dfrac{30}{10} = \mathbf{3\,\Omega}$, as above

Now try the following exercise

Exercise 27   Practical problems involving simple equations (Answers on page 255)

1. A formula used for calculating resistance of a cable is $R = (\rho l)/a$. Given $R = 1.25$, $l = 2500$ and $a = 2 \times 10^{-4}$ find the value of ρ.

2. Force F newtons is given by $F = ma$, where m is the mass in kilograms and a is the acceleration in metres per second squared. Find the acceleration when a force of 4 kN is applied to a mass of 500 kg.

3. $PV = mRT$ is the characteristic gas equation. Find the value of m when $P = 100 \times 10^3$, $V = 3.00$, $R = 288$ and $T = 300$

4. When three resistors R_1, R_2 and R_3 are connected in parallel the total resistance R_T is determined from
$$\frac{1}{R_T} = \frac{1}{R_1} + \frac{1}{R_2} + \frac{1}{R_3}$$

   (a) Find the total resistance when $R_1 = 3\,\Omega$, $R_2 = 6\,\Omega$ and $R_3 = 18\,\Omega$.

   (b) Find the value of R_3 given that $R_T = 3\,\Omega$, $R_1 = 5\,\Omega$ and $R_2 = 10\,\Omega$.

5. Five pens and two rulers cost 94p. If a ruler costs 5p more than a pen, find the cost of each.

6. Ohm's law may be represented by $I = V/R$, where I is the current in amperes, V is the voltage in volts and R is the resistance in ohms. A soldering iron takes a current of 0.30 A from a 240 V supply. Find the resistance of the element.

7.5 Further practical problems involving simple equations

Problem 22. The extension x m of an aluminium tie bar of length l m and cross-sectional area A m^2 when carrying a load of F newtons is given by the modulus of elasticity $E = Fl/Ax$. Find the extension of the tie bar (in mm) if $E = 70 \times 10^9$ N/m^2, $F = 20 \times 10^6$ N, $A = 0.1$ m^2 and $l = 1.4$ m

$E = Fl/Ax$, hence $70 \times 10^9 \dfrac{\text{N}}{\text{m}^2} = \dfrac{(20 \times 10^6\,\text{N})(1.4\,\text{m})}{(0.1\,\text{m}^2)(x)}$
$$\text{(the unit of } x \text{ is thus metres)}$$

$$70 \times 10^9 \times 0.1 \times x = 20 \times 10^6 \times 1.4$$
$$x = \frac{20 \times 10^6 \times 1.4}{70 \times 10^9 \times 0.1}$$

Cancelling gives:
$$x = \frac{2 \times 1.4}{7 \times 100}\,\text{m} = \frac{2 \times 1.4}{7 \times 100} \times 1000\,\text{mm}$$

Hence the extension of the tie bar, $x = 4$ mm

Problem 23. Power in a d.c. circuit is given by $P = \dfrac{V^2}{R}$ where V is the supply voltage and R is the circuit resistance. Find the supply voltage if the circuit resistance is 1.25 Ω and the power measured is 320 W

Since $P = \dfrac{V^2}{R}$ then $320 = \dfrac{V^2}{1.25}$
$$(320)(1.25) = V^2$$
i.e. $V^2 = 400$
Supply voltage, $V = \sqrt{400} = \pm 20\,\text{V}$

Problem 24. A painter is paid £4.20 per hour for a basic 36 hour week, and overtime is paid at one and a third times this rate. Determine how many hours the painter has to work in a week to earn £212.80

Basic rate per hour = £4.20;

overtime rate per hour = $1\frac{1}{3} \times £4.20 = £5.60$

Let the number of overtime hours worked = x

Then $(36)(4.20) + (x)(5.60) = 212.80$
$$151.20 + 5.60x = 212.80$$
$$5.60x = 212.80 - 151.20 = 61.60$$
$$x = \frac{61.60}{5.60} = 11$$

Thus 11 hours overtime would have to be worked to earn £212.80 per week.

Hence the total number of hours worked is $36 + 11$, i.e. **47 hours**.

Problem 25. A formula relating initial and final states of pressures, P_1 and P_2, volumes V_1 and V_2, and absolute temperatures, T_1 and T_2, of an ideal gas is $\dfrac{P_1 V_1}{T_1} = \dfrac{P_2 V_2}{T_2}$. Find the value of P_2 given $P_1 = 100 \times 10^3$, $V_1 = 1.0$, $V_2 = 0.266$, $T_1 = 423$ and $T_2 = 293$

Since $\dfrac{P_1 V_1}{T_1} = \dfrac{P_2 V_2}{T_2}$ then $\dfrac{(100 \times 10^3)(1.0)}{423} = \dfrac{P_2(0.266)}{293}$

'Cross-multiplying' gives:
$$(100 \times 10^3)(1.0)(293) = P_2(0.266)(423)$$
$$P_2 = \frac{(100 \times 10^3)(1.0)(293)}{(0.266)(423)}$$

Hence $P_2 = 260 \times 10^3$ or **2.6×10^5**

Problem 26. The stress f in a material of a thick cylinder can be obtained from $\dfrac{D}{d} = \sqrt{\left(\dfrac{f + p}{f - p}\right)}$. Calculate the stress, given that $D = 21.5$, $d = 10.75$ and $p = 1800$

Since $\dfrac{D}{d} = \sqrt{\left(\dfrac{f + p}{f - p}\right)}$ then $\dfrac{21.5}{10.75} = \sqrt{\left(\dfrac{f + 1800}{f - 1800}\right)}$

i.e. $2 = \sqrt{\left(\dfrac{f + 1800}{f - 1800}\right)}$

Squaring both sides gives:

$$4 = \frac{f + 1800}{f - 1800}$$

$$4(f - 1800) = f + 1800$$

$$4f - 7200 = f + 1800$$

$$4f - f = 1800 + 7200$$

$$3f = 9000$$

$$f = \frac{9000}{3} = 3000$$

Hence   **stress, $f = 3000$**

Problem 27.   12 workmen employed on a building site earn between them a total of £2415 per week. Labourers are paid £175 per week and craftsmen are paid £220 per week. How many craftsmen and how many labourers are employed?.

Let the number of craftsmen be c. The number of labourers is therefore $(12 - c)$ The wage bill equation is:

$$220c + 175(12 - c) = 2415$$

$$220c + 2100 - 175c = 2415$$

$$220c - 175c = 2415 - 2100$$

$$45c = 315$$

$$c = \frac{315}{45} = 7$$

Hence there are 7 craftsmen and $(12 - 7)$, i.e. 5 labourers on the site.

Now try the following exercise

Exercise 28   Practical problems involving simple equations (Answers on page 255)

1. A rectangle has a length of 20 cm and a width b cm. When its width is reduced by 4 cm its area becomes 160 cm². Find the original width and area of the rectangle.

2. Given $R_2 = R_1(1 + \alpha t)$, find α given $R_1 = 5.0$, $R_2 = 6.03$ and $t = 51.5$

3. If $v^2 = u^2 + 2as$, find u given $v = 24$, $a = -40$ and $s = 4.05$

4. The relationship between the temperature on a Fahrenheit scale and that on a Celsius scale is given by $F = \frac{9}{5}C + 32$. Express 113°F in degrees Celsius.

5. If $t = 2\pi\sqrt{(w/Sg)}$, find the value of S given $w = 1.219$, $g = 9.81$ and $t = 0.3132$

6. Two joiners and five mates earn £1216 between them for a particular job. If a joiner earns £48 more than a mate, calculate the earnings for a joiner and for a mate.

7. An alloy contains 60% by weight of copper, the remainder being zinc. How much copper must be mixed with 50 kg of this alloy to give an alloy containing 75% copper?

8. A rectangular laboratory has a length equal to one and a half times its width and a perimeter of 40 m. Find its length and width.

8

Transposition of formulae

8.1 Introduction to transposition of formulae

When a symbol other than the subject is required to be calculated it is usual to rearrange the formula to make a new subject. This rearranging process is called **transposing the formula** or **transposition**.

The rules used for transposition of formulae are the same as those used for the solution of simple equations (see Chapter 7) – basically, **that the equality of an equation must be maintained**.

8.2 Worked problems on transposition of formulae

> *Problem 1.* Transpose $p = q + r + s$ to make r the subject.

The aim is to obtain r on its own on the left-hand side (LHS) of the equation. Changing the equation around so that r is on the LHS gives:

$$q + r + s = p \qquad (1)$$

Subtracting $(q + s)$ from both sides of the equation gives:

$$q + r + s - (q + s) = p - (q + s)$$

Thus $\quad q + r + s - q - s = p - q - s$

i.e. $\qquad\qquad\qquad r = p - q - s \qquad (2)$

It is shown with simple equations, that a quantity can be moved from one side of an equation to the other with an appropriate change of sign. Thus equation (2) follows immediately from equation (1) above.

> *Problem 2.* If $a + b = w - x + y$, express x as the subject.

Rearranging gives:

$$w - x + y = a + b \quad \text{and} \quad -x = a + b - w - y$$

Multiplying both sides by -1 gives:

$$(-1)(-x) = (-1)(a + b - w - y)$$

i.e. $\qquad\qquad x = -a - b + w + y$

The result of multiplying each side of the equation by -1 is to change all the signs in the equation.

It is conventional to express answers with positive quantities first. Hence rather than $x = -a - b + w + y$, $x = w + y - a - b$, since the order of terms connected by $+$ and $-$ signs is immaterial.

> *Problem 3.* Transpose $v = f\lambda$ to make λ the subject.

Rearranging gives: $f\lambda = v$

Dividing both sides by f gives: $\dfrac{f\lambda}{f} = \dfrac{v}{f}$, i.e. $\lambda = \dfrac{v}{f}$

> *Problem 4.* When a body falls freely through a height h, the velocity v is given by $v^2 = 2gh$. Express this formula with h as the subject.

Rearranging gives: $2gh = v^2$

Dividing both sides by $2g$ gives: $\dfrac{2gh}{2g} = \dfrac{v^2}{2g}$, i.e. $h = \dfrac{v^2}{2g}$

> *Problem 5.* If $I = \dfrac{V}{R}$, rearrange to make V the subject.

Rearranging gives: $\dfrac{V}{R} = I$

Multiplying both sides by R gives:

$$R\left(\frac{V}{R}\right) = R(I)$$

Hence $V = IR$

Problem 6. Transpose $a = \dfrac{F}{m}$ for m

Rearranging gives: $\dfrac{F}{m} = a$

Multiplying both sides by m gives:

$$m\left(\frac{F}{m}\right) = m(a) \quad \text{i.e.} \quad F = ma$$

Rearranging gives: $ma = F$

Dividing both sides by a gives: $\dfrac{ma}{a} = \dfrac{F}{a}$

i.e. $m = \dfrac{F}{a}$

Problem 7. Rearrange the formula $R = \dfrac{\rho l}{a}$ to make (i) a the subject, and (ii) l the subject.

(i) Rearranging gives: $\dfrac{\rho l}{a} = R$

Multiplying both sides by a gives:

$$a\left(\frac{\rho l}{a}\right) = a(R) \quad \text{i.e.} \quad \rho l = aR$$

Rearranging gives: $aR = \rho l$

Dividing both sides by R gives:

$$\frac{aR}{R} = \frac{\rho l}{R}$$

i.e. $a = \dfrac{\rho l}{R}$

(ii) $\dfrac{\rho l}{a} = R$

Multiplying both sides by a gives:

$$\rho l = aR$$

Dividing both sides by ρ gives:

$$\frac{\rho l}{\rho} = \frac{aR}{\rho}$$

i.e. $l = \dfrac{aR}{\rho}$

Now try the following exercise

Exercise 29   Further problems on transposition of formulae (Answers on page 256)

Make the symbol indicated the subject of each of the formulae shown and express each in its simplest form.

1. $a + b = c - d - e$ (d)

2. $x + 3y = t$ (y)

3. $c = 2\pi r$ (r)

4. $y = mx + c$ (x)

5. $I = PRT$ (T)

6. $I = \dfrac{E}{R}$ (R)

7. $S = \dfrac{a}{1 - r}$ (r)

8. $F = \dfrac{9}{5}C + 32$ (C)

8.3   Further worked problems on transposition of formulae

Problem 8. Transpose the formula $v = u + \dfrac{ft}{m}$, to make f the subject.

Rearranging gives: $u + \dfrac{ft}{m} = v$ and $\dfrac{ft}{m} = v - u$

Multiplying each side by m gives:

$$m\left(\frac{ft}{m}\right) = m(v - u) \quad \text{i.e.} \quad ft = m(v - u)$$

Dividing both sides by t gives:

$$\frac{ft}{t} = \frac{m}{t}(v - u) \quad \text{i.e.} \quad f = \frac{m}{t}(v - u)$$

Problem 9. The final length, l_2 of a piece of wire heated through $\theta°C$ is given by the formula $l_2 = l_1(1 + \alpha\theta)$. Make the coefficient of expansion, α, the subject.

Rearranging gives: $l_1(1 + \alpha\theta) = l_2$

Removing the bracket gives: $l_1 + l_1\alpha\theta = l_2$

Rearranging gives: $l_1\alpha\theta = l_2 - l_1$

Dividing both sides by $l_1\theta$ gives:

$$\frac{l_1\alpha\theta}{l_1\theta} = \frac{l_2 - l_1}{l_1\theta} \quad \text{i.e.} \quad \alpha = \frac{l_2 - l_1}{l_1\theta}$$

Problem 10. A formula for the distance moved by a body is given by $s = \frac{1}{2}(v + u)t$. Rearrange the formula to make u the subject.

Rearranging gives:
$$\frac{1}{2}(v + u)t = s$$

Multiplying both sides by 2 gives: $(v + u)t = 2s$

Dividing both sides by t gives:
$$\frac{(v + u)t}{t} = \frac{2s}{t}$$

i.e.
$$v + u = \frac{2s}{t}$$

Hence
$$u = \frac{2s}{t} - v \quad \text{or} \quad u = \frac{2s - vt}{t}$$

Problem 11. A formula for kinetic energy is $k = \frac{1}{2}mv^2$. Transpose the formula to make v the subject.

Rearranging gives: $\frac{1}{2}mv^2 = k$

Whenever the prospective new subject is a squared term, that term is isolated on the LHS, and then the square root of both sides of the equation is taken.

Multiplying both sides by 2 gives: $mv^2 = 2k$

Dividing both sides by m gives:
$$\frac{mv^2}{m} = \frac{2k}{m}$$

i.e.
$$v^2 = \frac{2k}{m}$$

Taking the square root of both sides gives:

$$\sqrt{v^2} = \sqrt{\left(\frac{2k}{m}\right)}$$

i.e.
$$v = \sqrt{\left(\frac{2k}{m}\right)}$$

Problem 12. In a right angled triangle having sides x, y and hypotenuse z, Pythagoras' theorem states $z^2 = x^2 + y^2$. Transpose the formula to find x.

Rearranging gives: $x^2 + y^2 = z^2$

and
$$x^2 = z^2 - y^2$$

Taking the square root of both sides gives:

$$x = \sqrt{z^2 - y^2}$$

Problem 13. Given $t = 2\pi\sqrt{\dfrac{l}{g}}$, find g in terms of t, l and π.

Whenever the prospective new subject is within a square root sign, it is best to isolate that term on the LHS and then to square both sides of the equation.

Rearranging gives: $2\pi\sqrt{\dfrac{l}{g}} = t$

Dividing both sides by 2π gives: $\sqrt{\dfrac{l}{g}} = \dfrac{t}{2\pi}$

Squaring both sides gives: $\dfrac{l}{g} = \left(\dfrac{t}{2\pi}\right)^2 = \dfrac{t^2}{4\pi^2}$

Cross-multiplying, i.e. multiplying each term by $4\pi^2 g$, gives:
$$4\pi^2 l = gt^2$$

or
$$gt^2 = 4\pi^2 l$$

Dividing both sides by t^2 gives: $\dfrac{gt^2}{t^2} = \dfrac{4\pi^2 l}{t^2}$

i.e.
$$g = \frac{4\pi^2 l}{t^2}$$

Problem 14. The impedance of an a.c. circuit is given by $Z = \sqrt{R^2 + X^2}$. Make the reactance, X, the subject.

Rearranging gives: $\sqrt{R^2 + X^2} = Z$

Squaring both sides gives: $R^2 + X^2 = Z^2$

Rearranging gives: $X^2 = Z^2 - R^2$

Taking the square root of both sides gives:

$$X = \sqrt{Z^2 - R^2}$$

Problem 15. The volume V of a hemisphere is given by $V = \frac{2}{3}\pi r^3$. Find r in terms of V.

Rearranging gives:
$$\frac{2}{3}\pi r^3 = V$$

Multiplying both sides by 3 gives: $2\pi r^3 = 3V$

Dividing both sides by 2π gives:
$$\frac{2\pi r^3}{2\pi} = \frac{3V}{2\pi} \quad \text{i.e.} \quad r^3 = \frac{3V}{2\pi}$$

Taking the cube root of both sides gives:

$$\sqrt[3]{r^3} = \sqrt[3]{\left(\frac{3V}{2\pi}\right)} \quad \text{i.e.} \quad r = \sqrt[3]{\left(\frac{3V}{2\pi}\right)}$$

Now try the following exercise

Exercise 30   Further problems on transposition of formulae (Answers on page 256)

Make the symbol indicated the subject of each of the formulae shown and express each in its simplest form.

1. $y = \dfrac{\lambda(x - d)}{d}$ (x)

2. $A = \dfrac{3(F - f)}{L}$ (f)

3. $y = \dfrac{Ml^2}{8EI}$ (E)

4. $R = R_0(1 + \alpha t)$ (t)

5. $\dfrac{1}{R} = \dfrac{1}{R_1} + \dfrac{1}{R_2}$ (R_2)

6. $I = \dfrac{E - e}{R + r}$ (R)

7. $y = 4ab^2c^2$ (b)

8. $\dfrac{a^2}{x^2} + \dfrac{b^2}{y^2} = 1$ (x)

9. $t = 2\pi\sqrt{\dfrac{l}{g}}$ (l)

10. $v^2 = u^2 + 2as$ (u)

11. $A = \dfrac{\pi R^2 \theta}{360}$ (R)

12. $N = \sqrt{\left(\dfrac{a + x}{y}\right)}$ (a)

13. $Z = \sqrt{R^2 + (2\pi f L)^2}$ (L)

8.4   Harder worked problems on transposition of formulae

Problem 16.   Transpose the formula $p = \dfrac{a^2 x + a^2 y}{r}$ to make a the subject.

Rearranging gives: $\dfrac{a^2 x + a^2 y}{r} = p$

Multiplying both sides by r gives: $a^2 x + a^2 y = rp$

Factorizing the LHS gives: $\qquad a^2(x + y) = rp$

Dividing both sides by $(x + y)$ gives:

$$\frac{a^2(x + y)}{(x + y)} = \frac{rp}{(x + y)} \quad \text{i.e.} \quad a^2 = \frac{rp}{(x + y)}$$

Taking the square root of both sides gives:

$$a = \sqrt{\left(\frac{rp}{x + y}\right)}$$

Problem 17.   Make b the subject of the formula $a = \dfrac{x - y}{\sqrt{bd + be}}$

Rearranging gives: $\dfrac{x - y}{\sqrt{bd + be}} = a$

Multiplying both sides by $\sqrt{bd + be}$ gives:

$$x - y = a\sqrt{bd + be}$$

or $\quad a\sqrt{bd + be} = x - y$

Dividing both sides by a gives: $\sqrt{bd + be} = \dfrac{x - y}{a}$

Squaring both sides gives: $\qquad bd + be = \left(\dfrac{x - y}{a}\right)^2$

Factorizing the LHS gives: $\qquad b(d + e) = \left(\dfrac{x - y}{a}\right)^2$

Dividing both sides by $(d + e)$ gives: $\quad b = \dfrac{\left(\dfrac{x - y}{a}\right)^2}{(d + e)}$

i.e. $\qquad\qquad\qquad\qquad b = \dfrac{(x - y)^2}{a^2(d + e)}$

Problem 18.   If $cd = 3d + e - ad$, express d in terms of a, c and e.

Rearranging to obtain the terms in d on the LHS gives:

$$cd - 3d + ad = e$$

Factorizing the LHS gives:

$$d(c - 3 + a) = e$$

Dividing both sides by $(c - 3 + a)$ gives:

$$d = \frac{e}{c - 3 + a}$$

Problem 19.   If $a = \dfrac{b}{1 + b}$, make b the subject of the formula.

Rearranging gives:

$$\frac{b}{1+b} = a$$

Multiplying both sides by $(1 + b)$ gives: $b = a(1 + b)$

Removing the bracket gives: $b = a + ab$

Rearranging to obtain terms in b on the LHS gives:

$$b - ab = a$$

Factorizing the LHS gives: $b(1 - a) = a$

Dividing both sides by $(1 - a)$ gives:

$$b = \frac{a}{1-a}$$

Problem 20. Transpose the formula $V = \dfrac{Er}{R+r}$ to make r the subject.

Rearranging gives:

$$\frac{Er}{R+r} = V$$

Multiplying both sides by $(R + r)$ gives: $Er = V(R + r)$

Removing the bracket gives: $Er = VR + Vr$

Rearranging to obtain terms in r on the LHS gives:

$$Er - Vr = VR$$

Factorizing gives: $r(E - V) = VR$

Dividing both sides by $(E - V)$ gives:

$$r = \frac{VR}{E - V}$$

Problem 21. Transpose the formula $y = \dfrac{pq^2}{r+q^2} - t$ to make q the subject

---

Rearranging gives:

$$\frac{pq^2}{r+q^2} - t = y$$

and

$$\frac{pq^2}{r+q^2} = y + t$$

Multiplying both sides by $(r + q^2)$ gives:

$$pq^2 = (r + q^2)(y + t)$$

Removing brackets gives: $pq^2 = ry + rt + q^2 y + q^2 t$

Rearranging to obtain terms in q on the LHS gives:

$$pq^2 - q^2 y - q^2 t = ry + rt$$

Factorizing gives: $q^2(p - y - t) = r(y + t)$

Dividing both sides by $(p - y - t)$ gives:

$$q^2 = \frac{r(y + t)}{(p - y - t)}$$

Taking the square root of both sides gives:

$$q = \sqrt{\left(\frac{r(y + t)}{p - y - t}\right)}$$

Problem 22. Given that $\dfrac{D}{d} = \sqrt{\left(\dfrac{f+p}{f-p}\right)}$, express p in terms of D, d and f.

Rearranging gives:

$$\sqrt{\left(\frac{f+p}{f-p}\right)} = \frac{D}{d}$$

Squaring both sides gives:

$$\left(\frac{f+p}{f-p}\right) = \frac{D^2}{d^2}$$

Cross-multiplying, i.e. multiplying each term by $d^2(f - p)$, gives:

$$d^2(f + p) = D^2(f - p)$$

Removing brackets gives: $d^2 f + d^2 p = D^2 f - D^2 p$

Rearranging, to obtain terms in p on the LHS gives:

$$d^2 p + D^2 p = D^2 f - d^2 f$$

Factorizing gives: $p(d^2 + D^2) = f(D^2 - d^2)$

Dividing both sides by $(d^2 + D^2)$ gives:

$$p = \frac{f(D^2 - d^2)}{(d^2 - D^2)}$$

Now try the following exercise

Exercise 31 Further problems on transposition of formulae (Answers on page 256)

Make the symbol indicated the subject of each of the formulae shown in Problems 1 to 7, and express each in its simplest form.

1. $y = \dfrac{a^2 m - a^2 n}{x}$ (a)

2. $M = \pi(R^4 - r^4)$ (R)

3. $x + y = \dfrac{r}{3 + r}$ (r)

4. $m = \dfrac{\mu L}{L + rCR}$ (L)

5. $a^2 = \dfrac{b^2 - c^2}{b^2}$ (b)

6. $\dfrac{x}{y} = \dfrac{1 + r^2}{1 - r^2}$ (r)

7. $\dfrac{p}{q} = \sqrt{\left(\dfrac{a + 2b}{a - 2b}\right)}$ (b)

8. A formula for the focal length, f, of a convex lens is $\dfrac{1}{f} = \dfrac{1}{u} + \dfrac{1}{v}$. Transpose the formula to make v the subject and evaluate v when $f = 5$ and $u = 6$.

9. The quantity of heat, Q, is given by the formula $Q = mc(t_2 - t_1)$. Make t_2 the subject of the formula and evaluate t_2 when $m = 10$, $t_1 = 15$, $c = 4$ and $Q = 1600$.

10. The velocity, v, of water in a pipe appears in the formula $h = \dfrac{0.03Lv^2}{2dg}$. Express v as the subject of the formula and evaluate v when $h = 0.712$, $L = 150$, $d = 0.30$ and $g = 9.81$

11. The sag S at the centre of a wire is given by the formula: $S = \sqrt{\left(\dfrac{3d(l - d)}{8}\right)}$. Make l the subject of the formula and evaluate l when $d = 1.75$ and $S = 0.80$

12. In an electrical alternating current circuit the impedance Z is given by: $Z = \sqrt{\left\{R^2 + \left(\omega L - \dfrac{1}{\omega C}\right)^2\right\}}$.

Transpose the formula to make C the subject and hence evaluate C when $Z = 130$, $R = 120$, $\omega = 314$ and $L = 0.32$

Assignment 4

This assignment covers the material contained in Chapters 7 and 8. The marks for each question are shown in brackets at the end of each question.

1. Solve the following equations:
   (a) $3t - 2 = 5t - 4$
   (b) $4(k - 1) - 2(3k + 2) + 14 = 0$
   (c) $\dfrac{a}{2} - \dfrac{2a}{5} = 1$
   (d) $3\sqrt{y} = 2$
   (e) $\sqrt{\left(\dfrac{s + 1}{s - 1}\right)} = 2$ (20)

2. Distance $s = ut + \frac{1}{2}at^2$. Find the value of acceleration a when $s = 17.5$, time $t = 2.5$ and $u = 10$ (4)

3. A rectangular football pitch has its length equal to twice its width and a perimeter of 360 m. Find its length and width.   (4)

4. Transpose the following equations:
   (a) $y = mx + c$ for m
   (b) $x = \dfrac{2(y - z)}{t}$ for z
   (c) $\dfrac{1}{R_T} = \dfrac{1}{R_A} + \dfrac{1}{R_B}$ for R_A
   (d) $x^2 - y^2 = 3ab$ for y
   (e) $K = \dfrac{p - q}{1 + pq}$ for q (18)

5. The passage of sound waves through walls is governed by the equation:

$$v = \sqrt{\dfrac{K + \frac{4}{3}G}{\rho}}$$

Make the shear modulus G the subject of the formula.   (4)

9

Simultaneous equations

9.1 Introduction to simultaneous equations

Only one equation is necessary when finding the value of a **single unknown quantity** (as with simple equations in Chapter 7). However, when an equation contains **two unknown quantities** it has an infinite number of solutions. When two equations are available connecting the same two unknown values then a unique solution is possible. Similarly, for three unknown quantities it is necessary to have three equations in order to solve for a particular value of each of the unknown quantities, and so on.

Equations which have to be solved together to find the unique values of the unknown quantities, which are true for each of the equations, are called **simultaneous equations**.

Two methods of solving simultaneous equations analytically are:

(a) by **substitution**, and (b) by **elimination**.

(A graphical solution of simultaneous equations is shown in Chapter 12.)

9.2 Worked problems on simultaneous equations in two unknowns

Problem 1. Solve the following equations for x and y, (a) by substitution, and (b) by elimination:

$$x + 2y = -1 \quad\quad\quad (1)$$
$$4x - 3y = 18 \quad\quad\quad (2)$$

(a) **By substitution**

From equation (1): $x = -1 - 2y$

Substituting this expression for x into equation (2) gives:

$$4(-1 - 2y) - 3y = 18$$

This is now a simple equation in y.

Removing the bracket gives:

$$-4 - 8y - 3y = 18$$
$$-11y = 18 + 4 = 22$$
$$y = \frac{22}{-11} = -2$$

Substituting $y = -2$ into equation (1) gives:

$$x + 2(-2) = -1$$
$$x - 4 = -1$$
$$x = -1 + 4 = 3$$

Thus $x = 3$ and $y = -2$ is the solution to the simultaneous equations.

(Check: In equation (2), since $x = 3$ and $y = -2$, LHS $= 4(3) - 3(-2) = 12 + 6 = 18 =$ RHS.)

(b) **By elimination**

$$x + 2y = -1 \quad\quad\quad (1)$$
$$4x - 3y = 18 \quad\quad\quad (2)$$

If equation (1) is multiplied throughout by 4 the coefficient of x will be the same as in equation (2), giving:

$$4x + 8y = -4 \quad\quad\quad (3)$$

Subtracting equation (3) from equation (2) gives:

$$\begin{array}{rcl} 4x - 3y &=& 18 \quad\quad (2) \\ 4x + 8y &=& -4 \quad\quad (3) \\ \hline 0 - 11y &=& 22 \end{array}$$

Hence $y = \dfrac{22}{-11} = -2$

(Note, in the above subtraction, $18 - -4 = 18 + 4 = 22$)

Substituting $y = -2$ into either equation (1) or equation (2) will give $x = 3$ as in method (a). The solution $x = 3$, $y = -2$ is the only pair of values that satisfies both of the original equations.

Problem 2. Solve, by a substitution method, the simultaneous equations

$$3x - 2y = 12 \tag{1}$$
$$x + 3y = -7 \tag{2}$$

From equation (2), $x = -7 - 3y$

Substituting for x in equation (1) gives:

$$3(-7 - 3y) - 2y = 12$$

i.e. $\quad -21 - 9y - 2y = 12$

$$-11y = 12 + 21 = 33$$

Hence $\quad y = \dfrac{33}{-11} = -3$

Substituting $y = -3$ in equation (2) gives:

$$x + 3(-3) = -7$$

i.e. $\quad x - 9 = -7$

Hence $\quad x = -7 + 9 = 2$

Thus $x = 2$, $y = -3$ is the solution of the simultaneous equations.

(Such solutions should always be checked by substituting values into each of the original two equations.)

Problem 3. Use an elimination method to solve the simultaneous equations

$$3x + 4y = 5 \tag{1}$$
$$2x - 5y = -12 \tag{2}$$

If equation (1) is multiplied throughout by 2 and equation (2) by 3, then the coefficient of x will be the same in the newly formed equations. Thus

$2 \times$ equation (1) gives: $\qquad 6x + 8y = 10 \tag{3}$

$3 \times$ equation (2) gives: $\qquad 6x - 15y = -36 \tag{4}$

Equation (3) − equation (4) gives: $\quad 0 + 23y = 46$

i.e. $\qquad\qquad\qquad\qquad y = \dfrac{46}{23} = 2$

(Note $+8y - -15y = 8y + 15y = 23y$ and $10 - -36 = 10 + 36 = 46$. Alternatively, 'change the signs of the bottom line and add'.)

Substituting $y = 2$ in equation (1) gives:

$$3x + 4(2) = 5$$

from which $3x = 5 - 8 = -3$

and $\qquad\qquad x = -1$

Checking in equation (2), left-hand side $= 2(-1) - 5(2) = -2 - 10 = -12 =$ right-hand side.

Hence $x = -1$ and $y = 2$ is the solution of the simultaneous equations.

The elimination method is the most common method of solving simultaneous equations.

Problem 4. Solve

$$7x - 2y = 26 \tag{1}$$
$$6x + 5y = 29 \tag{2}$$

When equation (1) is multiplied by 5 and equation (2) by 2 the coefficients of y in each equation are numerically the same, i.e. 10, but are of opposite sign.

$5 \times$ equation (1) gives: $\qquad\qquad 35x - 10y = 130 \tag{3}$

$2 \times$ equation (2) gives: $\qquad\qquad 12x + 10y = 58 \tag{4}$

Adding equation (3) and (4) gives: $\quad 47x + 0 = 188$

Hence $x = \dfrac{188}{47} = 4$

[Note that when the signs of common coefficients are **different** the two equations are **added**, and when the signs of common coefficients are the **same** the two equations are **subtracted** (as in Problems 1 and 3)]

Substituting $x = 4$ in equation (1) gives:

$$7(4) - 2y = 26$$
$$28 - 2y = 26$$
$$28 - 26 = 2y$$
$$2 = 2y$$

Hence $y = 1$

Checking, by substituting $x = 4$ and $y = 1$ in equation (2), gives:

$$\text{LHS} = 6(4) + 5(1) = 24 + 5 = 29 = \text{RHS}$$

Thus the solution is $x = 4$, $y = 1$, since these values maintain the equality when substituted in both equations.

Now try the following exercise

Exercise 32   Further problems on simultaneous equations (Answers on page 256)

Solve the following simultaneous equations and verify the results.

1. $a + b = 7$
   $a - b = 3$

2. $2x + 5y = 7$
   $x + 3y = 4$

3. $3s + 2t = 12$
   $4s - t = 5$

4. $3x - 2y = 13$
   $2x + 5y = -4$

5. $5m - 3n = 11$
   $3m + n = 8$

6. $8a - 3b = 51$
   $3a + 4b = 14$

7. $5x = 2y$
   $3x + 7y = 41$

8. $5c = 1 - 3d$
   $2d + c + 4 = 0$

9.3 Further worked problems on simultaneous equations

Problem 5.   Solve

$$3p = 2q \tag{1}$$

$$4p + q + 11 = 0 \tag{2}$$

Rearranging gives:

$$3p - 2q = 0 \tag{3}$$

$$4p + q = -11 \tag{4}$$

Multiplying equation (4) by 2 gives:

$$8p + 2q = -22 \tag{5}$$

Adding equations (3) and (5) gives:

$$11p + 0 = -22$$

$$p = \frac{-22}{11} = -2$$

Substituting $p = -2$ into equation (1) gives:

$$3(-2) = 2q$$

$$-6 = 2q$$

$$q = \frac{-6}{2} = -3$$

Checking, by substituting $p = -2$ and $q = -3$ into equation (2) gives:

$$\text{LHS} = 4(-2) + (-3) + 11 = -8 - 3 + 11 = 0 = \text{RHS}$$

Hence the solution is $p = -2$, $q = -3$

Problem 6.   Solve

$$\frac{x}{8} + \frac{5}{2} = y \tag{1}$$

$$13 - \frac{y}{3} = 3x \tag{2}$$

Whenever fractions are involved in simultaneous equations it is usual to firstly remove them. Thus, multiplying equation (1) by 8 gives:

$$8\left(\frac{x}{8}\right) + 8\left(\frac{5}{2}\right) = 8y$$

i.e.

$$x + 20 = 8y \tag{3}$$

Multiplying equation (2) by 3 gives:

$$39 - y = 9x \tag{4}$$

Rearranging equations (3) and (4) gives:

$$x - 8y = -20 \tag{5}$$

$$9x + y = 39 \tag{6}$$

Multiplying equation (6) by 8 gives:

$$72x + 8y = 312 \tag{7}$$

Adding equations (5) and (7) gives:

$$73x + 0 = 292$$

$$x = \frac{292}{73} = 4$$

Substituting $x = 4$ into equation (5) gives:

$$4 - 8y = -20$$

$$4 + 20 = 8y$$

$$24 = 8y$$

$$y = \frac{24}{8} = 3$$

Checking: substituting $x = 4$, $y = 3$ in the original equations, gives:

Equation (1): $\text{LHS} = \dfrac{4}{8} + \dfrac{5}{2} = \dfrac{1}{2} + 2\dfrac{1}{2} = 3 = y = \text{RHS}$

Equation (2): $\text{LHS} = 13 - \dfrac{3}{3} = 13 - 1 = 12$

$$\text{RHS} = 3x = 3(4) = 12$$

Hence the solution is $x = 4$, $y = 3$

Problem 7. Solve

$$2.5x + 0.75 - 3y = 0$$

$$1.6x = 1.08 - 1.2y$$

It is often easier to remove decimal fractions. Thus multiplying equations (1) and (2) by 100 gives:

$$250x + 75 - 300y = 0 \tag{1}$$

$$160x = 108 - 120y \tag{2}$$

Rearranging gives:

$$250x - 300y = -75 \tag{3}$$

$$160x + 120y = 108 \tag{4}$$

Multiplying equation (3) by 2 gives:

$$500x - 600y = -150 \tag{5}$$

Multiplying equation (4) by 5 gives:

$$800x + 600y = 540 \tag{6}$$

Adding equations (5) and (6) gives:

$$1300x + 0 = 390$$

$$x = \frac{390}{1300} = \frac{39}{130} = \frac{3}{10} = \mathbf{0.3}$$

Substituting $x = 0.3$ into equation (1) gives:

$$250(0.3) + 75 - 300y = 0$$

$$75 + 75 = 300y$$

$$150 = 300y$$

$$y = \frac{150}{300} = \mathbf{0.5}$$

Checking $x = 0.3$, $y = 0.5$ in equation (2) gives:

$$\text{LHS} = 160(0.3) = 48$$

$$\text{RHS} = 108 - 120(0.5) = 108 - 60 = 48$$

Hence the solution is $x = 0.3$, $y = 0.5$

Now try the following exercise

Exercise 33   Further problems on simultaneous equations (Answers on page 256)

Solve the following simultaneous equations and verify the results.

1. $7p + 11 + 2q = 0$

    $-1 = 3q - 5p$

2. $\dfrac{x}{2} + \dfrac{y}{3} = 4$

    $\dfrac{x}{6} - \dfrac{y}{9} = 0$

3. $\dfrac{a}{2} - 7 = -2b$

    $12 = 5a + \dfrac{2}{3}b$

4. $\dfrac{3}{2}s - 2t = 8$

    $\dfrac{s}{4} + 3y = -2$

5. $\dfrac{x}{5} + \dfrac{2y}{3} = \dfrac{49}{15}$

    $\dfrac{3x}{7} - \dfrac{y}{2} + \dfrac{5}{7} = 0$

6. $v - 1 = \dfrac{u}{12}$

    $u + \dfrac{v}{4} - \dfrac{25}{2} = 0$

7. $1.5x - 2.2y = -18$

    $2.4x + 0.6y = 33$

8. $3b - 2.5a = 0.45$

    $1.6a + 0.8b = 0.8$

9.4 More difficult worked problems on simultaneous equations

Problem 8. Solve

$$\frac{2}{x} + \frac{3}{y} = 7 \tag{1}$$

$$\frac{1}{x} - \frac{4}{y} = -2 \tag{2}$$

In this type of equation the solution is easier if a substitution is initially made.

Let $\dfrac{1}{x} = a$ and $\dfrac{1}{y} = b$

Thus equation (1) becomes: $\qquad 2a + 3b = 7 \tag{3}$

and equation (2) becomes: $\qquad a - 4b = -2 \tag{4}$

Multiplying equation (4) by 2 gives: $2a - 8b = -4 \tag{5}$

Subtracting equation (5) from equation (3) gives:

$$0 + 11b = 11$$

i.e.

$$b = 1$$

Substituting $b = 1$ in equation (3) gives:

$$2a + 3 = 7$$

$$2a = 7 - 3 = 4$$

i.e.

$$a = 2$$

Checking, substituting $a = 2$ and $b = 1$ in equation (4) gives:

$$\text{LHS} = 2 - 4(1) = 2 - 4 = -2 = \text{RHS}$$

Hence $a = 2$ and $b = 1$

However, since $\dfrac{1}{x} = a$ then $x = \dfrac{1}{a} = \dfrac{1}{2}$

and since $\dfrac{1}{y} = b$ then $y = \dfrac{1}{b} = \dfrac{1}{1} = 1$

Hence the solution is $x = \frac{1}{2}$, $y = 1$, which may be checked in the original equations.

Problem 9. Solve

$$\frac{1}{2a} + \frac{3}{5b} = 4 \tag{1}$$

$$\frac{4}{a} + \frac{1}{2b} = 10.5 \tag{2}$$

Let $\dfrac{1}{a} = x$ and $\dfrac{1}{b} = y$

then

$$\frac{x}{2} + \frac{3}{5}y = 4 \tag{3}$$

$$4x + \frac{1}{2}y = 10.5 \tag{4}$$

To remove fractions, equation (3) is multiplied by 10 giving:

$$10\left(\frac{x}{2}\right) + 10\left(\frac{3}{5}y\right) = 10(4)$$

i.e. $\qquad\qquad 5x + 6y = 40 \tag{5}$

Multiplying equation (4) by 2 gives:

$$8x + y = 21 \tag{6}$$

Multiplying equation (6) by 6 gives:

$$48x + 6y = 126 \tag{7}$$

Subtracting equation (5) from equation (7) gives:

$$43x + 0 = 86$$

$$x = \frac{86}{43} = 2$$

Substituting $x = 2$ into equation (3) gives:

$$\frac{2}{2} + \frac{3}{5}y = 4$$

$$\frac{3}{5}y = 4 - 1 = 3$$

$$y = \frac{5}{3}(3) = 5$$

Since $\quad \dfrac{1}{a} = x \quad$ then $\quad a = \dfrac{1}{x} = \dfrac{1}{2}$

and since $\dfrac{1}{b} = y \quad$ then $\quad b = \dfrac{1}{y} = \dfrac{1}{5}$

Hence the solution is $a = \frac{1}{2}$, $b = \frac{1}{5}$, which may be checked in the original equations.

Problem 10. Solve

$$\frac{1}{x+y} = \frac{4}{27} \tag{1}$$

$$\frac{1}{2x-y} = \frac{4}{33} \tag{2}$$

To eliminate fractions, both sides of equation (1) are multiplied by $27(x+y)$ giving:

$$27(x+y)\left(\frac{1}{x+y}\right) = 27(x+y)\left(\frac{4}{27}\right)$$

i.e. $\qquad\qquad 27(1) = 4(x+y)$

$$27 = 4x + 4y \tag{3}$$

Similarly, in equation (2): $\quad 33 = 4(2x - y)$

i.e. $\qquad\qquad 33 = 8x - 4y \tag{4}$

Equation (3) + equation (4) gives:

$$60 = 12x, \quad \text{i.e.} \quad x = \frac{60}{12} = 5$$

Substituting $x = 5$ in equation (3) gives:

$$27 = 4(5) + 4y$$

from which $\qquad 4y = 27 - 20 = 7$

and $\qquad\qquad y = \frac{7}{4} = 1\frac{3}{4}$

Hence $x = 5$, $y = 1\frac{3}{4}$ is the required solution, which may be checked in the original equations.

Problem 11. Solve

$$\frac{x-1}{3} + \frac{y+2}{5} = \frac{2}{15} \tag{1}$$

$$\frac{1-x}{6} + \frac{5+y}{2} = \frac{5}{6} \tag{2}$$

Before equations (1) and (2) can be simultaneously solved, the fractions need to be removed and the equations rearranged.

Multiplying equation (1) by 15 gives:

$$15\left(\frac{x-1}{3}\right) + 15\left(\frac{y+2}{5}\right) = 15\left(\frac{2}{15}\right)$$

i.e. $\qquad 5(x-1) + 3(y+2) = 2$

$$5x - 5 + 3y + 6 = 2$$

$$5x + 3y = 2 + 5 - 6$$

Hence $\qquad\qquad 5x + 3y = 1 \tag{3}$

Multiplying equation (2) by 6 gives:

$$6\left(\frac{1-x}{6}\right) + 6\left(\frac{5+y}{2}\right) = 6\left(\frac{5}{6}\right)$$

i.e.

$$(1-x) + 3(5+y) = 5$$

$$1 - x + 15 + 3y = 5$$

$$-x + 3y = 5 - 1 - 15$$

Hence

$$-x + 3y = -11 \qquad (4)$$

Thus the initial problem containing fractions can be expressed as:

$$5x + 3y = 1 \qquad (3)$$

$$-x + 3y = -11 \qquad (4)$$

Subtracting equation (4) from equation (3) gives:

$$6x + 0 = 12$$

$$x = \frac{12}{6} = 2$$

Substituting $x = 2$ into equation (3) gives:

$$5(2) + 3y = 1$$

$$10 + 3y = 1$$

$$3y = 1 - 10 = -9$$

$$y = \frac{-9}{3} = -3$$

Checking, substituting $x = 2$, $y = -3$ in equation (4) gives:

$$\text{LHS} = -2 + 3(-3) = -2 - 9 = -11 = \text{RHS}$$

Hence the solution is $x = 2$, $y = -3$, which may be checked in the original equations.

Now try the following exercise

Exercise 34   Further more difficult problems on simultaneous equations (Answers on page 256)

In problems 1 to 7, solve the simultaneous equations and verify the results

1. $\dfrac{3}{x} + \dfrac{2}{y} = 14$

   $\dfrac{5}{x} - \dfrac{3}{y} = -2$

2. $\dfrac{4}{a} - \dfrac{3}{b} = 18$

   $\dfrac{2}{a} + \dfrac{5}{b} = -4$

3. $\dfrac{1}{2p} + \dfrac{3}{5q} = 5$

   $\dfrac{5}{p} - \dfrac{1}{2p} = \dfrac{35}{2}$

4. $\dfrac{5}{x} + \dfrac{3}{y} = 1.1$

   $\dfrac{3}{x} - \dfrac{7}{y} = -1.1$

5. $\dfrac{c+1}{4} - \dfrac{d+2}{3} + 1 = 0$

   $\dfrac{1-c}{5} + \dfrac{3-d}{4} + \dfrac{13}{20} = 0$

6. $\dfrac{3r+2}{5} - \dfrac{2s-1}{4} = \dfrac{11}{5}$

   $\dfrac{3+2r}{4} + \dfrac{5-s}{3} = \dfrac{15}{4}$

7. $\dfrac{5}{x+y} = \dfrac{20}{27}$

   $\dfrac{4}{2x-y} = \dfrac{16}{33}$

8. If $5x - \dfrac{3}{y} = 1$ and $x + \dfrac{4}{y} = \dfrac{5}{2}$ find the value of $\dfrac{xy+1}{y}$

9.5   Practical problems involving simultaneous equations

There are a number of situations in engineering and science where the solution of simultaneous equations is required. Some are demonstrated in the following worked problems.

Problem 12.   The law connecting friction F and load L for an experiment is of the form $F = aL + b$, where a and b are constants. When $F = 5.6$, $L = 8.0$ and when $F = 4.4$, $L = 2.0$. Find the values of a and b and the value of F when $L = 6.5$

Substituting $F = 5.6$, $L = 8.0$ into $F = aL + b$ gives:

$$5.6 = 8.0a + b \qquad (1)$$

Substituting $F = 4.4$, $L = 2.0$ into $F = aL + b$ gives:

$$4.4 = 2.0a + b \qquad (2)$$

Subtracting equation (2) from equation (1) gives:

$$1.2 = 6.0a$$

$$a = \frac{1.2}{6.0} = \frac{1}{5}$$

Substituting $a = \frac{1}{5}$ into equation (1) gives:

$$5.6 = 8.0\left(\tfrac{1}{5}\right) + b$$

$$5.6 = 1.6 + b$$

$$5.6 - 1.6 = b$$

i.e.

$$b = 4$$

Checking, substituting $a = \frac{1}{5}$ and $b = 4$ in equation (2), gives:

$$RHS = 2.0 \left(\tfrac{1}{5}\right) + 4 = 0.4 + 4 = 4.4 = LHS$$

Hence $a = \frac{1}{5}$ **and** $b = 4$

When $L = 6.5$, $F = aL + b = \frac{1}{5}(6.5) + 4 = 1.3 + 4$, i.e. $F = 5.30$

Problem 13. The equation of a straight line, of gradient m and intercept on the y-axis c, is $y = mx + c$. If a straight line passes through the point where $x = 1$ and $y = -2$, and also through the point where $x = 3\frac{1}{2}$ and $y = 10\frac{1}{2}$, find the values of the gradient and the y-axis intercept.

Substituting $x = 1$ and $y = -2$ into $y = mx + c$ gives:

$$-2 = m + c \tag{1}$$

Substituting $x = 3\frac{1}{2}$ and $y = 10\frac{1}{2}$ into $y = mx + c$ gives:

$$10\tfrac{1}{2} = 3\tfrac{1}{2}m + c \tag{2}$$

Subtracting equation (1) from equation (2) gives:

$$12\tfrac{1}{2} = 2\tfrac{1}{2}m \quad \text{from which,} \quad \mathbf{\mathit{m} = \dfrac{12\tfrac{1}{2}}{2\tfrac{1}{2}} = 5}$$

Substituting $m = 5$ into equation (1) gives:

$$-2 = 5 + c$$
$$\mathbf{\mathit{c} = -2 - 5 = -7}$$

Checking, substituting $m = 5$ and $c = -7$ in equation (2), gives:

$$RHS = \left(3\tfrac{1}{2}\right)(5) + (-7) = 17\tfrac{1}{2} - 7 = 10\tfrac{1}{2} = LHS$$

Hence the gradient, $m = 5$ and the y-axis intercept, $c = -7$

Problem 14. When Kirchhoff's laws are applied to the electrical circuit shown in Figure 9.1 the currents I_1 and I_2 are connected by the equations:

$$27 = 1.5I_1 + 8(I_1 - I_2) \tag{1}$$
$$-26 = 2I_2 - 8(I_1 - I_2) \tag{2}$$

Solve the equations to find the values of currents I_1 and I_2

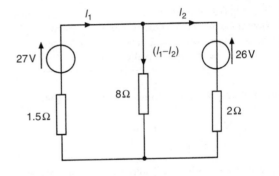

Fig. 9.1

Removing the brackets from equation (1) gives:

$$27 = 1.5I_1 + 8I_1 - 8I_2$$

Rearranging gives:

$$9.5I_1 - 8I_2 = 27 \tag{3}$$

Removing the brackets from equation (2) gives:

$$-26 = 2I_2 - 8I_1 + 8I_2$$

Rearranging gives:

$$-8I_1 + 10I_2 = -26 \tag{4}$$

Multiplying equation (3) by 5 gives:

$$47.5I_1 - 40I_2 = 135 \tag{5}$$

Multiplying equation (4) by 4 gives:

$$-32I_1 + 40I_2 = -104 \tag{6}$$

Adding equations (5) and (6) gives:

$$15.5I_1 + 0 = 31$$
$$I_2 = \dfrac{31}{15.5} = 2$$

Substituting $I_1 = 2$ into equation (3) gives:

$$9.5(2) - 8I_1 = 27$$
$$19 - 8I_2 = 27$$
$$19 - 27 = 8I_2$$
$$-8 = 8I_2$$
$$I_2 = -1$$

Hence the solution is $I_1 = 2$ **and** $I_2 = -1$ (which may be checked in the original equations.)

Problem 15. The distance s metres from a fixed point of a vehicle travelling in a straight line with constant acceleration, a m/s^2, is given by $s = ut + \frac{1}{2}at^2$, where u

is the initial velocity in m/s and t the time in seconds. Determine the initial velocity and the acceleration given that $s = 42$ m when $t = 2$ s and $s = 144$ m when $t = 4$ s. Find also the distance travelled after 3 s.

Substituting $s = 42$, $t = 2$ into $s = ut + \frac{1}{2}at^2$ gives:

$$42 = 2u + \frac{1}{2}a(2)^2$$

i.e. $42 = 2u + 2a$ (1)

Substituting $s = 144$, $t = 4$ into $s = ut + \frac{1}{2}at^2$ gives:

$$144 = 4u + \frac{1}{2}a(4)^2$$

i.e. $144 = 4u + 8a$ (2)

Multiplying equation (1) by 2 gives:

$$84 = 4u + 4a$$ (3)

Subtracting equation (3) from equation (2) gives:

$$60 = 0 + 4a$$

$$a = \frac{60}{4} = 15$$

Substituting $a = 15$ into equation (1) gives:

$$42 = 2u + 2(15)$$

$$42 - 30 = 2u$$

$$u = \frac{12}{2} = 6$$

Substituting $a = 15$, $u = 6$ in equation (2) gives:

$$\text{RHS} = 4(6) + 8(15) = 24 + 120 = 144 = \text{LHS}$$

Hence the initial velocity, $u = 6$ m/s and the acceleration, $a = 15$ m/s^2

Distance travelled after 3 s is given by $s = ut + \frac{1}{2}at^2$ where $t = 3$, $u = 6$ and $a = 15$

Hence $s = (6)(3) + \frac{1}{2}(15)(3)^2 = 18 + 67\frac{1}{2}$

i.e. distance travelled after 3 s $= 85\frac{1}{2}$ m

Problem 16. A craftsman and 4 labourers together earn £865 per week, whilst 4 craftsmen and 9 labourers earn £2340 basic per week. Determine the basic weekly wage of a craftsman and a labourer.

Let C represent the wage of a craftsman and L that of a labourer. Thus

$$C + 4L = 865$$ (1)

$$4C + 9L = 2340$$ (2)

Multiplying equation (1) by 4 gives:

$$4C + 16L = 3460$$ (3)

Subtracting equation (2) from equation (3) gives:

$$7L = 1120$$

$$L = \frac{1120}{7} = 160$$

Substituting $L = 138$ into equation (1) gives:

$$C + 4(160) = 865$$

$$C + 640 = 865$$

$$C = 865 - 640 = 225$$

Checking, substituting $C = 225$ and $L = 160$ into equation (2), gives:

$$\text{LHS} = 4(225) + 9(160) = 900 + 1440$$

$$= 2340 = \text{RHS}$$

Thus the basic weekly wage of a craftsman is £225 and that of a labourer is £160

Problem 17. The resistance $R\,\Omega$ of a length of wire at $t°$C is given by $R = R_0(1 + \alpha t)$, where R_0 is the resistance at $0°$C and α is the temperature coefficient of resistance in /$°$C. Find the values of α and R_0 if $R = 30\,\Omega$ at $50°$C and $R = 35\,\Omega$ at $100°$C.

Substituting $R = 30$, $t = 50$ into $R = R_0(1 + \alpha t)$ gives:

$$30 = R_0(1 + 50\alpha)$$ (1)

Substituting $R = 35$, $t = 100$ into $R = R_0(1 + \alpha t)$ gives:

$$35 = R_0(1 + 100\alpha)$$ (2)

Although these equations may be solved by the conventional substitution method, an easier way is to eliminate R_0 by division. Thus, dividing equation (1) by equation (2) gives:

$$\frac{30}{35} = \frac{R_0\,(1 + 50\alpha)}{R_0\,(1 + 100\alpha)} = \frac{1 + 50\alpha}{1 + 100\alpha}$$

'Cross-multiplying' gives:

$$30(1 + 100\alpha) = 35(1 + 50\alpha)$$

$$30 + 3000\alpha = 35 + 1750\alpha$$

$$3000\alpha - 1750\alpha = 35 - 30$$

$$1250\alpha = 5$$

i.e. $\alpha = \dfrac{5}{1250} = \dfrac{1}{250}$ or   **0.004**

Substituting $\alpha = \dfrac{1}{250}$ into equation (1) gives:

$$30 = R_0 \left\{ 1 + (50) \left(\dfrac{1}{250} \right) \right\}$$

$$30 = R_0(1.2)$$

$$R_0 = \dfrac{30}{1.2} = 25$$

Checking, substituting $\alpha = \dfrac{1}{250}$ and $R_0 = 25$ in equation (2) gives:

$$\text{RHS} = 25 \left\{ 1 + (100) \left(\dfrac{1}{250} \right) \right\}$$

$$= 25(1.4) = 35 = \text{LHS}$$

Thus the solution is $\alpha = 0.004/°C$ and $R_0 = 25\,\Omega$

Problem 18. The molar heat capacity of a solid compound is given by the equation $c = a + bT$, where a and b are constants. When $c = 52$, $T = 100$ and when $c = 172$, $T = 400$. Determine the values of a and b.

When $c = 52$, $T = 100$, hence

$$52 = a + 100b \qquad (1)$$

When $c = 172$, $T = 400$, hence

$$172 = a + 400b \qquad (2)$$

Equation (2) − equation (1) gives:

$$120 = 300b$$

from which, $\quad b = \dfrac{120}{300} = 0.4$

Substituting $b = 0.4$ in equation (1) gives:

$$52 = a + 100(0.4)$$

$$a = 52 - 40 = 12$$

Hence $\quad a = 12 \quad$ and $\quad b = 0.4$

Now try the following exercise

Exercise 35 Further practical problems involving simultaneous equations (Answers on page 256)

1. In a system of pulleys, the effort P required to raise a load W is given by $P = aW + b$, where a and b are constants. If $W = 40$ when $P = 12$ and $W = 90$ when $P = 22$, find the values of a and b.

2. Applying Kirchhoff's laws to an electrical circuit produces the following equations:

$$5 = 0.2I_1 + 2(I_1 - I_2)$$

$$12 = 3I_2 + 0.4I_2 - 2(I_1 - I_2)$$

Determine the values of currents I_1 and I_2

3. Velocity v is given by the formula $v = u + at$. If $v = 20$ when $t = 2$ and $v = 40$ when $t = 7$ find the values of u and a. Hence find the velocity when $t = 3.5$

4. Three new cars and 4 new vans supplied to a dealer together cost £97 700 and 5 new cars and 2 new vans of the same models cost £1 03 100. Find the cost of a car and a van.

5. $y = mx + c$ is the equation of a straight line of slope m and y-axis intercept c. If the line passes through the point where $x = 2$ and $y = 2$, and also through the point where $x = 5$ and $y = \frac{1}{2}$, find the slope and y-axis intercept of the straight line.

6. The resistance R ohms of copper wire at $t°C$ is given by $R = R_0(1 + \alpha t)$, where R_0 is the resistance at $0°C$ and α is the temperature coefficient of resistance. If $R = 25.44\,\Omega$ at $30°C$ and $R = 32.17\,\Omega$ at $100°C$, find α and R_0

7. The molar heat capacity of a solid compound is given by the equation $c = a + bT$. When $c = 52$, $T = 100$ and when $c = 172$, $T = 400$. Find the values of a and b

8. In an engineering process two variables p and q are related by: $q = ap + b/p$, where a and b are constants. Evaluate a and b if $q = 13$ when $p = 2$ and $q = 22$ when $p = 5$

10

Quadratic equations

10.1 Introduction to quadratic equations

As stated in Chapter 7, an **equation** is a statement that two quantities are equal and to '**solve an equation**' means 'to find the value of the unknown'. The value of the unknown is called the **root** of the equation.

A **quadratic equation** is one in which the highest power of the unknown quantity is 2. For example, $x^2 - 3x + 1 = 0$ is a quadratic equation.

There are four methods of **solving quadratic equations**.

These are: (i) by factorization (where possible)

(ii) by 'completing the square'

(iii) by using the 'quadratic formula'

or (iv) graphically (see Chapter 12).

10.2 Solution of quadratic equations by factorization

Multiplying out $(2x + 1)(x - 3)$ gives $2x^2 - 6x + x - 3$, i.e. $2x^2 - 5x - 3$. The reverse process of moving from $2x^2 - 5x - 3$ to $(2x + 1)(x - 3)$ is called **factorizing**.

If the quadratic expression can be factorized this provides the simplest method of solving a quadratic equation.

For example, if $2x^2 - 5x - 3 = 0$, then, by factorizing:

$$(2x + 1)(x - 3) = 0$$

Hence either $(2x + 1) = 0$ i.e. $x = -\frac{1}{2}$

or $(x - 3) = 0$ i.e. $x = 3$

The technique of factorizing is often one of 'trial and error'.

Problem 1. Solve the equations (a) $x^2 + 2x - 8 = 0$ (b) $3x^2 - 11x - 4 = 0$ by factorization.

(a) $x^2 + 2x - 8 = 0$. The factors of x^2 are x and x. These are placed in brackets thus: $(x\ )(x\ )$

The factors of -8 are $+8$ and -1, or -8 and $+1$, or $+4$ and -2, or -4 and $+2$. The only combination to give a middle term of $+2x$ is $+4$ and -2, i.e.

$$x^2 + 2x - 8 = (x + 4)(x - 2)$$

(Note that the product of the two inner terms added to the product of the two outer terms must equal the middle term, $+2x$ in this case.)

The quadratic equation $x^2 + 2x - 8 = 0$ thus becomes $(x + 4)(x - 2) = 0$.

Since the only way that this can be true is for either the first or the second, or both factors to be zero, then

either $(x + 4) = 0$ i.e. $x = -4$

or $(x - 2) = 0$ i.e. $x = 2$

Hence the roots of $x^2 + 2x - 8 = 0$ are $x = -4$ and 2

(b) $3x^2 - 11x - 4 = 0$

The factors of $3x^2$ are $3x$ and x. These are placed in brackets thus: $(3x\ )(x\ )$

The factors of -4 are -4 and $+1$, or $+4$ and -1, or -2 and 2.

Remembering that the product of the two inner terms added to the product of the two outer terms must equal $-11x$, the only combination to give this is $+1$ and -4, i.e.

$$3x^2 - 11x - 4 = (3x + 1)(x - 4)$$

The quadratic equation $3x^2 - 11x - 4 = 0$ thus becomes
$(3x + 1)(x - 4) = 0$

Hence, either $(3x + 1) = 0$ i.e. $x = -\frac{1}{3}$

or $(x - 4) = 0$ i.e. $x = 4$

and both solutions may be checked in the original equation.

Problem 2. Determine the roots of (a) $x^2 - 6x + 9 = 0$, and (b) $4x^2 - 25 = 0$, by factorization.

(a) $x^2 - 6x + 9 = 0$. Hence $(x-3)(x-3) = 0$, i.e. $(x-3)^2 = 0$ (the left-hand side is known as **a perfect square**). Hence $x = 3$ is the only root of the equation $x^2 - 6x + 9 = 0$.

(b) $4x^2 - 25 = 0$ (the left-hand side is **the difference of two squares**, $(2x)^2$ and $(5)^2$). Thus $(2x + 5)(2x - 5) = 0$

Hence either $(2x + 5) = 0$ i.e. $x = -\frac{5}{2}$

or $(2x - 5) = 0$ i.e. $x = \frac{5}{2}$

Problem 3. Solve the following quadratic equations by factorizing: (a) $4x^2 + 8x + 3 = 0$ (b) $15x^2 + 2x - 8 = 0$.

(a) $4x^2 + 8x + 3 = 0$. The factors of $4x^2$ are $4x$ and x or $2x$ and $2x$. The factors of 3 are 3 and 1, or -3 and -1. Remembering that the product of the inner terms added to the product of the two outer terms must equal $+8x$, the only combination that is true (by trial and error) is

$$(4x^2 + 8x + 3) = (2x + 3)(2x + 1)$$

Hence $(2x + 3)(2x + 1) = 0$ from which, either

$$(2x + 3) = 0 \text{ or } (2x + 1) = 0$$

Thus $2x = -3$, from which $x = -\frac{3}{2}$

or $2x = -1$, from which $x = -\frac{1}{2}$

which may be checked in the original equation.

(b) $15x^2 + 2x - 8 = 0$. The factors of $15x^2$ are $15x$ and x or $5x$ and $3x$. The factors of -8 are -4 and $+2$, or 4 and -2, or -8 and $+1$, or 8 and -1. By trial and error the only combination that works is

$$15x^2 + 2x - 8 = (5x + 4)(3x - 2)$$

Hence $(5x + 4)(3x - 2) = 0$ from which

either $5x + 4 = 0$

or $3x - 2 = 0$

Hence $x = -\frac{4}{5}$ or $x = \frac{2}{3}$
which may be checked in the original equation.

Problem 4. The roots of a quadratic equation are $\frac{1}{3}$ and -2. Determine the equation.

If the roots of a quadratic equation are α and β then $(x - \alpha)(x - \beta) = 0$

Hence if $\alpha = \frac{1}{3}$ and $\beta = -2$, then

$$\left(x - \frac{1}{3}\right)(x - (-2)) = 0$$

$$\left(x - \frac{1}{3}\right)(x + 2) = 0$$

$$x^2 - \frac{1}{3}x + 2x - \frac{2}{3} = 0$$

$$x^2 + \frac{5}{3}x - \frac{2}{3} = 0$$

Hence $3x^2 + 5x - 2 = 0$

Problem 5. Find the equations in x whose roots are (a) 5 and -5 (b) 1.2 and -0.4

(a) If 5 and -5 are the roots of a quadratic equation then

$$(x - 5)(x + 5) = 0$$

i.e. $x^2 - 5x + 5x - 25 = 0$

i.e. $x^2 - 25 = 0$

(b) If 1.2 and -0.4 are the roots of a quadratic equation then

$$(x - 1.2)(x + 0.4) = 0$$

i.e. $x^2 - 1.2x + 0.4x - 0.48 = 0$

i.e. $x^2 - 0.8x - 0.48 = 0$

Now try the following exercise

Exercise 36 Further problems on solving quadratic equations by factorization (Answers on page 256)

In Problems 1 to 12, solve the given equations by factorization.

1. $x^2 + 4x - 32 = 0$ 2. $x^2 - 16 = 0$

3. $(x + 2)^2 = 16$ 4. $2x^2 - x - 3 = 0$

5. $6x^2 - 5x + 1 = 0$ 6. $10x^2 + 3x - 4 = 0$

7. $x^2 - 4x + 4 = 0$ 8. $21x^2 - 25x = 4$

9. $8x^2 + 13x - 6 = 0$ 10. $5x^2 + 13x - 6 = 0$

11. $6x^2 - 5x - 4 = 0$ 12. $8x^2 + 2x - 15 = 0$

In Problems 13 to 18, determine the quadratic equations in x whose roots are

13. 3 and 1   14. 2 and -5 15. -1 and -4

16. $2\frac{1}{2}$ and $-\frac{1}{2}$ 17. 6 and -6 18. 2.4 and -0.7

10.3   Solution of quadratic equations by 'completing the square'

An expression such as x^2 or $(x+2)^2$ or $(x-3)^2$ is called a perfect square.

If $x^2 = 3$ then $x = \pm\sqrt{3}$

If $(x+2)^2 = 5$ then $x+2 = \pm\sqrt{5}$ and $x = -2 \pm \sqrt{5}$

If $(x-3)^2 = 8$ then $x-3 = \pm\sqrt{8}$ and $x = 3 \pm \sqrt{8}$

Hence if a quadratic equation can be rearranged so that one side of the equation is a perfect square and the other side of the equation is a number, then the solution of the equation is readily obtained by taking the square roots of each side as in the above examples. The process of rearranging one side of a quadratic equation into a perfect square before solving is called 'completing the square'.

$$(x+a)^2 = x^2 + 2ax + a^2$$

Thus in order to make the quadratic expression $x^2 + 2ax$ into a perfect square it is necessary to add (half the coefficient of x)^2 i.e. $\left(\dfrac{2a}{2}\right)^2$ or a^2

For example, $x^2 + 3x$ becomes a perfect square by adding $\left(\dfrac{3}{2}\right)^2$, i.e.

$$x^2 + 3x + \left(\frac{3}{2}\right)^2 = \left(x + \frac{3}{2}\right)^2$$

The method is demonstrated in the following worked problems.

Problem 6.   Solve $2x^2 + 5x = 3$ by 'completing the square'.

The procedure is as follows:

1. Rearrange the equation so that all terms are on the same side of the equals sign (and the coefficient of the x^2 term is positive).

Hence $2x^2 + 5x - 3 = 0$

2. Make the coefficient of the x^2 term unity. In this case this is achieved by dividing throughout by 2. Hence

$$\frac{2x^2}{2} + \frac{5x}{2} - \frac{3}{2} = 0$$

i.e. $x^2 + \dfrac{5}{2}x - \dfrac{3}{2} = 0$

3. Rearrange the equations so that the x^2 and x terms are on one side of the equals sign and the constant is on the other side. Hence

$$x^2 + \frac{5}{2}x = \frac{3}{2}$$

4. Add to both sides of the equation (half the coefficient of x)^2. In this case the coefficient of x is $\dfrac{5}{2}$. Half the coefficient squared is therefore $\left(\dfrac{5}{4}\right)^2$. Thus

$$x^2 + \frac{5}{2}x + \left(\frac{5}{4}\right)^2 = \frac{3}{2} + \left(\frac{5}{4}\right)^2$$

The LHS is now a perfect square, i.e.

$$\left(x + \frac{5}{4}\right)^2 = \frac{3}{2} + \left(\frac{5}{4}\right)^2$$

5. Evaluate the RHS. Thus

$$\left(x + \frac{5}{4}\right)^2 = \frac{3}{2} + \frac{25}{16} = \frac{24 + 25}{16} = \frac{49}{16}$$

6. Taking the square root of both sides of the equation (remembering that the square root of a number gives a $\pm$ answer). Thus

$$\sqrt{\left(x + \frac{5}{4}\right)^2} = \sqrt{\left(\frac{49}{16}\right)}$$

i.e. $x + \dfrac{5}{4} = \pm\dfrac{7}{4}$

7. Solve the simple equation. Thus

$$x = -\frac{5}{4} \pm \frac{7}{4}$$

i.e. $x = -\dfrac{5}{4} + \dfrac{7}{4} = \dfrac{2}{4} = \dfrac{1}{2}$

and $x = -\dfrac{5}{4} - \dfrac{7}{4} = -\dfrac{12}{4} = -3$

Hence $x = \frac{1}{2}$ or -3 are the roots of the equation $2x^2 + 5x = 3$

Problem 7.   Solve $2x^2 + 9x + 8 = 0$, correct to 3 significant figures, by 'completing the square'.

Making the coefficient of x^2 unity gives:

$$x^2 + \frac{9}{2}x + 4 = 0$$

and rearranging gives: $x^2 + \frac{9}{2}x = -4$

Adding to both sides (half the coefficient of $x)^2$ gives:

$$x^2 + \frac{9}{2}x + \left(\frac{9}{4}\right)^2 = \left(\frac{9}{4}\right)^2 - 4$$

The LHS is now a perfect square, thus

$$\left(x + \frac{9}{4}\right)^2 = \frac{81}{16} - 4 = \frac{17}{16}$$

Taking the square root of both sides gives:

$$x + \frac{9}{4} = \sqrt{\left(\frac{17}{16}\right)} = \pm 1.031$$

Hence $\qquad x = -\frac{9}{4} \pm 1.031$

i.e. $x = -1.22$ or -3.28, correct to 3 significant figures.

Problem 8. By 'completing the square', solve the quadratic equation $4.6y^2 + 3.5y - 1.75 = 0$, correct to 3 decimal places.

$$4.6y^2 + 3.5y - 1.75 = 0$$

Making the coefficient of y^2 unity gives:

$$y^2 + \frac{3.5}{4.6}y - \frac{1.75}{4.6} = 0$$

and rearranging gives:

$$y^2 + \frac{3.5}{4.6}y = \frac{1.75}{4.6}$$

Adding to both sides (half the coefficient of $y)^2$ gives:

$$y^2 + \frac{3.5}{4.6}y + \left(\frac{3.5}{9.2}\right)^2 = \frac{1.75}{4.6} + \left(\frac{3.5}{9.2}\right)^2$$

The LHS is now a perfect square, thus

$$\left(y + \frac{3.5}{9.2}\right)^2 = 0.5251654$$

Taking the square root of both sides gives:

$$y + \frac{3.5}{9.2} = \sqrt{0.5251654} = \pm 0.7246830$$

Hence $\qquad y = -\frac{3.5}{9.2} \pm 0.7246830$

i.e. $\qquad y = 0.344$ or -1.105

Now try the following exercise

Exercise 37 Further problems on solving quadratic equations by 'completing the square' (Answers on page 257)

In Problems 1 to 6, solve the given equations by completing the square, each correct to 3 decimal places.

1. $x^2 + 4x + 1 = 0$ $\qquad$ 2. $2x^2 + 5x - 4 = 0$

3. $3x^2 - x - 5 = 0$ $\qquad$ 4. $5x^2 - 8x + 2 = 0$

5. $4x^2 - 11x + 3 = 0$ $\qquad$ 6. $2x^2 + 5x = 2$

10.4 Solution of quadratic equations by formula

Let the general form of a quadratic equation be given by:

$$ax^2 + bx + c = 0 \quad \text{where } a, b \text{ and } c \text{ are constants.}$$

Dividing $ax^2 + bx + c = 0$ by a gives:

$$x^2 + \frac{b}{a}x + \frac{c}{a} = 0$$

Rearranging gives:

$$x^2 + \frac{b}{a}x = -\frac{c}{a}$$

Adding to each side of the equation the square of half the coefficient of the term in x to make the LHS a perfect square gives:

$$x^2 + \frac{b}{a}x + \left(\frac{b}{2a}\right)^2 = \left(\frac{b}{2a}\right)^2 - \frac{c}{a}$$

Rearranging gives:

$$\left(x + \frac{b}{a}\right)^2 = \frac{b^2}{4a^2} - \frac{c}{a} = \frac{b^2 - 4ac}{4a^2}$$

Taking the square root of both sides gives:

$$x + \frac{b}{2a} = \sqrt{\left(\frac{b^2 - 4ac}{4a^2}\right)} = \frac{\pm\sqrt{b^2 - 4ac}}{2a}$$

Hence $\qquad x = -\frac{b}{2a} \pm \frac{\sqrt{b^2 - 4ac}}{2a}$

i.e. the quadratic formula is $x = \dfrac{-b \pm \sqrt{b^2 - 4ac}}{2a}$

(This method of solution is 'completing the square' – as shown in Section 10.3.)

Summarizing:

if $ax^2 + bx + c = 0$ then
$$x = \frac{-b \pm \sqrt{b^2 - 4ac}}{2a}$$

This is known as the **quadratic formula**.

Problem 9. Solve (a) $x^2 + 2x - 8 = 0$ and (b) $3x^2 - 11x - 4 = 0$ by using the quadratic formula.

(a) Comparing $x^2 + 2x - 8 = 0$ with $ax^2 + bx + c = 0$ gives $a = 1$, $b = 2$ and $c = -8$

Substituting these values into the quadratic formula

$$x = \frac{-b \pm \sqrt{b^2 - 4ac}}{2a} \text{ gives:}$$

$$x = \frac{-2 \pm \sqrt{2^2 - 4(1)(-8)}}{2(1)} = \frac{-2 \pm \sqrt{4 + 32}}{2}$$

$$= \frac{-2 \pm \sqrt{36}}{2} = \frac{-2 \pm 6}{2}$$

$$= \frac{-2 + 6}{2} \text{ or } \frac{-2 - 6}{2}$$

Hence $x = \frac{4}{2} = \mathbf{2}$ or $\frac{-8}{2} = \mathbf{-4}$ (as in Problem 1(a)).

(b) Comparing $3x^2 - 11x - 4 = 0$ with $ax^2 + bx + c = 0$ gives $a = 3$, $b = -11$ and $c = -4$. Hence

$$x = \frac{-(-11) \pm \sqrt{(-11)^2 - 4(3)(-4)}}{2(3)}$$

$$= \frac{+11 \pm \sqrt{121 + 48}}{6} = \frac{11 \pm \sqrt{169}}{6}$$

$$= \frac{11 \pm 13}{6} = \frac{11 + 13}{6} \text{ or } \frac{11 - 13}{6}$$

Hence $x = \frac{24}{6} = \mathbf{4}$ or $\frac{-2}{6} = -\frac{1}{3}$ (as in Problem 1(b)).

Problem 10. Solve $4x^2 + 7x + 2 = 0$ giving the roots correct to 2 decimal places.

Comparing $4x^2 + 7x + 2 = 0$ with $ax^2 + bx + c$ gives $a = 4$, $b = 7$ and $c = 2$.

Hence

$$x = \frac{-7 \pm \sqrt{[(7)^2 - 4(4)(2)]}}{2(4)} = \frac{-7 \pm \sqrt{17}}{8}$$

$$= \frac{-7 \pm 4.123}{8} = \frac{-7 + 4.123}{8} \text{ or } \frac{-7 - 4.123}{8}$$

Hence $x = \mathbf{-0.36}$ or $\mathbf{-1.39}$, **correct to 2 decimal places.**

Problem 11. Use the quadratic formula to solve

$$\frac{x + 2}{4} + \frac{3}{x - 1} = 7 \text{ correct to 4 significant figures.}$$

Multiplying throughout by $4(x - 1)$ gives:

$$4(x - 1)\frac{(x + 2)}{4} + 4(x - 1)\frac{3}{(x - 1)} = 4(x - 1)(7)$$

i.e.
$$(x - 1)(x + 2) + (4)(3) = 28(x - 1)$$

$$x^2 + x - 2 + 12 = 28x - 28$$

Hence
$$x^2 - 27x + 38 = 0$$

Using the quadratic formula:

$$x = \frac{-(-27) \pm \sqrt{(-27)^2 - 4(1)(38)}}{2}$$

$$= \frac{27 \pm \sqrt{577}}{2} = \frac{27 \pm 24.0208}{2}$$

Hence $x = \frac{27 + 24.0208}{2} = 25.5104$

or $x = \frac{27 - 24.0208}{2} = 1.4896$

Hence $x = \mathbf{25.51}$ or $\mathbf{1.490}$, correct to 4 significant figures.

Now try the following exercise

Exercise 38   **Further problems on solving quadratic equations by formula (Answers on page 257)**

In Problems 1 to 6 solve the given equations by using the quadratic formula, correct to 3 decimal places.

1. $2x^2 + 5x - 4 = 0$ 2. $5.76x^2 + 2.86x - 1.35 = 0$

3. $2x^2 - 7x + 4 = 0$ 4. $4x + 5 = \dfrac{3}{x}$

5. $(2x + 1) = \dfrac{5}{x - 3}$ 6. $\dfrac{x + 1}{x - 1} = x - 3$

10.5 Practical problems involving quadratic equations

There are many **practical problems** where a quadratic equation has first to be obtained, from given information, before it is solved.

Problem 12. The area of a rectangle is $23.6\,\text{cm}^2$ and its width is 3.10 cm shorter than its length. Determine the dimensions of the rectangle, correct to 3 significant figures.

Let the length of the rectangle be $x\,\text{cm}$. Then the width is $(x - 3.10)\,\text{cm}$.

Area $=$ length $\times$ width $= x(x - 3.10) = 23.6$

i.e. $x^2 - 3.10x - 23.6 = 0$

Using the quadratic formula,

$$x = \frac{-(-3.10) \pm \sqrt{(-3.10)^2 - 4(1)(-23.6)}}{2(1)}$$

$$= \frac{3.10 \pm \sqrt{9.61 + 94.4}}{2} = \frac{3.10 \pm 10.20}{2}$$

$$= \frac{13.30}{2} \quad \text{or} \quad \frac{-7.10}{2}$$

Hence $x = 6.65\,\text{cm}$ or $-3.55\,\text{cm}$. The latter solution is neglected since length cannot be negative.

Thus length $x = 6.65\,\text{cm}$ and

width $= x - 3.10 = 6.65 - 3.10 = 3.55\,\text{cm}$

Hence the dimensions of the rectangle are 6.65 cm by 3.55 cm.

(Check: Area $= 6.65 \times 3.55 = 23.6\,\text{cm}^2$, correct to 3 significant figures.)

Problem 13. Calculate the diameter of a solid cylinder which has a height of 82.0 cm and a total surface area of $2.0\,\text{m}^2$

Total surface area of a cylinder

$=$ curved surface area $+ 2$ circular ends

$= 2\pi rh + 2\pi r^2$ (where $r =$ radius and $h =$ height)

Since the total surface area $= 2.0\,\text{m}^2$ and the height $h = 82\,\text{cm}$ or 0.82 m, then

$2.0 = 2\pi r(0.82) + 2\pi r^2$ i.e. $2\pi r^2 + 2\pi r(0.82) - 2.0 = 0$

Dividing throughout by 2π gives: $r^2 + 0.82r - \dfrac{1}{\pi} = 0$

Using the quadratic formula:

$$r = \frac{-0.82 \pm \sqrt{(0.82)^2 - 4(1)\left(-\dfrac{1}{\pi}\right)}}{2(1)}$$

$$= \frac{-0.82 \pm \sqrt{1.9456}}{2} = \frac{-0.82 \pm 1.3948}{2}$$

$$= 0.2874 \quad \text{or} \quad -1.1074$$

Thus the radius r of the cylinder is 0.2874 m (the negative solution being neglected).

Hence the diameter of the cylinder $= 2 \times 0.2874$

$$= \mathbf{0.5748\,m} \quad \text{or}$$

57.5 cm correct to 3 significant figures

Problem 14. The height s metres of a mass projected vertically upwards at time t seconds is $s = ut - \frac{1}{2}gt^2$. Determine how long the mass will take after being projected to reach a height of 16 m (a) on the ascent and (b) on the descent, when $u = 30\,\text{m/s}$ and $g = 9.81\,\text{m/s}^2$.

When height $s = 16\,\text{m}$, $16 = 30t - \frac{1}{2}(9.81)t^2$

i.e. $4.905t^2 - 30t + 16 = 0$

Using the quadratic formula:

$$t = \frac{-(-30) \pm \sqrt{(-30)^2 - 4(4.905)(16)}}{2(4.905)}$$

$$= \frac{30 \pm \sqrt{586.1}}{9.81} = \frac{30 \pm 24.21}{9.81} = 5.53 \quad \text{or} \quad 0.59$$

Hence the mass will reach a height of 16 m after 0.59 s on the ascent and after 5.53 s on the descent.

Problem 15. A shed is 4.0 m long and 2.0 m wide. A concrete path of constant width is laid all the way around the shed. If the area of the path is $9.50\,\text{m}^2$ calculate its width to the nearest centimetre.

Figure 10.1 shows a plan view of the shed with its surrounding path of width t metres.

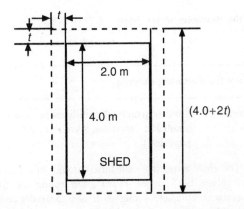

Fig. 10.1

Area of path $= 2(2.0 \times t) + 2t(4.0 + 2t)$

i.e. $9.50 = 4.0t + 8.0t + 4t^2$

or $4t^2 + 12.0t - 9.50 = 0$

Hence $t = \dfrac{-(12.0) \pm \sqrt{(12.0)^2 - 4(4)(-9.50)}}{2(4)}$

$= \dfrac{-12.0 \pm \sqrt{296.0}}{8} = \dfrac{-12.0 \pm 17.20465}{8}$

Hence $t = 0.6506$ m or -3.65058 m

Neglecting the negative result which is meaningless, the width of the path, $t = \mathbf{0.651\,m}$ or $\mathbf{65\,cm}$, correct to the nearest centimetre.

Problem 16.   If the total surface area of a solid cone is $486.2\,\text{cm}^2$ and its slant height is $15.3\,\text{cm}$, determine its base diameter.

From Chapter 22, page 166, the total surface area A of a solid cone is given by: $A = \pi r l + \pi r^2$ where l is the slant height and r the base radius.

If $A = 482.2$ and $l = 15.3$, then $482.2 = \pi r(15.3) + \pi r^2$

i.e. $\pi r^2 + 15.3\pi r - 482.2 = 0$

or $r^2 + 15.3r - \dfrac{482.2}{\pi} = 0$

Using the quadratic formula,

$r = \dfrac{-15.3 \pm \sqrt{\left[(15.3)^2 - 4\left(\dfrac{-482.2}{\pi}\right)\right]}}{2}$

$= \dfrac{-15.3 \pm \sqrt{848.0461}}{2} = \dfrac{-15.3 \pm 29.12123}{2}$

Hence radius $r = 6.9106$ cm (or -22.21 cm, which is meaningless, and is thus ignored).

Thus **the diameter of the base** $= 2r = 2(6.9106)$

$= \mathbf{13.82\,cm}$

Now try the following exercise

Exercise 39   Further practical problems involving quadratic equations (Answers on page 257)

1. The angle a rotating shaft turns through in t seconds is given by $\theta = \omega t + \frac{1}{2}\alpha t^2$. Determine the time taken to complete 4 radians if ω is 3.0 rad/s and α is 0.60 rad/s^2.

2. The power P developed in an electrical circuit is given by $P = 10I - 8I^2$, where I is the current in amperes. Determine the current necessary to produce a power of 2.5 watts in the circuit.

3. The area of a triangle is $47.6\,\text{cm}^2$ and its perpendicular height is $4.3\,\text{cm}$ more than its base length. Determine the length of the base correct to 3 significant figures.

4. The sag l metres in a cable stretched between two supports, distance x m apart is given by: $l = \dfrac{12}{x} + x$. Determine the distance between supports when the sag is 20 m.

5. The acid dissociation constant K_a of ethanoic acid is $1.8 \times 10^{-5}\,\text{mol dm}^{-3}$ for a particular solution. Using the Ostwald dilution law $K_a = \dfrac{x^2}{v(l - x)}$ determine x, the degree of ionization, given that $v = 10\,\text{dm}^3$.

6. A rectangular building is 15 m long by 11 m wide. A concrete path of constant width is laid all the way around the building. If the area of the path is $60.0\,\text{m}^2$, calculate its width correct to the nearest millimetre.

7. The total surface area of a closed cylindrical container is $20.0\,\text{m}^3$. Calculate the radius of the cylinder if its height is $2.80\,\text{m}^2$.

8. The bending moment M at a point in a beam is given by $M = \dfrac{3x(20 - x)}{2}$ where x metres is the distance from the point of support. Determine the value of x when the bending moment is 50 Nm.

9. A tennis court measures 24 m by 11 m. In the layout of a number of courts an area of ground must be allowed for at the ends and at the sides of each court. If a border of constant width is allowed around each court and the total area of the court and its border is $950\,\text{m}^2$, find the width of the borders.

10. Two resistors, when connected in series, have a total resistance of 40 ohms. When connected in parallel their total resistance is 8.4 ohms. If one of the resistors has a resistance R_x ohms:

    (a) show that $R_x^2 - 40R_x + 336 = 0$ and

    (b) calculate the resistance of each

10.6   The solution of linear and quadratic equations simultaneously

Sometimes a linear equation and a quadratic equation need to be solved simultaneously. An algebraic method of solution is shown in Problem 17; a graphical solution is shown in Chapter 12, page 92.

> *Problem 17.* Determine the values of x and y which simultaneously satisfy the equations:
> $y = 5x - 4 - 2x^2$ and $y = 6x - 7$

For a simultaneous solution the values of y must be equal, hence the RHS of each equation is equated. Thus

$$5x - 4 - 2x^2 = 6x - 7$$

Rearranging gives: $\quad 5x - 4 - 2x^2 - 6x + 7 = 0$

i.e. $\qquad\qquad\qquad -x + 3 - 2x^2 = 0$

or $\qquad\qquad\qquad 2x^2 + x - 3 = 0$

Factorizing gives: $\qquad (2x + 3)(x - 1) = 0$

i.e. $\qquad\qquad\qquad x = -\dfrac{3}{2} \quad \text{or} \quad x = 1$

In the equation $y = 6x - 7$,

when $x = -\dfrac{3}{2}$, $y = 6\left(\dfrac{-3}{2}\right) - 7 = -16$

and when $x = 1$, $y = 6 - 7 = -1$

[Checking the result in $y = 5x - 4 - 2x^2$:

when $x = -\dfrac{3}{2}$, $\quad y = 5\left(-\dfrac{3}{2}\right) - 4 - 2\left(-\dfrac{3}{2}\right)^2$

$$= -\dfrac{15}{2} - 4 - \dfrac{9}{2} = -16$$

as above; and when $x = 1$, $y = 5 - 4 - 2 = -1$ as above]

Hence the simultaneous solutions occur when $x = -\dfrac{3}{2}$, $y = -16$ and when $x = 1, y = -1$

Now try the following exercise

Exercise 40   Further problems on solving linear and quadratic equations simultaneously (Answers on page 257)

In Problems 1 to 3 determine the solutions of the simultaneous equations.

1. $y = x^2 + x + 1$
   $y = 4 - x$

2. $y = 15x^2 + 21x - 11$
   $y = 2x - 1$

3. $2x^2 + y = 4 + 5x$
   $x + y = 4$

Assignment 5

> This assignment covers the material contained in Chapters 9 and 10. The marks for each question are shown in brackets at the end of each question.

1. Solve the following pairs of simultaneous equations:

   (a) $7x - 3y = 23$
       $2x + 4y = -8$

   (b) $3a - 8 + \dfrac{b}{8} = 0$

       $b + \dfrac{a}{2} = \dfrac{21}{4}$

   (c) $\dfrac{2p + 1}{5} - \dfrac{1 - 4q}{2} = \dfrac{5}{2}$

       $\dfrac{1 - 3p}{7} + \dfrac{2q - 3}{5} + \dfrac{32}{35} = 0$ $\qquad$ (20)

2. In an engineering process two variables x and y are related by the equation $y = ax + \dfrac{b}{x}$ where a and b are constants. Evaluate a and b if $y = 15$ when $x = 1$ and $y = 13$ when $x = 3$ $\qquad$ (5)

3. Solve the following equations by factorization:

   (a) $x^2 - 9 = 0$ $\qquad$ (b) $2x^2 - 5x - 3 = 0$ $\qquad$ (6)

4. Determine the quadratic equation in x whose roots are 1 and -3 $\qquad$ (4)

5. Solve the equation $3x^2 - x - 4 = 0$ by completing the square. $\qquad$ (5)

6. Solve the equation $4x^2 - 9x + 3 = 0$ correct to 3 decimal places. $\qquad$ (5)

7. The current i flowing through an electronic device is given by:

   $$i = 0.005\,v^2 + 0.014\,v$$

   where v is the voltage. Calculate the values of v when $i = 3 \times 10^{-3}$ $\qquad$ (5)

11

Straight line graphs

11.1 Introduction to graphs

A **graph** is a pictorial representation of information showing how one quantity varies with another related quantity.

The most common method of showing a relationship between two sets of data is to use **Cartesian** or **rectangular axes** as shown in Fig. 11.1.

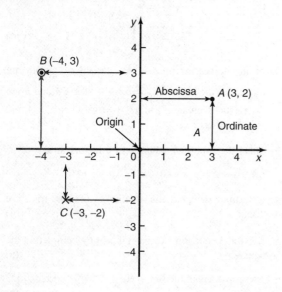

Fig. 11.1

The points on a graph are called **co-ordinates**. Point A in Fig. 11.1 has the co-ordinates $(3, 2)$, i.e. 3 units in the x direction and 2 units in the y direction. Similarly, point B has co-ordinates $(-4, 3)$ and C has co-ordinates $(-3, -2)$. The origin has co-ordinates $(0, 0)$.

The horizontal distance of a point from the vertical axis is called the **abscissa** and the vertical distance from the horizontal axis is called the **ordinate**.

11.2 The straight line graph

Let a relationship between two variables x and y be $y = 3x + 2$.

When $x = 0$, $y = 3(0) + 2 = 2$.

When $x = 1$, $y = 3(1) + 2 = 5$.

When $x = 2$, $y = 3(2) + 2 = 8$, and so on.

Thus co-ordinates $(0, 2)$, $(1, 5)$ and $(2, 8)$ have been produced from the equation by selecting arbitrary values of x, and are shown plotted in Fig. 11.2. When the points are joined together a **straight-line graph** results.

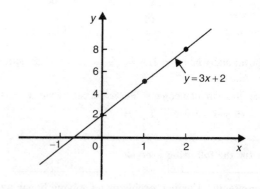

Fig. 11.2

The **gradient** or **slope** of a straight line is the ratio of the change in the value of y to the change in the value of x between any two points on the line. If, as x increases, $(\rightarrow)$, y also increases $(\uparrow)$, then the gradient is positive.

In Fig. 11.3(a), the gradient of AC

$$= \frac{\text{change in } y}{\text{change in } x} = \frac{CB}{BA} = \frac{7 - 3}{3 - 1} = \frac{4}{2} = 2$$

Straight line graphs 77

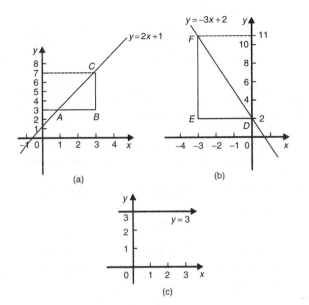

(a)

(b)

(c)

Fig. 11.3

If as x increases ($\rightarrow$), y decreases ($\downarrow$), then the gradient is negative.

In Fig. 11.3(b), the gradient of DF

$$= \frac{\text{change in } y}{\text{change in } x} = \frac{FE}{ED} = \frac{11-2}{-3-0}$$

$$= \frac{9}{-3} = -3$$

Figure 11.3(c) shows a straight line graph $y = 3$. Since the straight line is horizontal the gradient is zero.

The value of y when $x = 0$ is called the **y-axis intercept**. In Fig. 11.3(a) the y-axis intercept is 1 and in Fig. 11.3(b) is 2.

If the equation of a graph is of the form $y = mx + c$, where m and c are constants, **the graph will always be a straight line**, m representing the gradient and c the y-axis intercept. Thus $y = 5x + 2$ represents a straight line of gradient 5 and y-axis intercept 2. Similarly, $y = -3x - 4$ represents a straight line of gradient -3 and y-axis intercept -4

Summary of general rules to be applied when drawing graphs

(i) Give the graph a title clearly explaining what is being illustrated.

(ii) Choose scales such that the graph occupies as much space as possible on the graph paper being used.

(iii) Choose scales so that interpolation is made as easy as possible. Usually scales such as 1 cm = 1 unit, or 1 cm = 2 units, or 1 cm = 10 units are used. Awkward scales such as 1 cm = 3 units or 1 cm = 7 units should not be used.

(iv) The scales need not start at zero, particularly when starting at zero produces an accumulation of points within a small area of the graph paper.

(v) The co-ordinates, or points, should be clearly marked. This may be done either by a cross, or a dot and circle, or just by a dot (see Fig. 11.1).

(vi) A statement should be made next to each axis explaining the numbers represented with their appropriate units.

(vii) Sufficient numbers should be written next to each axis without cramping.

Problem 1. Plot the graph $y = 4x + 3$ in the range $x = -3$ to $x = +4$. From the graph, find (a) the value of y when $x = 2.2$, and (b) the value of x when $y = -3$

Whenever an equation is given and a graph is required, a table giving corresponding values of the variable is necessary. The table is achieved as follows:

When $x = -3$, $y = 4x + 3 = 4(-3) + 3 = -12 + 3 = -9$.

When $x = -2$, $y = 4(-2) + 3 = -8 + 3 = -5$, and so on.

Such a table is shown below:

x	-3	-2	-1	0	1	2	3	4
y	-9	-5	-1	3	7	11	15	19

The co-ordinates $(-3, -9)$, $(-2, -5)$, $(-1, -1)$, and so on, are plotted and joined together to produce the straight line shown in Fig. 11.4. (Note that the scales used on the x and y axes do not have to be the same) From the graph:

(a) when $x = 2.2$, $y = $ **11.8**, and

(b) when $y = -3$, $x = $ **-1.5**

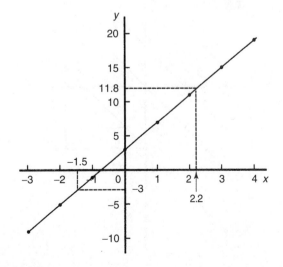

Fig. 11.4

Problem 2. Plot the following graphs on the same axes between the range $x = -4$ to $x = +4$, and determine the gradient of each.

(a) $y = x$ (b) $y = x + 2$

(c) $y = x + 5$ (d) $y = x - 3$

A table of co-ordinates is produced for each graph.

(a) $y = x$

x	−4	−3	−2	−1	0	1	2	3	4
y	−4	−3	−2	−1	0	1	2	3	4

(b) $y = x + 2$

x	−4	−3	−2	−1	0	1	2	3	4
y	−2	−1	0	1	2	3	4	5	6

(c) $y = x + 5$

x	−4	−3	−2	−1	0	1	2	3	4
y	1	2	3	4	5	6	7	8	9

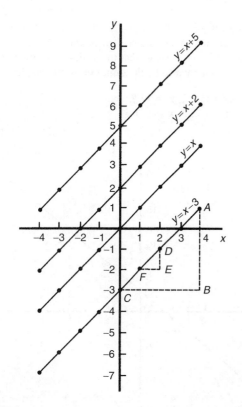

Fig. 11.5

(d) $y = x - 3$

x	−4	−3	−2	−1	0	1	2	3	4
y	−7	−6	−5	−4	−3	−2	−1	0	1

The co-ordinates are plotted and joined for each graph. The results are shown in Fig. 11.5. Each of the straight lines produced are parallel to each other, i.e. the slope or gradient is the same for each.

To find the gradient of any straight line, say, $y = x - 3$ a horizontal and vertical component needs to be constructed. In Fig. 11.5, AB is constructed vertically at $x = 4$ and BC constructed horizontally at $y = -3$. The gradient of AC

$$= \frac{AB}{BC} = \frac{1 - (-3)}{4 - 0} = \frac{4}{4} = 1$$

i.e. the gradient of the straight line $y = x - 3$ is 1. The actual positioning of AB and BC is unimportant for the gradient is also given by

$$\frac{DE}{EF} = \frac{-1 - (-2)}{2 - 1} = \frac{1}{1} = 1$$

The slope or gradient of each of the straight lines in Fig. 11.5 is thus 1 since they are all parallel to each other.

Problem 3. Plot the following graphs on the same axes between the values $x = -3$ to $x = +3$ and determine the gradient and y-axis intercept of each.

(a) $y = 3x$ (b) $y = 3x + 7$

(c) $y = -4x + 4$ (d) $y = -4x - 5$

A table of co-ordinates is drawn up for each equation.

(a) $y = 3x$

x	−3	−2	−1	0	1	2	3
y	−9	−6	−3	0	3	6	9

(b) $y = 3x + 7$

x	−3	−2	−1	0	1	2	3
y	−2	1	4	7	10	13	16

(c) $y = -4x + 4$

x	−3	−2	−1	0	1	2	3
y	16	12	8	4	0	−4	−8

(d) $y = -4x - 5$

x	−3	−2	−1	0	1	2	3
y	7	3	−1	−5	−9	−13	−17

Each of the graphs is plotted as shown in Fig. 11.6, and each is a straight line. $y = 3x$ and $y = 3x + 7$ are parallel to each other and thus have the same gradient. The gradient of AC is given by

$$\frac{CB}{BA} = \frac{16 - 7}{3 - 0} = \frac{9}{3} = 3$$

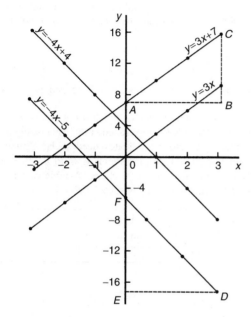

Fig. 11.6

Hence the gradient of both $y = 3x$ and $y = 3x + 7$ is 3.
$y = -4x + 4$ and $y = -4x - 5$ are parallel to each other and thus have the same gradient. The gradient of DF is given by

$$\frac{FE}{ED} = \frac{-5 - (-17)}{0 - 3} = \frac{12}{-3} = -4$$

Hence the gradient of both $y = -4x + 4$ and $y = -4x - 5$ is -4

The y-axis intercept means the value of y where the straight line cuts the y-axis. From Fig. 11.6,

$y = 3x$ cuts the y-axis at $y = 0$

$y = 3x + 7$ cuts the y-axis at $y = +7$

$y = -4x + 4$ cuts the y-axis at $y = +4$

and $y = -4x - 5$ cuts the y-axis at $y = -5$

Some general conclusions can be drawn from the graphs shown in Figs. 11.4, 11.5 and 11.6.
When an equation is of the form $y = mx + c$, where m and c are constants, then

(i) a graph of y against x produces a straight line,
(ii) m represents the slope or gradient of the line, and
(iii) c represents the y-axis intercept.

Thus, given an equation such as $y = 3x + 7$, it may be deduced 'on sight' that its gradient is $+3$ and its y-axis intercept is $+7$, as shown in Fig. 11.6. Similarly, if $y = -4x - 5$, then the gradient is -4 and the y-axis intercept is -5, as shown in Fig. 11.6.
When plotting a graph of the form $y = mx + c$, only two co-ordinates need be determined. When the co-ordinates are plotted a straight line is drawn between the two points. Normally, three co-ordinates are determined, the third one acting as a check.

Problem 4. The following equations represent straight lines. Determine, without plotting graphs, the gradient and y-axis intercept for each.

(a) $y = 3$ (b) $y = 2x$

(c) $y = 5x - 1$ (d) $2x + 3y = 3$

(a) $y = 3$ (which is of the form $y = 0x + 3$) represents a horizontal straight line intercepting the y-axis at **3**. Since the line is horizontal its **gradient is zero.**

(b) $y = 2x$ is of the form $y = mx + c$, where c is zero. Hence **gradient $= 2$** and **y-axis intercept $= 0$** (i.e. the origin).

(c) $y = 5x - 1$ is of the form $y = mx + c$. Hence **gradient $= 5$** and **y-axis intercept $= -1$**

(d) $2x + 3y = 3$ is not in the form $y = mx + c$ as it stands. Transposing to make y the subject gives $3y = 3 - 2x$, i.e.

$$y = \frac{3 - 2x}{3} = \frac{3}{3} - \frac{2x}{3}$$

i.e. $y = -\dfrac{2x}{3} + 1$

which is of the form $y = mx + c$.
Hence **gradient $= -\frac{2}{3}$** and **y-axis intercept $= +1$**

Problem 5. Without plotting graphs, determine the gradient and y-axis intercept values of the following equations:

(a) $y = 7x - 3$ (b) $3y = -6x + 2$

(c) $y - 2 = 4x + 9$ (d) $\dfrac{y}{3} = \dfrac{x}{3} - \dfrac{1}{5}$

(e) $2x + 9y + 1 = 0$

(a) $y = 7x - 3$ is of the form $y = mx + c$, hence **gradient, $m = 7$** and **y-axis intercept, $c = -3$**.

(b) Rearranging $3y = -6x + 2$ gives

$$y = -\frac{6x}{3} + \frac{2}{3}$$

i.e. $y = -2x + \frac{2}{3}$

which is of the form $y = mx + c$. Hence **gradient** $m = -2$ and **y-axis intercept,** $c = \frac{2}{3}$

(c) Rearranging $y - 2 = 4x + 9$ gives $y = 4x + 11$, hence **gradient** $= 4$ and **y-axis intercept** $= 11$

(d) Rearranging $\frac{y}{3} = \frac{x}{2} - \frac{1}{5}$ gives

$$y = 3\left(\frac{x}{2} - \frac{1}{5}\right) = \frac{3}{2}x - \frac{3}{5}$$

Hence **gradient** $= \frac{3}{2}$ and **y-axis intercept** $= -\frac{3}{5}$

(e) Rearranging $2x + 9y + 1 = 0$ gives

$$9y = -2x - 1,$$

i.e. $y = -\frac{2}{9}x - \frac{1}{9}$

Hence **gradient** $= -\frac{2}{9}$ and **y-axis intercept** $= -\frac{1}{9}$

Problem 6. Determine the gradient of the straight line graph passing through the co-ordinates (a) $(-2, 5)$ and $(3, 4)$, and (b) $(-2, -3)$ and $(-1, 3)$

A straight line graph passing through co-ordinates (x_1, y_1) and (x_2, y_2) has a gradient given by:

$$m = \frac{y_2 - y_1}{x_2 - x_1} \quad \text{(see Fig. 11.7)}$$

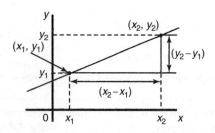

Fig. 11.7

(a) A straight line passes through $(-2, 5)$ and $(3, 4)$, hence $x_1 = -2$, $y_1 = 5$, $x_2 = 3$ and $y_2 = 4$,

hence gradient $m = \frac{y_2 - y_1}{x_2 - x_1}$

$= \frac{4 - 5}{3 - (-2)} = -\frac{1}{5}$

(b) A straight line passes through $(-2, -3)$ and $(-1, 3)$, hence $x_1 = -2$, $y_1 = -3$, $x_2 = -1$ and $y_2 = 3$,

hence gradient, $m = \frac{y_2 - y_1}{x_2 - x_1} = \frac{3 - (-3)}{-1 - (-2)}$

$= \frac{3 + 3}{-1 + 2} = \frac{6}{1} = 6$

Problem 7. Plot the graph $3x + y + 1 = 0$ and $2y - 5 = x$ on the same axes and find their point of intersection.

Rearranging $3x + y + 1 = 0$ gives $y = -3x - 1$

Rearranging $2y - 5 = x$ gives $2y = x + 5$ and $y = \frac{1}{2}x + 2\frac{1}{2}$.

Since both equations are of the form $y = mx + c$ both are straight lines. Knowing an equation is a straight line means that only two co-ordinates need to be plotted and a straight line drawn through them. A third co-ordinate is usually determined to act as a check. A table of values is produced for each equation as shown below.

x	1	0	−1
$-3x - 1$	−4	−1	2

x	2	0	−3
$\frac{1}{2}x + 2\frac{1}{2}$	$3\frac{1}{2}$	$2\frac{1}{2}$	1

The graphs are plotted as shown in Fig. 11.8

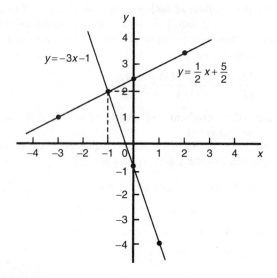

Fig. 11.8

The two straight lines are seen to intersect at $(-1, 2)$

Now try the following exercise

Exercise 41   Further problems on straight line graphs
(Answers on page 257)

1. Corresponding values obtained experimentally for two quantities are:

x	−2.0	−0.5	0	1.0	2.5	3.0	5.0
y	−13.0	−5.5	−3.0	2.0	9.5	12.0	22.0

Use a horizontal scale for x of $1\,\text{cm} = \frac{1}{2}$ unit and a vertical scale for y of $1\,\text{cm} = 2$ units and draw a graph of x against y. Label the graph and each of its axes. By interpolation, find from the graph the value of y when x is 3.5

2. The equation of a line is $4y = 2x + 5$. A table of corresponding values is produced and is shown below. Complete the table and plot a graph of y against x. Find the gradient of the graph.

x	−4	−3	−2	−1	0	1	2	3	4
y		−0.25		1.25				3.25	

3. Determine the gradient and intercept on the y-axis for each of the following equations:

   (a) $y = 4x - 2$ (b) $y = -x$

   (c) $y = -3x - 4$ (d) $y = 4$

4. Find the gradient and intercept on the y-axis for each of the following equations:

   (a) $2y - 1 = 4x$ (b) $6x - 2y = 5$

   (c) $3(2y - 1) = \dfrac{x}{4}$

Determine the gradient and y-axis intercept for each of the equations in Problems 5 and 6 and sketch the graphs.

5. (a) $y = 6x - 3$ (b) $y = -2x + 4$

   (c) $y = 3x$ (d) $y = 7$

6. (a) $2y + 1 = 4x$ (b) $2x + 3y + 5 = 0$

   (c) $3(2y - 4) = \dfrac{x}{3}$ (d) $5x - \dfrac{y}{2} - \dfrac{7}{3} = 0$

7. Determine the gradient of the straight line graphs passing through the co-ordinates:

   (a) $(2, 7)$ and $(-3, 4)$

   (b) $(-4, -1)$ and $(-5, 3)$

   (c) $\left(\dfrac{1}{4}, -\dfrac{3}{4}\right)$ and $\left(-\dfrac{1}{2}, \dfrac{5}{8}\right)$

8. State which of the following equations will produce graphs which are parallel to one another:

   (a) $y - 4 = 2x$ (b) $4x = -(y + 1)$

   (c) $x = \dfrac{1}{2}(y + 5)$ (d) $1 + \dfrac{1}{2}y = \dfrac{3}{2}x$

   (e) $2x = \dfrac{1}{2}(7 - y)$

9. Draw a graph of $y - 3x + 5 = 0$ over a range of $x = -3$ to $x = 4$. Hence determine (a) the value of y when $x = 1.3$ and (b) the value of x when $y = -9.2$

10. Draw on the same axes the graphs of $y = 3x - 5$ and $3y + 2x = 7$. Find the co-ordinates of the point of intersection. Check the result obtained by solving the two simultaneous equations algebraically.

11. Plot the graphs $y = 2x + 3$ and $2y = 15 - 2x$ on the same axes and determine their point of intersection.

11.3 Practical problems involving straight line graphs

When a set of co-ordinate values are given or are obtained experimentally and it is believed that they follow a law of the form $y = mx + c$, then if a straight line can be drawn reasonably close to most of the co-ordinate values when plotted, this verifies that a law of the form $y = mx + c$ exists. From the graph, constants m (i.e. gradient) and c (i.e. y-axis intercept) can be determined. This technique is called **determination of law** (see Chapter 15).

Problem 8.   The temperature in degrees Celsius and the corresponding values in degrees Fahrenheit are shown in the table below. Construct rectangular axes, choose a suitable scale and plot a graph of degrees Celsius (on the horizontal axis) against degrees Fahrenheit (on the vertical scale).

°C	10	20	40	60	80	100
°F	50	68	104	140	176	212

From the graph find (a) the temperature in degrees Fahrenheit at 55°C, (b) the temperature in degrees Celsius at 167°F, (c) the Fahrenheit temperature at 0°C, and (d) the Celsius temperature at 230°F.

The co-ordinates (10, 50), (20, 68), (40, 104), and so on are plotted as shown in Fig. 11.9. When the co-ordinates are joined, a straight line is produced. Since a straight line results there is a linear relationship between degrees Celsius and degrees Fahrenheit.

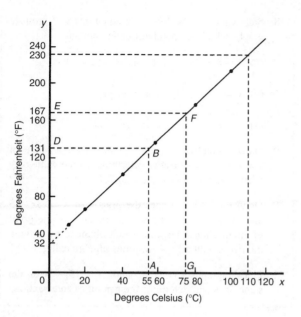

Fig. 11.9

(a) To find the Fahrenheit temperature at 55°C a vertical line *AB* is constructed from the horizontal axis to meet the straight line at *B*. The point where the horizontal line *BD* meets the vertical axis indicates the equivalent Fahrenheit temperature.

Hence 55°C is equivalent to 131°F.

This process of finding an equivalent value in between the given information in the above table is called **interpolation**.

(b) To find the Celsius temperature at 167°F, a horizontal line *EF* is constructed as shown in Fig. 11.9. The point where the vertical line *FG* cuts the horizontal axis indicates the equivalent Celsius temperature.

Hence 167°F is equivalent to 75°C.

(c) If the graph is assumed to be linear even outside of the given data, then the graph may be extended at both ends (shown by broken line in Fig. 11.9).

From Fig. 11.9, **0°C corresponds to 32°F.**

(d) **230°F is seen to correspond to 110°C.**

The process of finding equivalent values outside of the given range is called **extrapolation**.

Problem 9. In an experiment on Charles's law, the value of the volume of gas, $V\,\mathrm{m}^3$, was measured for various temperatures $T°$C. Results are shown below.

$V\,\mathrm{m}^3$	25.0	25.8	26.6	27.4	28.2	29.0
$T°$C	60	65	70	75	80	85

Plot a graph of volume (vertical) against temperature (horizontal) and from it find (a) the temperature when the volume is $28.6\,\mathrm{m}^3$, and (b) the volume when the temperature is 67°C.

If a graph is plotted with both the scales starting at zero then the result is as shown in Fig. 11.10. All of the points lie in the top right-hand corner of the graph, making interpolation difficult. A more accurate graph is obtained if the temperature axis starts at 55°C and the volume axis starts at $24.5\,\mathrm{m}^3$. The axes corresponding to these values is shown by the broken lines in Fig. 11.10 and are called **false axes**, since the origin is not now at zero. A magnified version of this relevant part of the graph is shown in Fig. 11.11. From the graph:

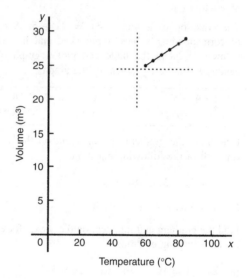

Fig. 11.10

(a) when the volume is $28.6\,\mathrm{m}^3$, the equivalent temperature is **82.5°C**, and

(b) when the temperature is 67°C, the equivalent volume is **$26.1\,\mathrm{m}^3$**

Problem 10. In an experiment demonstrating Hooke's law, the strain in an aluminium wire was measured for various stresses. The results were:

Stress N/mm^2	4.9	8.7	15.0
Strain	0.00007	0.00013	0.00021

Stress N/mm^2	18.4	24.2	27.3
Strain	0.00027	0.00034	0.00039

Plot a graph of stress (vertically) against strain (horizontally). Find:

(a) Young's Modulus of Elasticity for aluminium which is given by the gradient of the graph,

(b) the value of the strain at a stress of $20\,N/mm^2$, and

(c) the value of the stress when the strain is 0.00020

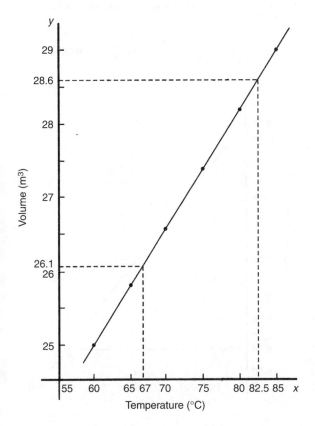

Fig. 11.11

The co-ordinates (0.00007, 4.9), (0.00013, 8.7), and so on, are plotted as shown in Fig. 11.12. The graph produced is the best straight line which can be drawn corresponding to these points. (With experimental results it is unlikely that all the points will lie exactly on a straight line.) The graph, and each of its axes, are labelled. Since the straight line passes through the origin, then stress is directly proportional to strain for the given range of values.

(a) The gradient of the straight line AC is given by

$$\frac{AB}{BC} = \frac{28 - 7}{0.00040 - 0.00010} = \frac{21}{0.00030}$$

$$= \frac{21}{3 \times 10^{-4}} = \frac{7}{10^{-4}}$$

$$= 7 \times 10^4 = 70\,000\,N/mm^2$$

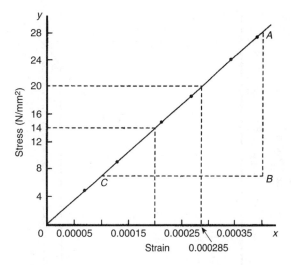

Fig. 11.12

Thus Young's Modulus of Elasticity for aluminium is $70\,000\,N/mm^2$

Since $1\,m^2 = 10^6\,mm^2$, $70\,000\,N/mm^2$ is equivalent to $70\,000 \times 10^6\,N/m^2$, i.e. $\mathbf{70 \times 10^9\,N/m^2}$ **(or Pascals)** From Fig. 11.12:

(b) the value of the strain at a stress of $20\,N/mm^2$ is **0.000285**, and

(c) the value of the stress when the strain is 0.00020 is **$14\,N/mm^2$**

Problem 11. The following values of resistance R ohms and corresponding voltage V volts are obtained from a test on a filament lamp.

R ohms	30	48.5	73	107	128
V volts	16	29	52	76	94

Choose suitable scales and plot a graph with R representing the vertical axis and V the horizontal axis. Determine (a) the gradient of the graph, (b) the R axis intercept value, (c) the equation of the graph, (d) the value of resistance when the voltage is $60\,V$, and (e) the value of the voltage when the resistance is 40 ohms. (f) If the graph were to continue in the same manner, what value of resistance would be obtained at $110\,V$?

The co-ordinates (16, 30), (29, 48.5), and so on, are shown plotted in Fig. 11.13 where the best straight line is drawn through the points.

(a) The slope or gradient of the straight line AC is given by

$$\frac{AB}{BC} = \frac{135 - 10}{100 - 0} = \frac{125}{100} = \mathbf{1.25}$$

(Note that the vertical line AB and the horizontal line BC may be constructed anywhere along the length of the

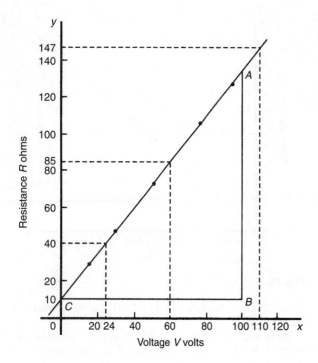

Fig. 11.13

straight line. However, calculations are made easier if the horizontal line *BC* is carefully chosen, in this case, 100).

(b) The *R*-axis intercept is at **R = 10 ohms** (by extrapolation).

(c) The equation of a straight line is $y = mx + c$, when y is plotted on the vertical axis and x on the horizontal axis. m represents the gradient and c the *y*-axis intercept. In this case, R corresponds to y, V corresponds to x, $m = 1.25$ and $c = 10$. Hence the equation of the graph is $R = (1.25\,V + 10)\,\Omega$.

From Fig. 11.13,

(d) when the voltage is 60 V, the resistance is **85 Ω**

(e) when the resistance is 40 ohms, the voltage is **24 V**, and

(f) by extrapolation, when the voltage is 110 V, the resistance is **147 Ω**.

Problem 12. Experimental tests to determine the breaking stress σ of rolled copper at various temperatures t gave the following results.

Stress σ N/cm^2	8.46	8.04	7.78
Temperature $t°$C	70	200	280

Stress σ N/cm^2	7.37	7.08	6.63
Temperature $t°$C	410	500	640

Show that the values obey the law $\sigma = at + b$, where a and b are constants and determine approximate values for a and b. Use the law to determine the stress at 250°C and the temperature when the stress is 7.54 N/cm^2.

The co-ordinates (70, 8.46), (200, 8.04), and so on, are plotted as shown in Fig. 11.14. Since the graph is a straight line then the values obey the law $\sigma = at + b$, and the gradient of the straight line is

$$a = \frac{AB}{BC} = \frac{8.36 - 6.76}{100 - 600} = \frac{1.60}{-500} = -0.0032$$

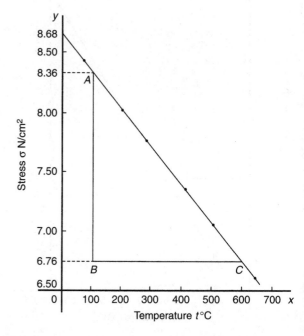

Fig. 11.14

Vertical axis intercept, $b = 8.68$

Hence the law of the graph is $\sigma = 0.0032t + 8.68$

When the temperature is 250°C, stress σ is given by

$$\sigma = -0.0032(250) + 8.68 = 7.88\,\text{N/cm}^2$$

Rearranging $\sigma = -0.0032t + 8.68$ gives

$$0.0032t = 8.68 - \sigma,$$

i.e. $$t = \frac{8.68 - \sigma}{0.0032}$$

Hence when the stress $\sigma = 7.54$ N/cm^2, temperature

$$t = \frac{8.68 - 7.54}{0.0032} = 356.3°\text{C}$$

Now try the following exercise

Exercise 42 Further practical problems involving straight line graphs (Answers on page 257)

1. The resistance R ohms of a copper winding is measured at various temperatures $t°C$ and the results are as follows:

R ohms	112	120	126	131	134
$t°C$	20	36	48	58	64

Plot a graph of R (vertically) against t (horizontally) and find from it (a) the temperature when the resistance is $122\,\Omega$ and (b) the resistance when temperature is $52°C$.

2. The speed of a motor varies with armature voltage as shown by the following experimental results:

n (rev/min)	285	517	615	750	917	1050
V volts	60	95	110	130	155	175

Plot a graph of speed (horizontally) against voltage (vertically) and draw the best straight line through the points. Find from the graph (a) the speed at a voltage of $145\,V$, and (b) the voltage at a speed of $400\,$rev/min.

3. The following table gives the force F newtons which, when applied to a lifting machine, overcomes a corresponding load of L newtons.

Force F newtons	25	47	64	120	149	187
Load L newtons	50	140	210	430	550	700

Choose suitable scales and plot a graph of F (vertically) against L (horizontally). Draw the best straight line through the points. Determine from the graph (a) the gradient, (b) the F-axis intercept, (c) the equation of the graph, (d) the force applied when the load is $310\,N$, and (e) the load that a force of $160\,N$ will overcome. (f) If the graph were to continue in the same manner, what value of force will be needed to overcome a $800\,N$ load?

4. The following table gives the results of tests carried out to determine the breaking stress σ of rolled copper at various temperatures, t:

Stress σ (N/cm^2)	8.51	8.07	7.80
Temperature $t(°C)$	75	220	310

Stress σ (N/cm^2)	7.47	7.23	6.78
Temperature $t(°C)$	420	500	650

Plot a graph of stress (vertically) against temperature (horizontally). Draw the best straight line through the plotted co-ordinates. Determine the slope of the graph and the vertical axis intercept.

5. The velocity v of a body after varying time intervals t was measured as follows:

t (seconds)	2	5	8	11	15	18
v (m/s)	16.9	19.0	21.1	23.2	26.0	28.1

Plot v vertically and t horizontally and draw a graph of velocity against time. Determine from the graph (a) the velocity after $10\,s$, (b) the time at $20\,m/s$ and (c) the equation of the graph.

6. The mass m of a steel joint varies with length L as follows:

mass, m (kg)	80	100	120	140	160
length, L (m)	3.00	3.74	4.48	5.23	5.97

Plot a graph of mass (vertically) against length (horizontally). Determine the equation of the graph.

7. The crushing strength of mortar varies with the percentage of water used in its preparation, as shown below.

Crushing strength, F (tonnes)	1.64	1.36	1.07	0.78	0.50	0.22
% of water used, $w\%$	6	9	12	15	18	21

Plot a graph of F (vertically) against w (horizontally).

(a) Interpolate and determine the crushing strength when 10% of water is used.
(b) Assuming the graph continues in the same manner extrapolate and determine the percentage of water used when the crushing strength is 0.15 tonnes.
(c) What is the equation of the graph?

8. In an experiment demonstrating Hooke's law, the strain in a copper wire was measured for various stresses. The results were:

Stress (pascals)	10.6×10^6	18.2×10^6	24.0×10^6
Strain	0.00011	0.00019	0.00025

Stress (pascals)	30.7×10^6	39.4×10^6
Strain	0.00032	0.00041

Plot a graph of stress (vertically) against strain (horizontally). Determine (a) Young's Modulus of Elasticity for copper, which is given by the gradient of the graph, (b) the value of strain at a stress of 21×10^6 Pa, (c) the value of stress when the strain is 0.00030

9. An experiment with a set of pulley blocks gave the following results:

Effort,						
E (newtons)	9.0	11.0	13.6	17.4	20.8	23.6
Load,						
L (newtons)	15	25	38	57	74	88

Plot a graph of effort (vertically) against load (horizontally) and determine (a) the gradient, (b) the vertical axis intercept, (c) the law of the graph, (d) the effort when the load is 30 N and (e) the load when the effort is 19 N

10. The variation of pressure p in a vessel with temperature T is believed to follow a law of the form $p = aT + b$, where a and b are constants. Verify this law for the results given below and determine the approximate values of a and b. Hence determine the pressures at temperatures of 285 K and 310 K and the temperature at a pressure of 250 kPa.

Pressure, p						
kPa	244	247	252	258	262	267
Temperature,						
T K	273	277	282	289	294	300

12

Graphical solution of equations

12.1 Graphical solution of simultaneous equations

Linear simultaneous equations in two unknowns may be solved graphically by:

 (i) plotting the two straight lines on the same axes, and

(ii) noting their point of intersection.

The co-ordinates of the point of intersection give the required solution.

Problem 1. Solve graphically the simultaneous equations

$$2x - y = 4$$

$$x + y = 5$$

Rearranging each equation into $y = mx + c$ form gives:

$$y = 2x - 4 \qquad (1)$$

$$y = -x + 5 \qquad (2)$$

Only three co-ordinates need be calculated for each graph since both are straight lines.

x	0	1	2
$y = 2x - 4$	−4	−2	0

x	0	1	2
$y = -x + 5$	5	4	3

Each of the graphs is plotted as shown in Fig. 12.1. The point of intersection is at (3,2) and since this is the only point which lies simultaneously on both lines then $x = 3$, $y = 2$ is the solution of the simultaneous equations.

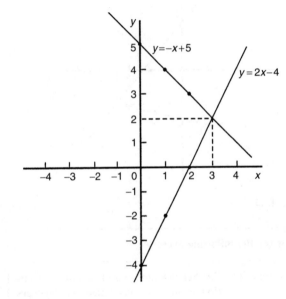

Fig. 12.1

Problem 2. Solve graphically the equations

$$1.20x + y = 1.80$$

$$x - 5.0y = 8.50$$

Rearranging each equation into $y = mx + c$ form gives:

$$y = -1.20x + 1.80 \qquad (1)$$

$$y = \frac{x}{5.0} - \frac{8.5}{5.0}$$

i.e. $y = 0.20x - 1.70 \qquad (2)$

Three co-ordinates are calculated for each equation as shown below.

x	0	1	2
$y = -1.20x + 1.80$	1.80	0.60	-0.60

x	0	1	2
$y = 0.20x - 1.70$	-1.70	-1.50	-1.30

The two lines are plotted as shown in Fig. 12.2. The point of intersection is $(2.50, -1.20)$. Hence the solution of the simultaneous equation is $x = \mathbf{2.50}$, $y = \mathbf{-1.20}$. (It is sometimes useful initially to sketch the two straight lines to determine the region where the point of intersection is. Then, for greater accuracy, a graph having a smaller range of values can be drawn to 'magnify' the point of intersection).

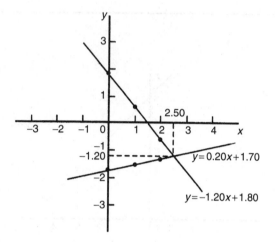

Fig. 12.2

Now try the following exercise

Exercise 43 **Further problems on the graphical solution of simultaneous equations (Answers on page 257)**

In Problems 1 to 5, solve the simultaneous equations graphically.

1. $x + y = 2$
   $3y - 2x = 1$

2. $y = 5 - x$
   $x - y = 2$

3. $3x + 4y = 5$
   $2x - 5y + 12 = 0$

4. $1.4x - 7.06 = 3.2y$
   $2.1x - 6.7y = 12.87$

5. $3x - 2y = 0$
   $4x + y + 11 = 0$

6. The friction force F newtons and load L newtons are connected by a law of the form $F = aL + b$, where a and b are constants. When $F = 4$ newtons, $L = 6$ newtons and when $F = 2.4$ newtons, $L = 2$ newtons. Determine graphically the values of a and b.

12.2 Graphical solutions of quadratic equations

A general **quadratic equation** is of the form

$$y = ax^2 + bx + c,$$

where a, b and c are constants and a is not equal to zero.

A graph of a quadratic equation always produces a shape called a **parabola**.

The gradient of the curve between 0 and A and between B and C in Fig. 12.3 is positive, whilst the gradient between A and B is negative. Points such as A and B are called **turning points**. At A the gradient is zero and, as x increases, the gradient of the curve changes from positive just before A to negative just after. Such a point is called a **maximum value**. At B the gradient is also zero, and, as x increases, the gradient of the curve changes from negative just before B to positive just after. Such a point is called a **minimum value**.

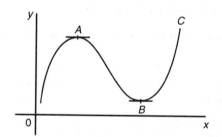

Fig. 12.3

Quadratic graphs

(i) $y = ax^2$

Graphs of $y = x^2$, $y = 3x^2$ and $y = \frac{1}{2}x^2$ are shown in Fig. 12.4.

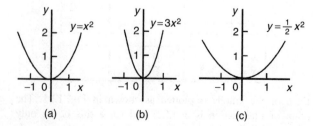

Fig. 12.4

All have minimum values at the origin (0,0).

Graphs of $y = -x^2$, $y = -3x^2$ and $y = -\frac{1}{2}x^2$ are shown in Fig. 12.5.

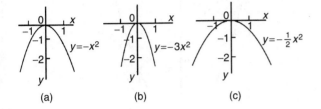

(a)　　　　　(b)　　　　　(c)

Fig. 12.5

All have maximum values at the origin (0,0).

When $y = ax^2$:

(a) curves are symmetrical about the y-axis,

(b) the magnitude of 'a' affects the gradient of the curve,

and (c) the sign of 'a' determines whether it has a maximum or minimum value.

(ii) $y = ax^2 + c$

Graphs of $y = x^2 + 3$, $y = x^2 - 2$, $y = -x^2 + 2$ and $y = -2x^2 - 1$ are shown in Fig. 12.6.

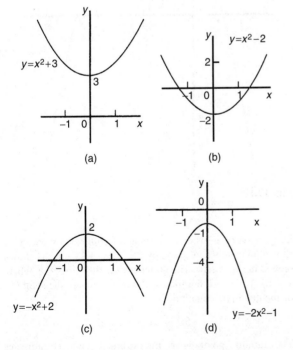

(a)　　　　　(b)

(c)　　　　　(d)

Fig. 12.6

When $y = ax^2 + c$:

(a) curves are symmetrical about the y-axis,

(b) the magnitude of 'a' affects the gradient of the curve,

and (c) the constant 'c' is the y-axis intercept.

(iii) $y = ax^2 + bx + c$

Whenever 'b' has a value other than zero the curve is displaced to the right or left of the y-axis. When b/a is positive, the curve is displaced $b/2a$ to the left of the y-axis, as shown in Fig. 12.7(a). When b/a is negative the curve is displaced $b/2a$ to the right of the y-axis, as shown in Fig. 12.7(b).

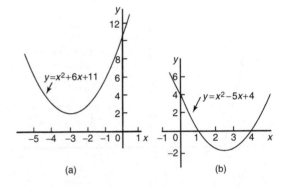

(a)　　　　　(b)

Fig. 12.7

Quadratic equations of the form $ax^2 + bx + c = 0$ may be solved graphically by:

(i) plotting the graph $y = ax^2 + bx + c$, and

(ii) noting the points of intersection on the x-axis (i.e. where $y = 0$).

The x values of the points of intersection give the required solutions since at these points both $y = 0$ and

$$ax^2 + bx + c = 0$$

The number of solutions, or roots of a quadratic equation, depends on how many times the curve cuts the x-axis and there can be no real roots (as in Fig. 12.7(a)) or one root (as in Figs. 12.4 and 12.5) or two roots (as in Fig. 12.7(b)).

> *Problem 3.* Solve the quadratic equation
>
> $$4x^2 + 4x - 15 = 0$$
>
> graphically given that the solutions lie in the range $x = -3$ to $x = 2$. Determine also the co-ordinates and nature of the turning point of the curve.

Let $y = 4x^2 + 4x - 15$. A table of values is drawn up as shown below.

x		-3	-2	-1	0	1	2
$4x^2$		36	16	4	0	4	16
$4x$		-12	-8	-4	0	4	8
-15		-15	-15	-15	-15	-15	-15
$y = 4x^2 + 4x - 15$		9	-7	-15	-15	-7	9

A graph of $y = 4x^2 + 4x - 15$ is shown in Fig. 12.8. The only points where $y = 4x^2 + 4x - 15$ and $y = 0$ are the points marked A and B. This occurs at $x = -2.5$ and $x = 1.5$ and these are the solutions of the quadratic equation $4x^2 + 4x - 15 = 0$. (By substituting $x = -2.5$ and $x = 1.5$ into the original equation the solutions may be checked.) The curve has a turning point at $(-0.5, -16)$ and the nature of the point is a **minimum**.

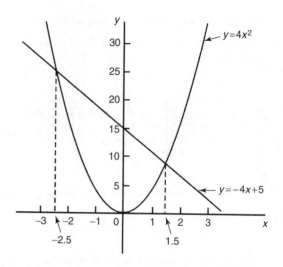

Fig. 12.9

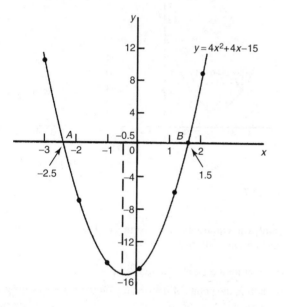

Fig. 12.8

An alternative graphical method of solving

$$4x^2 + 4x - 15 = 0$$

is to rearrange the equation as $4x^2 = -4x + 15$ and then plot two separate graphs – in this case $y = 4x^2$ and $y = -4x + 15$. Their points of intersection give the roots of equation

$$4x^2 = -4x + 15,$$

i.e. $4x^2 + 4x - 15 = 0$. This is shown in Fig. 12.9, where the roots are $x = -2.5$ and $x = 1.5$ as before.

Problem 4. Solve graphically the quadratic equation $-5x^2 + 9x + 7.2 = 0$ given that the solutions lie between $x = -1$ and $x = 3$. Determine also the co-ordinates of the turning point and state its nature.

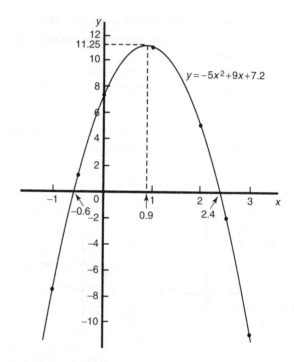

Fig. 12.10

Let $y = -5x^2 + 9x + 7.2$. A table of values is drawn up as shown on page 91. A graph of $y = -5x^2 + 9x + 7.2$ is shown plotted in Fig. 12.10. The graph crosses the x-axis (i.e. where $y = 0$) at $x = -0.6$ and $x = 2.4$ and these are the solutions of the quadratic equation

$$-5x^2 + 9x + 7.2 = 0$$

The turning point is a **maximum** having co-ordinates **(0.9, 11.25)**.

x	−1	−0.5	0	1
−5x²	−5	−1.25	0	−5
+9x	−9	−4.5	0	9
+7.2	7.2	7.2	7.2	7.2
y = −5x² + 9x + 7.2	−6.8	1.45	7.2	11.2

x	2	2.5	3
−5x²	−20	−31.25	−45
+9x	18	22.5	27
+7.2	7.2	7.2	7.2
y = −5x² + 9x + 7.2	5.2	−1.55	−10.8

Problem 5. Plot a graph of $y = 2x^2$ and hence solve the equations:

(a) $2x^2 - 8 = 0$ and (b) $2x^2 - x - 3 = 0$

A graph of $y = 2x^2$ is shown in Fig. 12.11.

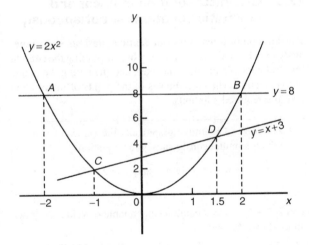

Fig. 12.11

(a) Rearranging $2x^2 - 8 = 0$ gives $2x^2 = 8$ and the solution of this equation is obtained from the points of intersection of $y = 2x^2$ and $y = 8$, i.e. at co-ordinates (−2,8) and (2,8), shown as A and B, respectively, in Fig. 12.11. Hence the solutions of $2x^2 - 8 = 0$ are **x = −2 and x = +2**

(b) Rearranging $2x^2 - x - 3 = 0$ gives $2x^2 = x + 3$ and the solution of this equation is obtained from the points of intersection of $y = 2x^2$ and $y = x + 3$, i.e. at C and D in Fig. 12.11. Hence the solutions of $2x^2 - x - 3 = 0$ are **x = −1 and x = 1.5**

Problem 6. Plot the graph of $y = -2x^2 + 3x + 6$ for values of x from $x = -2$ to $x = 4$. Use the graph to find the roots of the following equations:

(a) $-2x^2 + 3x + 6 = 0$ (b) $-2x^2 + 3x + 2 = 0$

(c) $-2x^2 + 3x + 9 = 0$ (d) $-2x^2 + x + 5 = 0$

A table of values is drawn up as shown below.

x	−2	−1	0	1	2	3	4
−2x²	−8	−2	0	−2	−8	−18	−32
+3x	−6	−3	0	3	6	9	12
+6	6	6	6	6	6	6	6
y	−8	1	6	7	4	−3	−14

A graph of $-2x^2 + 3x + 6$ is shown in Fig. 12.12.

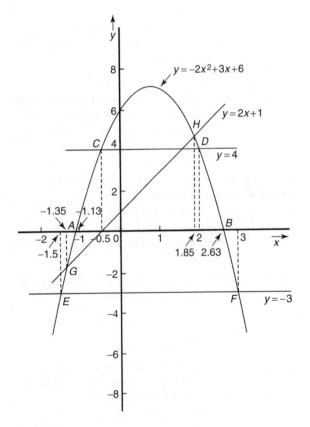

Fig. 12.12

(a) The parabola $y = -2x^2 + 3x + 6$ and the straight line $y = 0$ intersect at A and B, where **x = −1.13** **and x = 2.63** and these are the roots of the equation $-2x^2 + 3x + 6 = 0$

(b)   Comparing $y = -2x^2 + 3x + 6$ (1)

with $0 = -2x^2 + 3x + 2$ (2)

shows that if 4 is added to both sides of equation (2), the right-hand side of both equations will be the same. Hence $4 = -2x^2 + 3x + 6$. The solution of this equation is found from the points of intersection of the line $y = 4$ and the parabola $y = -2x^2 + 3x + 6$, i.e. points C and D in Fig. 12.12. Hence the roots of $-2x^2 + 3x + 2 = 0$ are $x = -0.5$ and $x = 2$

(c) $-2x^2 + 3x + 9 = 0$ may be rearranged as

$$-2x^2 + 3x + 6 = -3$$

and the solution of this equation is obtained from the points of intersection of the line $y = -3$ and the parabola $y = -2x^2 + 3x + 6x$, i.e. at points E and F in Fig. 12.12. Hence the roots of $-2x^2 + 3x + 9 = 0$ are $x = -1.5$ and $x = 3$

(d)   Comparing $y = -2x^2 + 3x + 6$ (3)

with $0 = -2x^2 + x + 5$ (4)

shows that if $2x + 1$ is added to both sides of equation (4) the right-hand side of both equations will be the same. Hence equation (4) may be written as

$$2x + 1 = -2x^2 + 3x + 6$$

The solution of this equation is found from the points of intersection of the line $y = 2x + 1$ and the parabola $y = -2x^2 + 3x + 6$, i.e. points G and H in Fig. 12.12. Hence the roots of $-2x^2 + x + 5 = 0$ are $x = -1.35$ and $x = 1.85$

Now try the following exercise

Exercise 44   Further problems on solving quadratic equations graphically (Answers on page 257)

1. Sketch the following graphs and state the nature and co-ordinates of their turning points

    (a) $y = 4x^2$ (b) $y = 2x^2 - 1$

    (c) $y = -x^2 + 3$ (d) $y = -\frac{1}{2}x^2 - 1$

    Solve graphically the quadratic equations in Problems 2 to 5 by plotting the curves between the given limits. Give answers correct to 1 decimal place.

2. $4x^2 - x - 1 = 0$; $x = -1$ to $x = 1$

3. $x^2 - 3x = 27$; $x = -5$ to $x = 8$

4. $2x^2 - 6x - 9 = 0$; $x = -2$ to $x = 5$

5. $2x(5x - 2) = 39.6$; $x = -2$ to $x = 3$

6. Solve the quadratic equation $2x^2 + 7x + 6 = 0$ graphically, given that the solutions lie in the range

$x = -3$ to $x = 1$. Determine also the nature and co-ordinates of its turning point.

7. Solve graphically the quadratic equation

$$10x^2 - 9x - 11.2 = 0$$

given that the roots lie between $x = -1$ and $x = 2$

8. Plot a graph of $y = 3x^2$ and hence solve the equations.

    (a) $3x^2 - 8 = 0$ and (b) $3x^2 - 2x - 1 = 0$

9. Plot the graphs $y = 2x^2$ and $y = 3 - 4x$ on the same axes and find the co-ordinates of the points of intersection. Hence determine the roots of the equation $2x^2 + 4x - 3 = 0$

10. Plot a graph of $y = 10x^2 - 13x - 30$ for values of x between $x = -2$ and $x = 3$.
Solve the equation $10x^2 - 13x - 30 = 0$ and from the graph determine (a) the value of y when x is 1.3, (b) the value of x when y is 10 and (c) the roots of the equation $10x^2 - 15x - 18 = 0$

12.3 Graphical solution of linear and quadratic equations simultaneously

The solution of **linear and quadratic equations simultaneously** may be achieved graphically by: (i) plotting the straight line and parabola on the same axes, and (ii) noting the points of intersection. The co-ordinates of the points of intersection give the required solutions.

Problem 7.   Determine graphically the values of x and y which simultaneously satisfy the equations

$$y = 2x^2 - 3x - 4 \text{ and } y = 2 - 4x$$

$y = 2x^2 - 3x - 4$ is a parabola and a table of values is drawn up as shown below:

x	-2	-1	0	1	2	3
$2x^2$	8	2	0	2	8	18
$-3x$	6	3	0	-3	-6	-9
-4	-4	-4	-4	-4	-4	-4
y	10	1	-4	-5	-2	5

$y = 2 - 4x$ is a straight line and only three co-ordinates need be calculated:

x	0	1	2
y	2	-2	-6

The two graphs are plotted in Fig. 12.13 and the points of intersection, shown as A and B, are at co-ordinates $(-2,10)$

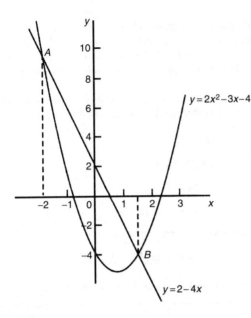

Fig. 12.13

and $(1\frac{1}{2} - 4)$. Hence the simultaneous solutions occur when $x = -2$, $y = 10$ and when $x = 1\frac{1}{2}$, $y = -4$

(These solutions may be checked by substituting into each of the original equations)

Now try the following exercise

Exercise 45 Further problems on solving linear and quadratic equations simultaneously (Answers on page 258)

1. Determine graphically the values of x and y which simultaneously satisfy the equations $y = 2(x^2 - 2x - 4)$ and $y + 4 = 3x$

2. Plot the graph of $y = 4x^2 - 8x - 21$ for values of x from -2 to $+4$. Use the graph to find the roots of the following equations:

   (a) $4x^2 - 8x - 21 = 0$ (b) $4x^2 - 8x - 16 = 0$

   (c) $4x^2 - 6x - 18 = 0$

12.4 Graphical solution of cubic equations

A **cubic equation** of the form $ax^3 + bx^2 + cx + d = 0$ may be solved graphically by: (i) plotting the graph

$$y = ax^3 + bx^2 + cx + d$$

and (ii) noting the points of intersection on the x-axis (i.e. where $y = 0$). The x-values of the points of intersection give

the required solution since at these points both $y = 0$ and $ax^3 + bx^2 + cx + d = 0$.

The number of solutions, or roots of a cubic equation depends on how many times the curve cuts the x-axis and there can be one, two or three possible roots, as shown in Fig. 12.14.

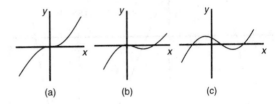

Fig. 12.14

Problem 8. Solve graphically the cubic equation

$$4x^3 - 8x^2 - 15x + 9 = 0$$

given that the roots lie between $x = -2$ and $x = 3$. Determine also the co-ordinates of the turning points and distinguish between them.

Let $y = 4x^3 - 8x^2 - 15x + 9$. A table of values is drawn up as shown below.

x	-2	-1	0	1	2	3
$4x^3$	-32	-4	0	4	32	108
$-8x^2$	-32	-8	0	-8	-32	-72
$-15x$	30	15	0	-15	-30	-45
$+9$	9	9	9	9	9	9
y	-25	12	9	-10	-21	0

A graph of $y = 4x^3 - 8x^2 - 15x + 9$ is shown in Fig. 12.15. The graph crosses the x-axis (where $y = 0$) at $x = -1\frac{1}{2}$, $x = \frac{1}{2}$ and $x = 3$ and these are the solutions to the cubic equation $4x^3 - 8x^2 - 15x + 9 = 0$. The turning points occur at $(-0.6, 14.2)$, which is a **maximum**, and $(2, -21)$, which is a **minimum**.

Problem 9. Plot the graph of $y = 2x^3 - 7x^2 + 4x + 4$ for values of x between $x = -1$ and $x = 3$. Hence determine the roots of the equation $2x^3 - 7x^2 + 4x + 4 = 0$

A table of values is drawn up as shown below.

x	-1	0	1	2	3
$2x^3$	-2	0	2	16	54
$-7x^2$	-7	0	-7	-28	-63
$+4x$	-4	0	4	8	12
$+4$	4	4	4	4	4
y	-9	4	3	0	7

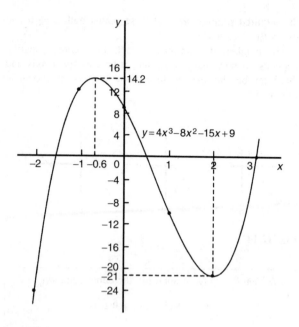

Fig. 12.15

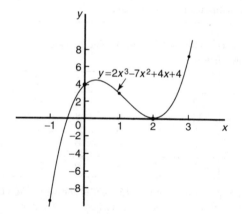

Fig. 12.16

A graph of $y = 2x^3 - 7x^2 + 4x + 4$ is shown in Fig. 12.16. The graph crosses the x-axis at $x = -0.5$ and touches the x-axis at $x = 2$.

Hence the solutions of the equation $2x^3 - 7x^2 + 4x + 4 = 0$ are $x = -0.5$ and $x = 2$

Now try the following exercise

Exercise 46 Further problems on solving cubic equations (Answers on page 258)

1. Plot the graph $y = 4x^3 + 4x^2 - 11x - 6$ between $x = -3$ and $x = 2$ and use the graph to solve the cubic equation $4x^3 + 4x^2 - 11x - 6 = 0$

2. By plotting a graph of $y = x^3 - 2x^2 - 5x + 6$ between $x = -3$ and $x = 4$ solve the equation $x^3 - 2x^2 - 5x + 6 = 0$. Determine also the co-ordinates of the turning points and distinguish between them.

In Problems 3 to 6, solve graphically the cubic equations given, each correct to 2 significant figures.

3. $x^3 - 1 = 0$

4. $x^3 - x^2 - 5x + 2 = 0$

5. $x^3 - 2x^2 = 2x - 2$

6. $2x^3 - x^2 - 9.08x + 8.28 = 0$

7. Show that the cubic equation

$$8x^3 + 36x^2 + 54x + 27 = 0$$

has only one real root and determine its value.

Assignment 6

This assignment covers the material contained in Chapters 11 and 12. The marks for each question are shown in brackets at the end of each question.

1. Determine the gradient and intercept on the y-axis for the following equations:
   (a) $y = -5x + 2$ (b) $3x + 2y + 1 = 0$ (5)

2. The equation of a line is $2y = 4x + 7$. A table of corresponding values is produced and is as shown below. Complete the table and plot a graph of y against x. Determine the gradient of the graph.   (6)

x	-3	-2	-1	0	1	2	3
y	-2.5					7.5	

3. Plot the graphs $y = 3x + 2$ and $\dfrac{y}{2} + x = 6$ on the same axes and determine the co-ordinates of their point of intersection.   (7)

4. The velocity v of a body over varying time intervals t was measured as follows:

t seconds	2	5	7	10	14	17
v m/s	15.5	17.3	18.5	20.3	22.7	24.5

Plot a graph with velocity vertical and time horizontal. Determine from the graph (a) the gradient, (b) the vertical axis intercept, (c) the equation of the graph, (d) the velocity after 12.5 s, and (e) the time when the velocity is 18 m/s. (9)

5. Solve, correct to 1 decimal place, the quadratic equation $2x^2 - 6x - 9 = 0$ by plotting values of x from $x = -2$ to $x = 5$ (7)

6. Plot the graph of $y = x^3 + 4x^2 + x - 6$ for values of x between $x = -4$ and $x = 2$. Hence determine the roots of the equation $x^3 + 4x^2 + x - 6 = 0$ (7)

7. Plot a graph of $y = 2x^2$ from $x = -3$ to $x = +3$ and hence solve the equations:
   (a) $2x^2 - 8 = 0$ (b) $2x^2 - 4x - 6 = 0$ (9)

13

Logarithms

13.1 Introduction to logarithms

With the use of calculators firmly established, logarithmic tables are now rarely used for calculation. However, the theory of logarithms is important, for there are several scientific and engineering laws that involve the rules of logarithms.

If a number y can be written in the form a^x, then the index x is called the 'logarithm of y to the base of a',

i.e. $\boxed{\text{if } y = a^x \quad \text{then} \quad x = \log_a y}$

Thus, since $1000 = 10^3$, then $3 = \log_{10} 1000$

Check this using the 'log' button on your calculator.

(a) Logarithms having a base of 10 are called **common logarithms** and $\log_{10}$ is usually abbreviated to lg. The following values may be checked by using a calculator:

$$\lg 17.9 = 1.2528\ldots, \lg 462.7 = 2.6652\ldots$$

and $\lg 0.0173 = -1.7619\ldots$

(b) Logarithms having a base of e (where 'e' is a mathematical constant approximately equal to 2.7183) are called **hyperbolic, Napierian** or **natural logarithms**, and $\log_e$ is usually abbreviated to ln. The following values may be checked by using a calculator:

$$\ln 3.15 = 1.1474\ldots, \ln 362.7 = 5.8935\ldots$$

and $\ln 0.156 = -1.8578\ldots$

For more on Napierian logarithms see Chapter 14.

13.2 Laws of logarithms

There are three laws of logarithms, which apply to any base:

(i) To multiply two numbers:

$$\boxed{\log (A \times B) = \log A + \log B}$$

The following may be checked by using a calculator:

$$\lg 10 = 1,$$

also $\lg 5 + \lg 2 = 0.69897\ldots + 0.301029\ldots = 1$

Hence $\lg(5 \times 2) = \lg 10 = \lg 5 + \lg 2$

(ii) To divide two numbers:

$$\boxed{\log \left(\frac{A}{B}\right) = \log A - \log B}$$

The following may be checked using a calculator:

$$\ln \left(\frac{5}{2}\right) = \ln 2.5 = 0.91629\ldots$$

Also $\ln 5 - \ln 2 = 1.60943\ldots - 0.69314\ldots$

$$= 0.91629\ldots$$

Hence $\ln \left(\frac{5}{2}\right) = \ln 5 - \ln 2$

(iii) To raise a number to a power:

$$\boxed{\lg A^n = n \log A}$$

The following may be checked using a calculator:

$$\lg 5^2 = \lg 25 = 1.39794\ldots$$

Also $2\lg 5 = 2 \times 0.69897\ldots = 1.39794\ldots$

Hence $\lg 5^2 = 2\lg 5$

Problem 1. Evaluate (a) $\log_3 9$ (b) $\log_{10} 10$
(c) $\log_{16} 8$

(a) Let $x = \log_3 9$ then $3^x = 9$ from the definition of a logarithm, i.e. $3^x = 3^2$, from which $x = 2$.

Hence $\log_3 9 = 2$

(b) Let $x = \log_{10} 10$ then $10^x = 10$ from the definition of a logarithm, i.e. $10^x = 10^1$, from which $x = 1$.

Hence $\log_{10} 10 = 1$ (which may be checked by a calculator).

(c) Let $x = \log_{16} 8$ then $16^x = 8$, from the definition of a logarithm, i.e. $(2^4)^x = 2^3$, i.e. $2^{4x} = 2^3$ from the laws of indices, from which, $4x = 3$ and $x = \frac{3}{4}$

Hence $\log_{16} 8 = \frac{3}{4}$

Problem 2. Evaluate (a) $\lg 0.001$ (b) $\ln e$
(c) $\log_3 \dfrac{1}{81}$

(a) Let $x = \lg 0.001 = \log_{10} 0.001$ then $10^x = 0.001$, i.e. $10^x = 10^{-3}$, from which $x = -3$

Hence $\lg 0.001 = -3$ (which may be checked by a calculator).

(b) Let $x = \ln e = \log_e e$ then $e^x = e$, i.e. $e^x = e^1$ from which $x = 1$

Hence $\ln e = 1$ (which may be checked by a calculator).

(c) Let $x = \log_3 \dfrac{1}{81}$ then $3^x = \dfrac{1}{81} = \dfrac{1}{3^4} = 3^{-4}$, from which $x = -4$

Hence $\log_3 \dfrac{1}{81} = -4$

Problem 3. Solve the following equations:
(a) $\lg x = 3$ (b) $\log_2 x = 3$ (c) $\log_5 x = -2$

(a) If $\lg x = 3$ then $\log_{10} x = 3$ and $x = 10^3$, i.e. $x = 1000$

(b) If $\log_2 x = 3$ then $x = 2^3 = 8$

(c) If $\log_5 x = -2$ then $x = 5^{-2} = \dfrac{1}{5^2} = \dfrac{1}{25}$

Problem 4. Write (a) $\log 30$ (b) $\log 450$ in terms of $\log 2$, $\log 3$ and $\log 5$ to any base.

(a) $\log 30 = \log(2 \times 15) = \log(2 \times 3 \times 5)$

$= \log 2 + \log 3 + \log 5$

by the first law of logarithms

(b) $\log 450 = \log(2 \times 225) = \log(2 \times 3 \times 75)$

$= \log(2 \times 3 \times 3 \times 25)$

$= \log(2 \times 3^2 \times 5^2)$

$= \log 2 + \log 3^2 + \log 5^2$

by the first law of logarithms

i.e $\log 450 = \log 2 + 2\log 3 + 2\log 5$

by the third law of logarithms

Problem 5. Write $\log\left(\dfrac{8 \times \sqrt[4]{5}}{81}\right)$ in terms of $\log 2$, $\log 3$ and $\log 5$ to any base.

$\log\left(\dfrac{8 \times \sqrt[4]{5}}{81}\right) = \log 8 + \log \sqrt[4]{5} - \log 81$, by the first and second laws of logarithms

$= \log 2^3 + \log 5^{(1/4)} - \log 3^4$

by the laws of indices

i.e. $\log\left(\dfrac{8 \times \sqrt[4]{5}}{81}\right) = 3\log 2 + \dfrac{1}{4}\log 5 - 4\log 3$

by the third law of logarithms

Problem 6. Simplify $\log 64 - \log 128 + \log 32$

$64 = 2^6$, $128 = 2^7$ and $32 = 2^5$

Hence $\log 64 - \log 128 + \log 32 = \log 2^6 - \log 2^7 + \log 2^5$

$= 6\log 2 - 7\log 2 + 5\log 2$

by the third law of logarithms

$= 4\ \log\ 2$

Problem 7. Evaluate $\dfrac{\log 25 - \log 125 + \frac{1}{2}\log 625}{3\log 5}$

$\dfrac{\log 25 - \log 125 + \frac{1}{2}\log 625}{3\log 5} = \dfrac{\log 5^2 - \log 5^3 + \frac{1}{2}\log 5^4}{3\log 5}$

$= \dfrac{2\log 5 - 3\log 5 + \frac{4}{2}\log 5}{3\log 5}$

$= \dfrac{1\log 5}{3\log 5} = \dfrac{1}{3}$

Problem 8. Solve the equation:

$\log(x - 1) + \log(x + 1) = 2\log(x + 2)$

$\log(x - 1) + \log(x + 1) = \log(x - 1)(x + 1)$ from the first law of logarithms

$= \log(x^2 - 1)$

$2\log(x + 2) = \log(x + 2)^2 = \log(x^2 + 4x + 4)$

Hence if $\log(x^2 - 1) = \log(x^2 + 4x + 4)$

then $x^2 - 1 = x^2 + 4x + 4$

i.e. $-1 = 4x + 4$

i.e. $-5 = 4x$

i.e. $x = -\dfrac{5}{4}$ or $-1\dfrac{1}{4}$

Now try the following exercise

Exercise 47   Further problems on the laws of logarithms (Answers on page 258)

In Problems 1 to 11, evaluate the given expression:

1. $\log_{10} 10\,000$ 2. $\log_2 16$ 3. $\log_5 125$

4. $\log_2 \dfrac{1}{8}$ 5. $\log_8 2$ 6. $\log_7 343$

7. $\lg 100$ 8. $\lg 0.01$ 9. $\log_4 8$

10. $\log_{27} 3$ 11. $\ln e^2$

In Problems 12 to 18 solve the equations:

12. $\log_{10} x = 4$ 13. $\log x = 5$

14. $\log_3 x = 2$ 15. $\log_4 x = -2\dfrac{1}{2}$

16. $\lg x = -2$ 17. $\log_8 x = -\dfrac{4}{3}$

18. $\ln x = 3$

In Problems 19 to 22 write the given expressions in terms of $\log 2$, $\log 3$ and $\log 5$ to any base:

19. $\log 60$ 20. $\log 300$

21. $\log \left(\dfrac{16 \times \sqrt[4]{5}}{27} \right)$ 22. $\log \left(\dfrac{125 \times \sqrt[4]{16}}{\sqrt[4]{81^3}} \right)$

Simplify the expressions given in Problems 23 to 25:

23. $\log 27 - \log 9 + \log 81$

24. $\log 64 + \log 32 - \log 128$

25. $\log 8 - \log 4 + \log 32$

Evaluate the expressions given in Problems 26 and 27:

26. $\dfrac{\frac{1}{2} \log 16 - \frac{1}{3} \log 8}{\log 4}$

27. $\dfrac{\log 9 - \log 3 + \frac{1}{2} \log 81}{2 \log 3}$

Solve the equations given in Problems 28 to 30:

28. $\log x^4 - \log x^3 = \log 5x - \log 2x$

29. $\log 2t^3 - \log t = \log 16 + \log t$

30. $2 \log b^2 - 3 \log b = \log 8b - \log 4b$

13.3   Indicial equations

The laws of logarithms may be used to solve certain equations involving powers – called **indicial equations**. For example, to solve, say, $3^x = 27$, logarithms to a base of 10 are taken of both sides,

i.e. $\log_{10} 3^x = \log_{10} 27$

and $x \log_{10} 3 = \log_{10} 27$ by the third law of logarithms.

Rearranging gives $x = \dfrac{\log_{10} 27}{\log_{10} 3} = \dfrac{1.43136\ldots}{0.4771\ldots} = 3$ which may be readily checked.

(Note, $(\log 8 / \log 2)$ is **not** equal to $\lg(8/2)$)

Problem 9. Solve the equation $2^x = 3$, correct to 4 significant figures.

Taking logarithms to base 10 of both sides of $2^x = 3$ gives:

$$\log_{10} 2^x = \log_{10} 3$$

i.e. $x \log_{10} 2 = \log_{10} 3$

Rearranging gives:

$$x = \frac{\log_{10} 3}{\log_{10} 2} = \frac{0.47712125\ldots}{0.30102999\ldots} = \mathbf{1.585}$$

correct to 4 significant figures.

Problem 10. Solve the equation $2^{x+1} = 3^{2x-5}$ correct to 2 decimal places.

Taking logarithms to base 10 of both sides gives:

$$\log_{10} 2^{x+1} = \log_{10} 3^{2x-5}$$

i.e. $(x + 1)\log_{10} 2 = (2x - 5)\log_{10} 3$

$x \log_{10} 2 + \log_{10} 2 = 2x \log_{10} 3 - 5 \log_{10} 3$

$x(0.3010) + (0.3010) = 2x(0.4771) - 5(0.4771)$

i.e. $0.3010x + 0.3010 = 0.9542x - 2.3855$

Hence $2.3855 + 0.3010 = 0.9542x - 0.3010x$

$2.6865 = 0.6532x$

from which $x = \dfrac{2.6865}{0.6532} = \mathbf{4.11}$

correct to 2 decimal places.

Problem 11. Solve the equation $x^{3.2} = 41.15$, correct to 4 significant figures.

Taking logarithms to base 10 of both sides gives:

$$\log_{10} x^{3.2} = \log_{10} 41.15$$

$$3.2 \log_{10} x = \log_{10} 41.15$$

Hence $\log_{10} x = \dfrac{\log_{10} 41.15}{3.2} = 0.50449$

Thus $x = $ antilog $0.50449 = 10^{0.50449} = \mathbf{3.195}$ correct to 4 significant figures.

Now try the following exercise

Exercise 48 Indicial equations (Answers on page 258)

Solve the following indicial equations for x, each correct to 4 significant figures:

1. $3^x = 6.4$ 2. $2^x = 9$ 3. $2^{x-1} = 3^{2x-1}$

4. $x^{1.5} = 14.91$ 5. $25.28 = 4.2^x$ 6. $4^{2x-1} = 5^{x+2}$

7. $x^{-0.25} = 0.792$ 8. $0.027^x = 3.26$

13.4 Graphs of logarithmic functions

A graph of $y = \log_{10} x$ is shown in Fig. 13.1 and a graph of $y = \log_e x$ is shown in Fig. 13.2. Both are seen to be of similar shape; in fact, the same general shape occurs for a logarithm to any base.

In general, with a logarithm to any base a, it is noted that:

(i) $\log_a 1 = 0$

Let $\log_a = x$, then $a^x = 1$ from the definition of the logarithm.

If $a^x = 1$ then $x = 0$ from the laws of logarithms.

Hence $\log_a 1 = 0$. In the above graphs it is seen that $\log_{10} 1 = 0$ and $\log_e 1 = 0$

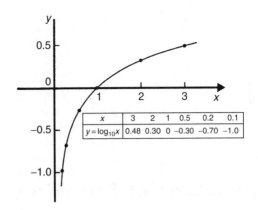

Fig. 13.1

x	3	2	1	0.5	0.2	0.1
$y = \log_{10}x$	0.48	0.30	0	−0.30	−0.70	−1.0

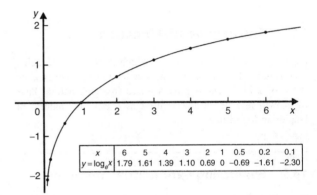

Fig. 13.2

x	6	5	4	3	2	1	0.5	0.2	0.1
$y = \log_e x$	1.79	1.61	1.39	1.10	0.69	0	−0.69	−1.61	−2.30

(ii) $\log_a a = 1$

Let $\log_a a = x$ then $a^x = a$ from the definition of a logarithm.

If $a^x = a$ then $x = 1$

Hence $\log_a a = 1$. (Check with a calculator that $\log_{10} 10 = 1$ and $\log_e e = 1$)

(iii) $\log_a 0 \to -\infty$

Let $\log_a 0 = x$ then $a^x = 0$ from the definition of a logarithm.

If $a^x = 0$, and a is a positive real number, then x must approach minus infinity. (For example, check with a calculator, $2^{-2} = 0.25$, $2^{-20} = 9.54 \times 10^{-7}$, $2^{-200} = 6.22 \times 10^{-61}$, and so on.)

Hence $\log_a 0 \to -\infty$

14

Exponential functions

14.1 The exponential function

An exponential function is one which contains e^x, e being a constant called the exponent and having an approximate value of 2.7183. The exponent arises from the natural laws of growth and decay and is used as a base for natural or Napierian logarithms.

14.2 Evaluating exponential functions

The value of e^x may be determined by using:

(a) a calculator, or

(b) the power series for e^x (see Section 14.3), or

(c) tables of exponential functions.

The most common method of evaluating an exponential function is by using a scientific notation **calculator**, this now having replaced the use of tables. Most scientific notation calculators contain an e^x function which enables all practical values of e^x and e^{-x} to be determined, correct to 8 or 9 significant figures. For example

$$e^1 = 2.7182818$$

$$e^{2.4} = 11.023176$$

$$e^{-1.618} = 0.19829489$$

correct to 8 significant figures.

In practical situations the degree of accuracy given by a calculator is often far greater than is appropriate. The accepted convention is that the final result is stated to one significant figure greater than the least significant measured value. Use your calculator to check the following values:

$$e^{0.12} = 1.1275, \text{ correct to 5 significant figures}$$

$$e^{-1.47} = 0.22993, \text{ correct to 5 decimal places}$$

$$e^{-0.431} = 0.6499, \text{ correct to 4 decimal places}$$

$$e^{9.32} = 11\,159, \text{ correct to 5 significant figures}$$

$$e^{-2.785} = 0.0617291, \text{ correct to 7 decimal places}$$

Problem 1. Using a calculator, evaluate, correct to 5 significant figures:

(a) $e^{2.731}$ (b) $e^{-3.162}$ (c) $\dfrac{5}{3}e^{5.253}$

(a) $e^{2.731} = 15.348227\ldots = \mathbf{15.348}$, correct to 5 significant figures.

(b) $e^{-3.162} = 0.04234097\ldots = \mathbf{0.042341}$, correct to 5 significant figures.

(c) $\dfrac{5}{3}e^{5.253} = \dfrac{5}{3}(191.138825\ldots) = \mathbf{318.56}$, correct to 5 significant figures.

Problem 2. Use a calculator to determine the following, each correct to 4 significant figures:

(a) $3.72e^{0.18}$ (b) $53.2e^{-1.4}$ (c) $\dfrac{5}{122}e^7$

(a) $3.72e^{0.18} = (3.72)(1.197217\ldots) = \mathbf{4.454}$, correct to 4 significant figures.

(b) $53.2e^{-1.4} = (53.2)(0.246596\ldots) = \mathbf{13.12}$, correct to 4 significant figures.

(c) $\dfrac{5}{122}e^7 = \dfrac{5}{122}(1096.6331\ldots) = \mathbf{44.94}$, correct to 4 significant figures.

Problem 3. Evaluate the following correct to 4 decimal places, using a calculator:

(a) $0.0256(e^{5.21} - e^{2.49})$ (b) $5\left(\dfrac{e^{0.25} - e^{-0.25}}{e^{0.25} + e^{-0.25}}\right)$

(a) $0.0256(e^{5.21} - e^{2.49})$

$\qquad = 0.0256(183.094058\ldots - 12.0612761\ldots)$

$\qquad = \mathbf{4.3784}$, correct to 4 decimal places

(b) $5\left(\dfrac{e^{0.25} - e^{-0.25}}{e^{0.25} + e^{-0.25}}\right)$

$\qquad = 5\left(\dfrac{1.28402541\ldots - 0.77880078\ldots}{1.28402541\ldots + 0.77880078\ldots}\right)$

$\qquad = 5\left(\dfrac{0.5052246\ldots}{2.0628261\ldots}\right)$

$\qquad = \mathbf{1.2246}$, correct to 4 decimal places

Problem 4. The instantaneous voltage v in a capacitive circuit is related to time t by the equation $v = Ve^{-t/CR}$ where V, C and R are constants. Determine v, correct to 4 significant figures, when $t = 30 \times 10^{-3}$ seconds, $C = 10 \times 10^{-6}$ farads, $R = 47 \times 10^3$ ohms and $V = 200$ volts.

$v = Ve^{-t/CR} = 200e^{(-30\times 10^{-3})/(10\times 10^{-6}\times 47\times 10^3)}$

Using a calculator, $v = 200e^{-0.0638297\ldots}$

$\qquad = 200(0.9381646\ldots)$

$\qquad = \mathbf{187.6\ volts}$

Now try the following exercise

Exercise 49 Further problems on evaluating exponential functions (Answers on page 258)

In Problems 1 and 2 use a calculator to evaluate the given functions correct to 4 significant figures:

1. (a) $e^{4.4}$ (b) $e^{-0.25}$ (c) $e^{0.92}$

2. (a) $e^{-1.8}$ (b) $e^{-0.78}$ (c) e^{10}

3. Evaluate, correct to 5 significant figures:

(a) $3.5e^{2.8}$ (b) $-\dfrac{6}{5}e^{-1.5}$ (c) $2.16e^{5.7}$

4. Use a calculator to evaluate the following, correct to 5 significant figures:

(a) $e^{1.629}$ (b) $e^{-2.7483}$ (c) $0.62e^{4.178}$

In Problems 5 and 6, evaluate correct to 5 decimal places.

5. (a) $\dfrac{1}{7}e^{3.4629}$ (b) $8.52e^{-1.2651}$ (c) $\dfrac{5e^{2.6921}}{3e^{1.1171}}$

6. (a) $\dfrac{5.6823}{e^{-2.1347}}$ (b) $\dfrac{e^{2.1127} - e^{-2.1127}}{2}$

(c) $\dfrac{4(e^{-1.7295} - 1)}{e^{3.6817}}$

7. The length of a bar, l, at a temperature θ is given by $l = l_0 e^{\alpha\theta}$, where l_0 and α are constants. Evaluate l, correct to 4 significant figures, when $l_0 = 2.587$, $\theta = 321.7$ and $\alpha = 1.771 \times 10^{-4}$

14.3 The power series for e^x

The value of e^x can be calculated to any required degree of accuracy since it is defined in terms of the following **power series**:

$$e^x = 1 + x + \frac{x^2}{2!} + \frac{x^3}{3!} + \frac{x^4}{4!} + \cdots \qquad (1)$$

(where $3! = 3 \times 2 \times 1$ and is called 'factorial 3')

The series is valid for all values of x.

The series is said to **converge**, i.e. if all the terms are added, an actual value for e^x (where x is a real number) is obtained. The more terms that are taken, the closer will be the value of e^x to its actual value. The value of the exponent e, correct to say 4 decimal places, may be determined by substituting $x = 1$ in the power series of equation (1). Thus

$$e^1 = 1 + 1 + \frac{(1)^2}{2!} + \frac{(1)^3}{3!} + \frac{(1)^4}{4!} + \frac{(1)^5}{5!} + \frac{(1)^6}{6!}$$

$$+ \frac{(1)^7}{7!} + \frac{(1)^8}{8!} + \cdots$$

$$= 1 + 1 + 0.5 + 0.16667 + 0.04167 + 0.00833$$

$$+ 0.00139 + 0.00020 + 0.00002 + \cdots$$

$$= 2.71828$$

i.e. e $= 2.7183$ correct to 4 decimal places

The value of $e^{0.05}$, correct to say 8 significant figures, is found by substituting $x = 0.05$ in the power series for e^x. Thus

$$e^{0.05} = 1 + 0.05 + \frac{(0.05)^2}{2!} + \frac{(0.05)^3}{3!}$$

$$+ \frac{(0.05)^4}{4!} + \frac{(0.05)^5}{5!} + \cdots$$

$$= 1 + 0.05 + 0.00125 + 0.000020833$$

$$+ 0.000000260 + 0.000000003$$

and by adding,

$$e^{0.05} = 1.0512711, \text{ correct to 8 significant figures}$$

In this example, successive terms in the series grow smaller very rapidly and it is relatively easy to determine the value of $e^{0.05}$ to a high degree of accuracy. However, when x is nearer to unity or larger than unity, a very large number of terms are required for an accurate result.

If in the series of equation (1), x is replaced by $-x$, then

$$e^{-x} = 1 + (-x) + \frac{(-x)^2}{2!} + \frac{(-x)^3}{3!} + \cdots$$

$$e^{-x} = 1 - x + \frac{x^2}{2!} - \frac{x^3}{3!} + \cdots$$

In a similar manner the power series for e^x may be used to evaluate any exponential function of the form ae^{kx}, where a and k are constants. In the series of equation (1), let x be replaced by kx. Then

$$ae^{kx} = a\left\{1 + (kx) + \frac{(kx)^2}{2!} + \frac{(kx)^3}{3!} + \cdots\right\}$$

Thus $5e^{2x} = 5\left\{1 + (2x) + \frac{(2x)^2}{2!} + \frac{(2x)^3}{3!} + \cdots\right\}$

$$= 5\left\{1 + 2x + \frac{4x^2}{2} + \frac{8x^3}{6} + \cdots\right\}$$

i.e. $5e^{2x} = 5\left\{1 + 2x + 2x^2 + \frac{4}{3}x^3 + \cdots\right\}$

Problem 5. Determine the value of $5e^{0.5}$, correct to 5 significant figures by using the power series for e^x.

$$e^x = 1 + x + \frac{x^2}{2!} + \frac{x^3}{3!} + \frac{x^4}{4!} + \cdots$$

Hence $e^{0.5} = 1 + 0.5 + \frac{(0.5)^2}{(2)(1)} + \frac{(0.5)^3}{(3)(2)(1)}$

$$+ \frac{(0.5)^4}{(4)(3)(2)(1)} + \frac{(0.5)^5}{(5)(4)(3)(2)(1)}$$

$$+ \frac{(0.5)^6}{(6)(5)(4)(3)(2)(1)}$$

$$= 1 + 0.5 + 0.125 + 0.020833$$

$$+ 0.0026042 + 0.0002604 + 0.0000217$$

i.e. $e^{0.5} = 1.64872$ correct to 6 significant figures

Hence $5e^{0.5} = 5(1.64872) = \mathbf{8.2436}$, correct to 5 significant figures.

Problem 6. Determine the value of $3e^{-1}$, correct to 4 decimal places, using the power series for e^x.

Substituting $x = -1$ in the power series

$$e^x = 1 + x + \frac{x^2}{2!} + \frac{x^3}{3!} + \frac{x^4}{4!} + \cdots$$

gives $e^{-1} = 1 + (-1) + \frac{(-1)^2}{2!} + \frac{(-1)^3}{3!} + \frac{(-1)^4}{4!} + \cdots$

$$= 1 - 1 + 0.5 - 0.166667 + 0.041667$$

$$- 0.008333 + 0.001389 - 0.000198 + \cdots$$

$$= 0.367858 \text{ correct to 6 decimal places}$$

Hence $3e^{-1} = (3)(0.367858) = \mathbf{1.1036}$ correct to 4 decimal places.

Problem 7. Expand $e^x(x^2 - 1)$ as far as the term in x^5

The power series for e^x is

$$e^x = 1 + x + \frac{x^2}{2!} + \frac{x^3}{3!} + \frac{x^4}{4!} + \frac{x^5}{5!} + \cdots$$

Hence

$$e^x(x^2 - 1) = \left(1 + x + \frac{x^2}{2!} + \frac{x^3}{3!} + \frac{x^4}{4!} + \frac{x^5}{5!} + \cdots\right)(x^2 - 1)$$

$$= \left(x^2 + x^3 + \frac{x^4}{2!} + \frac{x^5}{3!} + \cdots\right)$$

$$- \left(1 + x + \frac{x^2}{2!} + \frac{x^3}{3!} + \frac{x^4}{4!} + \frac{x^5}{5!} + \cdots\right)$$

Grouping like terms gives:

$$e^x(x^2 - 1) = -1 - x + \left(x^2 - \frac{x^2}{2!}\right) + \left(x^3 - \frac{x^3}{3!}\right)$$

$$+ \left(\frac{x^4}{2!} - \frac{x^4}{4!}\right) + \left(\frac{x^5}{3!} - \frac{x^5}{5!}\right) + \cdots$$

$$= \mathbf{-1 - x + \frac{1}{2}x^2 + \frac{5}{6}x^3 + \frac{11}{24}x^4 + \frac{19}{120}x^5}$$

when expanded as far as the term in x^5

Now try the following exercise

Exercise 50 **Further problems on the power series for e^x (Answers on page 258)**

1. Evaluate $5.6e^{-1}$, correct to 4 decimal places, using the power series for e^x.

2. Use the power series for e^x to determine, correct to 4 significant figures, (a) e^2 (b) $e^{-0.3}$ and check your result by using a calculator.

3. Expand $(1 - 2x)e^{2x}$ as far as the term in x^4.

4. Expand $(2e^{x^2})(x^{1/2})$ to six terms.

14.4 Graphs of exponential functions

Values of e^x and e^{-x} obtained from a calculator, correct to 2 decimal places, over a range $x = -3$ to $x = 3$, are shown in the following table.

x	-3.0	-2.5	-2.0	-1.5	-1.0	-0.5	0
e^x	0.05	0.08	0.14	0.22	0.37	0.61	1.00
e^{-x}	20.09	12.18	7.9	4.48	2.72	1.65	1.00

x	0.5	1.0	1.5	2.0	2.5	3.0
e^x	1.65	2.72	4.48	7.39	12.18	20.09
e^{-x}	0.61	0.37	0.22	0.14	0.08	0.05

Figure 14.1 shows graphs of $y = e^x$ and $y = e^{-x}$

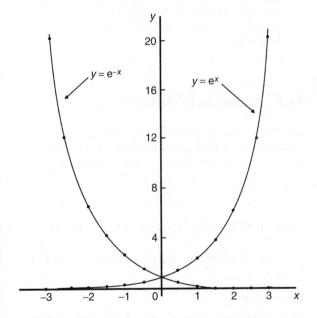

Fig. 14.1

Problem 8. Plot a graph of $y = 2e^{0.3x}$ over a range of $x = -2$ to $x = 3$. Hence determine the value of y when $x = 2.2$ and the value of x when $y = 1.6$

A table of values is drawn up as shown below.

x	-3	-2	-1	0	1	2	3
$0.3x$	-0.9	-0.6	-0.3	0	0.3	0.6	0.9
$e^{0.3x}$	0.407	0.549	0.741	1.000	1.350	1.822	2.460
$2e^{0.3x}$	0.81	1.10	1.48	2.00	2.70	3.64	4.92

A graph of $y = 2e^{0.3x}$ is shown plotted in Fig. 14.2.

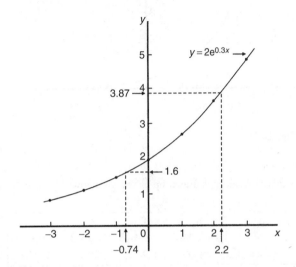

Fig. 14.2

From the graph, **when $x = 2.2$, $y = 3.87$** and **when $y = 1.6$, $x = -0.74$**

Problem 9. Plot a graph of $y = \frac{1}{3}e^{-2x}$ over the range $x = -1.5$ to $x = 1.5$. Determine from the graph the value of y when $x = -1.2$ and the value of x when $y = 1.4$

A table of values is drawn up as shown below.

x	-1.5	-1.0	-0.5	0	0.5	1.0	1.5
$-2x$	3	2	1	0	-1	-2	-3
e^{-2x}	20.086	7.389	2.718	1.00	0.368	0.135	0.050
$\frac{1}{3}e^{-2x}$	6.70	2.46	0.91	0.33	0.12	0.05	0.02

A graph of $\frac{1}{3}e^{-2x}$ is shown in Fig. 14.3.
From the graph, **when $x = -1.2$, $y = 3.67$** and **when $y = 1.4$, $x = -0.72$**

Problem 10. The decay of voltage, v volts, across a capacitor at time t seconds is given by $v = 250e^{-t/3}$. Draw a graph showing the natural decay curve over the first 6 seconds. From the graph, find (a) the voltage after 3.4 s, and (b) the time when the voltage is 150 V.

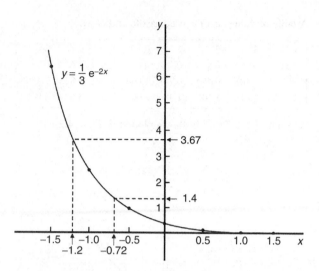

$y = \frac{1}{3} e^{-2x}$

Fig. 14.3

A table of values is drawn up as shown below.

t	0	1	2	3
$e^{-t/3}$	1.00	0.7165	0.5134	0.3679
$v = 250e^{-t/3}$	250.0	179.1	128.4	91.97

t	4	5	6
$e^{-t/3}$	0.2636	0.1889	0.1353
$v = 250e^{-t/3}$	65.90	47.22	33.83

The natural decay curve of $v = 250e^{-t/3}$ is shown in Fig. 14.4.

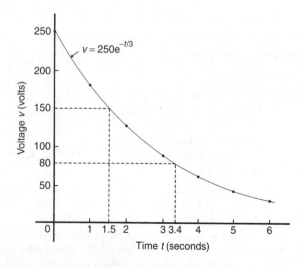

$v = 250e^{-t/3}$

Fig. 14.4

From the graph:

(a) **when time $t = 3.4$ s, voltage $v = 80$ volts**

and (b) **when voltage $v = 150$ volts, time $t = 1.5$ seconds.**

Now try the following exercise

Exercise 51　Further problems on exponential graphs (Answers on page 258)

1. Plot a graph of $y = 3e^{0.2x}$ over the range $x = -3$ to $x = 3$. Hence determine the value of y when $x = 1.4$ and the value of x when $y = 4.5$

2. Plot a graph of $y = \frac{1}{2}e^{-1.5x}$ over a range $x = -1.5$ to $x = 1.5$ and hence determine the value of y when $x = -0.8$ and the value of x when $y = 3.5$

3. In a chemical reaction the amount of starting material C cm^3 left after t minutes is given by $C = 40e^{-0.006t}$. Plot a graph of C against t and determine (a) the concentration C after 1 hour, and (b) the time taken for the concentration to decrease by half.

4. The rate at which a body cools is given by $\theta = 250e^{-0.05t}$ where the excess of temperature of a body above its surroundings at time t minutes is $\theta°$C. Plot a graph showing the natural decay curve for the first hour of cooling. Hence determine (a) the temperature after 25 minutes, and (b) the time when the temperature is 195°C

14.5　Napierian logarithms

Logarithms having a base of e are called **hyperbolic, Napierian** or **natural logarithms** and the Napierian logarithm of x is written as $\log_e x$, or more commonly, $\ln x$.

14.6　Evaluating Napierian logarithms

The value of a Napierian logarithm may be determined by using:

(a) a calculator, or

(b) a relationship between common and Napierian logarithms, or

(c) Napierian logarithm tables.

The most common method of evaluating a Napierian logarithm is by a scientific notation **calculator**, this now having replaced the use of four-figure tables, and also the relationship between common and Napierian logarithms,

$$\log_e y = 2.3026 \log_{10} y$$

Most scientific notation calculators contain a '$\ln x$' function which displays the value of the Napierian logarithm of a number when the appropriate key is pressed.

Using a calculator,

$$\ln 4.692 = 1.5458589\ldots = 1.5459,$$
correct to 4 decimal places

and $\ln 35.78 = 3.57738907\ldots = 3.5774,$
correct to 4 decimal places

Use your calculator to check the following values:

$\ln 1.732 = 0.54928$, correct to 5 significant figures

$\ln 1 = 0$

$\ln 0.52 = -0.6539$, correct to 4 decimal places

$\ln 593 = 6.3852$, correct to 5 significant figures

$\ln 1750 = 7.4674$, correct to 4 decimal places

$\ln 0.17 = -1.772$, correct to 4 significant figures

$\ln 0.00032 = -8.04719$, correct to 6 significant figures

$\ln e^3 = 3$

$\ln e^1 = 1$

From the last two examples we can conclude that

$$\log_e e^x = x$$

This is useful when solving equations involving exponential functions.

For example, to solve $e^{3x} = 8$, take Napierian logarithms of both sides, which gives

$$\ln e^{3x} = \ln 8$$

i.e. $\qquad\qquad 3x = \ln 8$

from which $\qquad x = \frac{1}{3}\ln 8 = \mathbf{0.6931},$
correct to 4 decimal places

Problem 11. Using a calculator evaluate correct to 5 significant figures:

(a) $\ln 47.291$ (b) $\ln 0.06213$ (c) $3.2 \ln 762.923$

(a) $\ln 47.291 = 3.8563200\ldots = \mathbf{3.8563}$, correct to 5 significant figures.

(b) $\ln 0.06213 = -2.7785263\ldots = \mathbf{-2.7785}$, correct to 5 significant figures.

(c) $3.2 \ln 762.923 = 3.2(6.6371571\ldots) = \mathbf{21.239}$, correct to 5 significant figures.

Problem 12. Use a calculator to evaluate the following, each correct to 5 significant figures:

(a) $\dfrac{1}{4}\ln 4.7291$ (b) $\dfrac{\ln 7.8693}{7.8693}$

(c) $\dfrac{5.29 \ln 24.07}{e^{-0.1762}}$

(a) $\dfrac{1}{4}\ln 4.7291 = \dfrac{1}{4}(1.5537349\ldots) = \mathbf{0.38843}$, correct to 5 significant figures.

(b) $\dfrac{\ln 7.8693}{7.8693} = \dfrac{2.06296911\ldots}{7.8693} = \mathbf{0.26215}$, correct to 5 significant figures.

(c) $\dfrac{5.29 \ln 24.07}{e^{-0.1762}} = \dfrac{5.29(3.18096625\ldots)}{0.83845027\ldots} = \mathbf{20.070}$, correct to 5 significant figures.

Problem 13. Evaluate the following:

(a) $\dfrac{\ln e^{2.5}}{\lg 10^{0.5}}$ (b) $\dfrac{4e^{2.23}\lg 2.23}{\ln 2.23}$ (correct to 3 decimal places)

(a) $\dfrac{\ln e^{2.5}}{\lg 10^{0.5}} = \dfrac{2.5}{0.5} = \mathbf{5}$

(b) $\dfrac{4e^{2.23}\lg 2.23}{\ln 2.23} = \dfrac{4(9.29986607\ldots)(0.34830486\ldots)}{0.80200158\ldots}$

$\qquad\qquad = \mathbf{16.156}$, correct to 3 decimal places

---

Problem 14. Solve the equation $7 = 4e^{-3x}$ to find x, correct to 4 significant figures.

Rearranging $7 = 4e^{-3x}$ gives:

$$\frac{7}{4} = e^{-3x}$$

Taking the reciprocal of both sides gives:

$$\frac{4}{7} = \frac{1}{e^{-3x}} = e^{3x}$$

Taking Napierian logarithms of both sides gives:

$$\ln\left(\frac{4}{7}\right) = \ln(e^{3x})$$

Since $\log_e e^\alpha = \alpha$, then $\ln\left(\dfrac{4}{7}\right) = 3x$

Hence $x = \dfrac{1}{3}\ln\left(\dfrac{4}{7}\right) = \dfrac{1}{3}(-0.55962) = \mathbf{-0.1865}$, correct to 4 significant figures.

Problem 15. Given $20 = 60(1 - e^{-t/2})$ determine the value of t, correct to 3 significant figures.

Rearranging $20 = 60(1 - e^{-t/2})$ gives:

$$\frac{20}{60} = 1 - e^{-1/2}$$

and

$$e^{-t/2} = 1 - \frac{20}{60} = \frac{2}{3}$$

Taking the reciprocal of both sides gives:

$$e^{t/2} = \frac{3}{2}$$

Taking Napierian logarithms of both sides gives:

$$\ln e^{t/2} = \ln \frac{3}{2}$$

i.e.

$$\frac{t}{2} = \ln \frac{3}{2}$$

from which, $t = 2 \ln \dfrac{3}{2} = \mathbf{0.881}$, correct to 3 significant figures.

Problem 16. Solve the equation $3.72 = \ln \left(\dfrac{5.14}{x} \right)$ to find x.

From the definition of a logarithm, since $3.72 = \left(\dfrac{5.14}{x} \right)$

then $e^{3.72} = \dfrac{5.14}{x}$

Rearranging gives: $x = \dfrac{5.14}{e^{3.72}} = 5.14 e^{-3.72}$

i.e. $x = \mathbf{0.1246}$,

correct to 4 significant figures

Now try the following exercise

Exercise 52   Further problems on evaluating Napierian logarithms (Answers on page 258)

In Problems 1 to 3 use a calculator to evaluate the given functions, correct to 4 decimal places.

1. (a) $\ln 1.73$ (b) $\ln 5.413$ (c) $\ln 9.412$

2. (a) $\ln 17.3$ (b) $\ln 541.3$ (c) $\ln 9412$

3. (a) $\ln 0.173$ (b) $\ln 0.005413$ (c) $\ln 0.09412$

In Problems 4 and 5, evaluate correct to 5 significant figures.

4. (a) $\dfrac{1}{6} \ln 5.2932$ (b) $\dfrac{\ln 82.473}{4.829}$ (c) $\dfrac{5.62 \ln 321.62}{e^{1.2942}}$

5. (a) $\dfrac{2.946 \ln e^{1.76}}{\lg 10^{1.41}}$ (b) $\dfrac{5 e^{-0.1629}}{2 \ln 0.00165}$

    (c) $\dfrac{\ln 4.8629 - \ln 2.4711}{5.173}$

In Problems 6 to 10 solve the given equations, each correct to 4 significant figures.

6. $1.5 = 4e^{2t}$ 7. $7.83 = 2.91e^{-1.7x}$

8. $16 = 24(1 - e^{-t/2})$ 9. $5.17 = \ln \left(\dfrac{x}{4.64} \right)$

10. $3.72 \ln \left(\dfrac{1.59}{x} \right) = 2.43$

14.7   Laws of growth and decay

The laws of exponential growth and decay are of the form $y = Ae^{-kx}$ and $y = A(1 - e^{-kx})$, where A and k are constants. When plotted, the form of each of these equations is as shown in Fig. 14.5. The laws occur frequently in engineering and science and examples of quantities related by a natural law include:

(i)   Linear expansion $l = l_0 e^{\alpha\theta}$

(ii) Change in electrical resistance with temperature

$$R_\theta = R_0 e^{\alpha\theta}$$

(iii) Tension in belts $T_1 = T_0 e^{\mu\theta}$

(iv) Newton's law of cooling $\theta = \theta_0 e^{-kt}$

(v)   Biological growth $y = y_0 e^{kt}$

(vi) Discharge of a capacitor $q = Q e^{-t/CR}$

(vii) Atmospheric pressure $p = p_0 e^{-h/c}$

(viii) Radioactive decay $N = N_0 e^{-\lambda t}$

(ix) Decay of current in an inductive circuit $i = Ie^{-Rt/L}$

(x)   Growth of current in a capacitive circuit

$$i = I(1 - e^{-t/CR})$$

Problem 17. The resistance R of an electrical conductor at temperature $\theta°C$ is given by $R = R_0 e^{\alpha\theta}$, where α is a constant and $R_0 = 5 \times 10^3$ ohms. Determine the value of α, correct to 4 significant figures, when $R = 6 \times 10^3$ ohms and $\theta = 1500°C$. Also, find the temperature, correct to the nearest degree, when the resistance R is 5.4×10^3 ohms.

Transposing $R = R_0 e^{\alpha\theta}$ gives $\dfrac{R}{R_0} = e^{\alpha\theta}$

Taking Napierian logarithms of both sides gives:

$$\ln \frac{R}{R_0} = \ln e^{\alpha\theta} = \alpha\theta$$

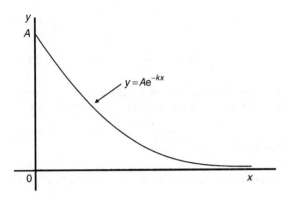

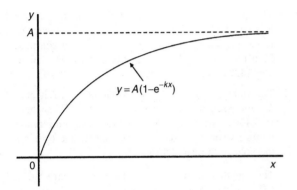

Fig. 14.5

Hence

$$\alpha = \frac{1}{\theta} \ln \frac{R}{R_0} = \frac{1}{1500} \ln \left(\frac{6 \times 10^3}{5 \times 10^3} \right)$$

$$= \frac{1}{1500}(0.1823215\ldots) = 1.215477\ldots \times 10^{-4}$$

Hence $\alpha = 1.215 \times 10^{-4}$, correct to 4 significant figures.

From above, $\ln \dfrac{R}{R_0} = \alpha\theta$ hence $\theta = \dfrac{1}{\alpha} \ln \dfrac{R}{R_0}$

When $R = 5.4 \times 10^3$, $\alpha = 1.215477\ldots \times 10^{-4}$ and $R_0 = 5 \times 10^3$

$$\theta = \frac{1}{1.215477\ldots \times 10^{-4}} \ln \left(\frac{5.4 \times 10^3}{5 \times 10^3} \right)$$

$$= \frac{10^4}{1.215477\ldots}(7.696104\ldots \times 10^{-2})$$

$$= 633°C \text{ correct to the nearest degree.}$$

Problem 18. In an experiment involving Newton's law of cooling, the temperature $\theta(°C)$ is given by $\theta = \theta_0 e^{-kt}$. Find the value of constant k when $\theta_0 = 56.6°C$, $\theta = 16.5°C$ and $t = 83.0$ seconds.

Transposing $\theta = \theta_0 e^{-kt}$ gives $\dfrac{\theta}{\theta_0} = e^{-kt}$ from which

$$\frac{\theta_0}{\theta} = \frac{1}{e^{-kt}} = e^{kt}$$

Taking Napierian logarithms of both sides gives:

$$\ln \frac{\theta_0}{\theta} = kt$$

from which,

$$k = \frac{1}{t} \ln \frac{\theta_0}{\theta} = \frac{1}{83.0} \ln \left(\frac{56.6}{16.5} \right)$$

$$= \frac{1}{83.0}(1.2326486\ldots)$$

Hence $k = 1.485 \times 10^{-2}$

Problem 19. The current i amperes flowing in a capacitor at time t seconds is given by $i = 8.0(1 - e^{-t/CR})$, where the circuit resistance R is 25×10^3 ohms and capacitance C is 16×10^{-6} farads. Determine (a) the current i after 0.5 seconds and (b) the time, to the nearest millisecond, for the current to reach 6.0 A. Sketch the graph of current against time.

(a) Current $i = 8.0(1 - e^{-t/CR})$

$$= 8.0[1 - e^{0.5/(16 \times 10^{-6})(25 \times 10^3)}]$$

$$= 8.0(1 - e^{-1.25})$$

$$= 8.0(1 - 0.2865047\ldots)$$

$$= 8.0(0.7134952\ldots)$$

$$= 5.71 \text{ amperes}$$

(b) Transposing $i = 8.0(1 - e^{-t/CR})$ gives $\dfrac{i}{8.0} = 1 - e^{-t/CR}$

from which, $e^{-t/CR} = 1 - \dfrac{i}{8.0} = \dfrac{8.0 - i}{8.0}$

Taking the reciprocal of both sides gives:

$$e^{t/CR} = \frac{8.0}{8.0 - i}$$

Taking Napierian logarithms of both sides gives:

$$\frac{t}{CR} = \ln \left(\frac{8.0}{8.0 - i} \right)$$

Hence $t = CR \ln \left(\dfrac{8.0}{8.0 - i} \right)$

$$= (16 \times 10^{-6})(25 \times 10^3) \ln \left(\frac{8.0}{8.0 - 6.0} \right)$$

when $i = 6.0$ amperes,

i.e. $t = \dfrac{400}{10^3} \ln\left(\dfrac{8.0}{2.0}\right) = 0.4\ln 4.0$

$\qquad\qquad = 0.4(1.3862943\ldots) = 0.5545\,\text{s}$

$\qquad\qquad = \mathbf{555\,ms}$, to the nearest millisecond

A graph of current against time is shown in Fig. 14.6.

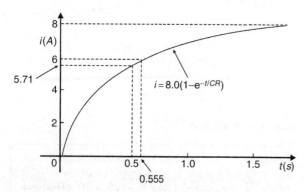

Fig. 14.6

Problem 20. The temperature θ_2 of a winding which
is being heated electrically at time t is given by:
$\theta_2 = \theta_1(1-e^{-t/\tau})$ where θ_1 is the temperature (in degrees
Celsius) at time $t=0$ and τ is a constant. Calculate.
(a) θ_1, correct to the nearest degree, when θ_2 is 50°C,
    t is 30 s and τ is 60 s
(b) the time t, correct to 1 decimal place, for θ_2 to be
    half the value of θ_1

(a) Transposing the formula to make θ_1 the subject gives:

$$\theta_1 = \frac{\theta_2}{(1 - e^{-t/\tau})} = \frac{50}{1 - e^{-30/60}}$$

$$= \frac{50}{1 - e^{-0.5}} = \frac{50}{0.393469\ldots}$$

i.e. $\theta_1 = \mathbf{127°C}$, correct to the nearest degree.

(b) Transposing to make t the subject of the formula gives:

$$\frac{\theta_2}{\theta_1} = 1 - e^{-t/\tau}$$

from which, $e^{-t/\tau} = 1 - \dfrac{\theta_2}{\theta_1}$

Hence $-\dfrac{t}{\tau} = \ln\left(1 - \dfrac{\theta_2}{\theta_1}\right)$

i.e. $t = -\tau\ln\left(1 - \dfrac{\theta_2}{\theta_1}\right)$

Since $\theta_2 = \dfrac{1}{2}\theta_1$

$$t = -60\ln\left(1 - \frac{1}{2}\right) = -60\ln 0.5 = 41.59\,\text{s}$$

Hence the time for the temperature θ_2 to be one half of
the value of θ_1 is 41.6 s, correct to 1 decimal place.

Now try the following exercise

**Exercise 53   Further problems on the laws of growth
and decay (Answers on page 258)**

1. The pressure p pascals at height h metres above
   ground level is given by $p = p_0e^{-h/C}$, where p_0
   is the pressure at ground level and C is a constant.
   Find pressure p when $p_0 = 1.012 \times 10^5$ Pa, height
   $h = 1420\,\text{m}$ and $C = 71500$.

2. The voltage drop, v volts, across an inductor L henrys
   at time t seconds is given by $v = 200e^{-Rt/L}$, where
   $R = 150\Omega$ and $L = 12.5 \times 10^{-3}$ H. Determine (a) the
   voltage when $t = 160 \times 10^{-6}$ s, and (b) the time for
   the voltage to reach 85 V.

3. The length l metres of a metal bar at temperature
   $t°C$ is given by $l = l_0e^{\alpha t}$, where l_0 and α are constants. Determine (a) the value of l when $l_0 = 1.894$,
   $\alpha = 2.038 \times 10^{-4}$ and $t = 250°C$, and (b) the value of
   l_0 when $l = 2.416$, $t = 310°C$ and $\alpha = 1.682 \times 10^{-4}$

4. The temperature $\theta_2°C$ of an electrical conductor at time
   t seconds is given by $\theta_2 = \theta_1(1 - e^{-t/T})$, where θ_1 is
   the initial temperature and T seconds is a constant.
   Determine (a) θ_2 when $\theta_1 = 159.9°C$, $t = 30$ s and
   $T = 80$ s, and (b) the time t for θ_2 to fall to half the
   value of θ_1 if T remains at 80 s.

5. A belt is in contact with a pulley for a sector of
   $\theta = 1.12$ radians and the coefficient of friction between
   these two surfaces is $\mu = 0.26$. Determine the tension
   on the taut side of the belt, T newtons, when tension
   on the slack side is given by $T_0 = 22.7$ newtons,
   given that these quantities are related by the law
   $T = T_0e^{\mu\theta}$

6. The instantaneous current i at time t is given by:

   $$i = 10e^{-t/CR}$$

   when a capacitor is being charged. The capacitance
   C is 7×10^{-6} farads and the resistance R is
   0.3×10^6 ohms. Determine:

   (a) the instantaneous current when t is 2.5 seconds,
       and
   (b) the time for the instantaneous current to fall to
       5 amperes.

Sketch a curve of current against time from $t = 0$ to $t = 6$ seconds.

7. The amount of product x (in mol/cm^3) found in a chemical reaction starting with 2.5 mol/cm^3 of reactant is given by $x = 2.5(1 - e^{-4t})$ where t is the time, in minutes, to form product x. Plot a graph at 30 second intervals up to 2.5 minutes and determine x after 1 minute.

8. The current i flowing in a capacitor at time t is given by:

$$i = 12.5(1 - e^{-t/CR})$$

where resistance R is 30 kilohms and the capacitance C is 20 microfarads. Determine

(a) the current flowing after 0.5 seconds, and

(b) the time for the current to reach 10 amperes.

9. The amount A after n years of a sum invested P is given by the compound interest law: $A = Pe^{-rn/100}$ when the per unit interest rate r is added continuously. Determine, correct to the nearest pound, the amount after 8 years for a sum of £1500 invested if the interest rate is 6% per annum.

Assignment 7

This assignment covers the material contained in Chapters 13 and 14. The marks for each question are shown in brackets at the end of each question.

1. Evaluate $\log_{16} 8$ (3)

2. Solve (a) $\log_3 x = -2$

(b) $\log 2x^2 + \log x = \log 32 - \log x$ (6)

3. Solve the following equations, each correct to 3 significant figures:

(a) $2^x = 5.5$

(b) $3^{2t-1} = 7^{t+2}$

(c) $3e^{2x} = 4.2$ (11)

4. Evaluate the following, each correct to 4 significant figures:

(a) $e^{-0.683}$ (b) $1.34e^{2.16}$ (c) $\dfrac{5(e^{-2.73} - 1)}{e^{1.68}}$ (6)

5. Use the power series for e^x to evaluate $e^{1.5}$, correct to 4 significant figures. (4)

6. Expand xe^{3x} to six terms. (5)

7. Plot a graph of $y = \frac{1}{2}e^{-1.2x}$ over the range $x = -2$ to $x = +1$ and hence determine, correct to 1 decimal place, (a) the value of y when $x = -0.75$, and (b) the value of x when $y = 4.0$ (6)

8. Evaluate the following, each correct to 3 decimal places:

(a) $\ln 462.9$ (b) $\ln 0.0753$ (c) $\dfrac{\ln 3.68 - \ln 2.91}{4.63}$ (6)

9. Two quantities x and y are related by the equation $y = ae^{-kx}$, where a and k are constants. Determine, correct to 1 decimal place, the value of y when $a = 2.114$, $k = -3.20$ and $x = 1.429$ (3)

15

Reduction of non-linear laws to linear form

15.1 Determination of law

Frequently, the relationship between two variables, say x and y, is not a linear one, i.e. when x is plotted against y a curve results. In such cases the non-linear equation may be modified to the linear form, $y = mx + c$, so that the constants, and thus the law relating the variables can be determined. This technique is called 'determination of law'.

Some examples of the reduction of equations to linear form include:

(i) $y = ax^2 + b$ compares with $Y = mX + c$, where $m = a$, $c = b$ and $X = x^2$.

Hence y is plotted vertically against x^2 horizontally to produce a straight line graph of gradient 'a' and y-axis intercept 'b'

(ii) $y = \dfrac{a}{x} + b$

y is plotted vertically against $\dfrac{1}{x}$ horizontally to produce a straight line graph of gradient 'a' and y-axis intercept 'b'

(iii) $y = ax^2 + bx$

Dividing both sides by x gives $\dfrac{y}{x} = ax + b$.

Comparing with $Y = mX + c$ shows that $\dfrac{y}{x}$ is plotted vertically against x horizontally to produce a straight line graph of gradient 'a' and $\dfrac{y}{x}$ axis intercept 'b'

Problem 1. Experimental values of x and y, shown below, are believed to be related by the law $y = ax^2 + b$. By plotting a suitable graph verify this law and determine approximate values of a and b.

x	1	2	3	4	5
y	9.8	15.2	24.2	36.5	53.0

If y is plotted against x a curve results and it is not possible to determine the values of constants a and b from the curve. Comparing $y = ax^2 + b$ with $Y = mX + c$ shows that y is to be plotted vertically against x^2 horizontally. A table of values is drawn up as shown below.

x	1	2	3	4	5
x^2	1	4	9	16	25
y	9.8	15.2	24.2	36.5	53.0

A graph of y against x^2 is shown in Fig. 15.1, with the best straight line drawn through the points. Since a straight line graph results, the law is verified.

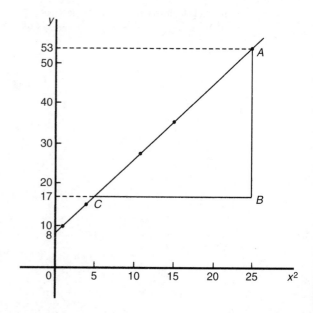

Fig. 15.1

From the graph, gradient $a = \dfrac{AB}{BC} = \dfrac{53 - 17}{25 - 5} = \dfrac{36}{20} = 1.8$

and the y-axis intercept, $b = 8.0$

Hence the law of the graph is $y = 1.8x^2 + 8.0$

Problem 2. Values of load L newtons and distance d metres obtained experimentally are shown in the following table.

Load, L N	32.3	29.6	27.0	23.2
distance, d m	0.75	0.37	0.24	0.17

Load, L N	18.3	12.8	10.0	6.4
distance, d m	0.12	0.09	0.08	0.07

Verify that load and distance are related by a law of the form $L = \dfrac{a}{d} + b$ and determine approximate values of a and b. Hence calculate the load when the distance is 0.20 m and the distance when the load is 20 N.

Comparing $L = \dfrac{a}{d} + b$ i.e. $L = a\left(\dfrac{1}{d}\right) + b$ with $Y = mX + c$

shows that L is to be plotted vertically against $\dfrac{1}{d}$ horizontally. Another table of values is drawn up as shown below.

L	32.3	29.6	27.0	23.2	18.3	12.8	10.0	6.4
d	0.75	0.37	0.24	0.17	0.12	0.09	0.08	0.07
$\dfrac{1}{d}$	1.33	2.70	4.17	5.88	8.33	11.11	12.50	14.29

A graph of L against $\dfrac{1}{d}$ is shown in Fig. 15.2. A straight line can be drawn through the points, which verifies that load and distance are related by a law of the form $L = \dfrac{a}{d} + b$.

Gradient of straight line, $a = \dfrac{AB}{BC} = \dfrac{31 - 11}{2 - 12}$

$\qquad\qquad = \dfrac{20}{-10} = -2$

L-axis intercept, $\qquad b = 35$

Hence the law of the graph is $L = -\dfrac{2}{d} + 35$

When the distance $d = 0.20$ m,

$\qquad$ load $L = \dfrac{-2}{0.20} + 35 = 25.0$ N

Rearranging $L = -\dfrac{2}{d} + 35$ gives

$\dfrac{2}{d} = 35 - L$ and $d = \dfrac{2}{35 - L}$

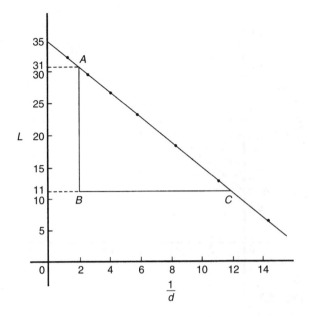

Fig. 15.2

Hence when the load $L = 20$ N,

$\qquad$ distance $d = \dfrac{2}{35 - 20} = \dfrac{2}{15} = 0.13$ m

---

Problem 3. The solubility s of potassium chlorate is shown by the following table:

$t°C$	10	20	30	40	50	60	80	100
s	4.9	7.6	11.1	15.4	20.4	26.4	40.6	58.0

The relationship between s and t is thought to be of the form $s = 3 + at + bt^2$. Plot a graph to test the supposition and use the graph to find approximate values of a and b. Hence calculate the solubility of potassium chlorate at 70°C.

Rearranging $s = 3 + at + bt^2$ gives $s - 3 = at + bt^2$ and

$\dfrac{s - 3}{t} = a + bt$ or $\dfrac{s - 3}{t} = bt + a$

which is of the form $Y = mX + c$, showing that $\dfrac{s - 3}{t}$ is to be plotted vertically and t horizontally. Another table of values is drawn up as shown below.

t	10	20	30	40	50	60	80	100
s	4.9	7.6	11.1	15.4	20.4	26.4	40.6	58.0
$\dfrac{s-3}{t}$	0.19	0.23	0.27	0.31	0.35	0.39	0.47	0.55

A graph of $\dfrac{s - 3}{t}$ against t is shown plotted in Fig. 15.3.

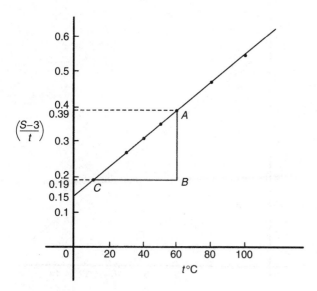

Fig. 15.3

A straight line fits the points which shows that s and t are related by $s = 3 + at + bt^2$

Gradient of straight line,

$$b = \frac{AB}{BC} = \frac{0.39 - 0.19}{60 - 10} = \frac{0.20}{50} = \mathbf{0.004}$$

Vertical axis intercept, $a = \mathbf{0.15}$

Hence the law of the graph is $s = 3 + 0.15t + 0.004t^2$

The solubility of potassium chlorate at 70°C is given by

$$s = 3 + 0.15(70) + 0.004(70)^2$$

$$= 3 + 10.5 + 19.6 = \mathbf{33.1}$$

___

Now try the following exercise

**Exercise 54   Further problems on reducing non-lin-
ear laws to linear form (Answers on
page 259)**

In Problems 1 to 5, x and y are two related variables and
all other letters denote constants. For the stated laws to be
verified it is necessary to plot graphs of the variables in a
modified form. State for each (a) what should be plotted
on the vertical axis, (b) what should be plotted on the
horizontal axis, (c) the gradient and (d) the vertical axis
intercept.

1. $y = d + cx^2$

2. $y - a = b\sqrt{x}$

3. $y - e = \dfrac{f}{x}$

4. $y - cx = bx^2$

5. $y = \dfrac{a}{x} + bx$

6. In an experiment the resistance of wire is measured for
   wires of different diameters with the following results.

R ohms	1.64	1.14	0.89	0.76	0.63
d mm	1.10	1.42	1.75	2.04	2.56

   It is thought that R is related to d by the law
   $R = (a/d^2) + b$, where a and b are constants. Verify this
   and find the approximate values for a and b. Determine
   the cross-sectional area needed for a resistance reading
   of 0.50 ohms.

7. Corresponding experimental values of two quantities x
   and y are given below.

x	1.5	3.0	4.5	6.0	7.5	9.0
y	11.5	25.0	47.5	79.0	119.5	169.0

   By plotting a suitable graph verify that y and x are
   connected by a law of the form $y = kx^2 + c$, where
   k and c are constants. Determine the law of the graph
   and hence find the value of x when y is 60.0

8. Experimental results of the safe load L kN, applied to
   girders of varying spans, d m, are shown below.

Span, d m	2.0	2.8	3.6	4.2	4.8
Load, L kN	475	339	264	226	198

   It is believed that the relationship between load and
   span is $L = c/d$, where c is a constant. Determine (a)
   the value of constant c and (b) the safe load for a span
   of 3.0 m.

9. The following results give corresponding values of two
   quantities x and y which are believed to be related by
   a law of the form $y = ax^2 + bx$ where a and b are
   constants.

y	33.86	55.54	72.80	84.10	111.4	168.1
x	3.4	5.2	6.5	7.3	9.1	12.4

   Verify the law and determine approximate values of a
   and b.

   Hence determine (i) the value of y when x is 8.0 and
   (ii) the value of x when y is 146.5

15.2   Determination of law involving logarithms

Examples of reduction of equations to linear form involving
logarithms include:

(i) $y = ax^n$

Taking logarithms to a base of 10 of both sides gives:

$$\lg y = \lg(ax^n) = \lg a + \lg x^n$$

i.e. $\lg y = n \lg x + \lg a$ by the laws of logarithms

which compares with $Y = mX + c$

and shows that $\lg y$ is plotted vertically against $\lg x$ horizontally to produce a straight line graph of gradient n and $\lg y$-axis intercept $\lg a$

(ii) $y = ab^x$

Taking logarithms to a base of 10 of the both sides gives:

$$\lg y = \lg(ab^x)$$

i.e. $\lg y = \lg a + \lg b^x$

i.e. $\lg y = x \lg b + \lg a$ by the laws of logarithms

or $\lg y = (\lg b)x + \lg a$

which compares with $Y = mX + c$

and shows that $\lg y$ is plotted vertically against x horizontally to produce a straight line graph of gradient $\lg b$ and $\lg y$-axis intercept $\lg a$

(iii) $y = ae^{bx}$

Taking logarithms to a base of e of both sides gives:

$$\ln y = \ln(ae^{bx})$$

i.e. $\ln y = \ln a + \ln e^{bx}$

i.e. $\ln y = \ln a + bx \ln e$

i.e. $\ln y = bx + \ln a$ since $\ln e = 1$

which compares with $Y = mX + c$

and shows that $\ln y$ is plotted vertically against x horizontally to produce a straight line graph of gradient b and $\ln y$-axis intercept $\ln a$

Problem 4. The current flowing in, and the power dissipated by, a resistor are measured experimentally for various values and the results are as shown below.

Current, I amperes	2.2	3.6	4.1	5.6	6.8
Power, P watts	116	311	403	753	1110

Show that the law relating current and power is of the form $P = RI^n$, where R and n are constants, and determine the law.

Taking logarithms to a base of 10 of both sides of $P = RI^n$ gives:

$$\lg P = \lg(RI^n) = \lg R + \lg I^n = \lg R + n \lg I$$

by the laws of logarithms.

i.e. $\lg P = n \lg I + \lg R$, which is of the form $Y = mX + c$, showing that $\lg P$ is to be plotted vertically against $\lg I$ horizontally.

A table of values for $\lg I$ and $\lg P$ is drawn up as shown below.

I	2.2	3.6	4.1	5.6	6.8
$\lg I$	0.342	0.556	0.613	0.748	0.833
P	116	311	403	753	1110
$\lg P$	2.064	2.493	2.605	2.877	3.045

A graph of $\lg P$ against $\lg I$ is shown in Fig. 15.4 and since a straight line results the law $P = RI^n$ is verified.

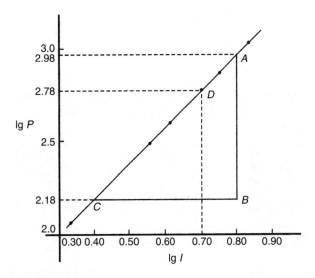

Fig. 15.4

Gradient of straight line,

$$n = \frac{AB}{BC} = \frac{2.98 - 2.18}{0.8 - 0.4} = \frac{0.80}{0.4} = 2$$

It is not possible to determine the vertical axis intercept on sight since the horizontal axis scale does not start at zero. Selecting any point from the graph, say point D, where $\lg I = 0.70$ and $\lg P = 2.78$, and substituting values into $\lg P = n \lg I + \lg R$ gives

$$2.78 = (2)(0.70) + \lg R$$

from which $\lg R = 2.78 - 1.40 = 1.38$

Hence $R = \text{antilog } 1.38(= 10^{1.38}) = 24.0$

Hence the law of the graph is $P = 24.0I^2$

Problem 5. The periodic time, T, of oscillation of a pendulum is believed to be related to its length, l, by a law of the form $T = kl^n$, where k and n

are constants. Values of T were measured for various lengths of the pendulum and the results are as shown below.

Periodic time, T s	1.0	1.3	1.5	1.8	2.0	2.3
Length, l m	0.25	0.42	0.56	0.81	1.0	1.32

Show that the law is true and determine the approximate values of k and n. Hence find the periodic time when the length of the pendulum is 0.75 m.

From para (i), if $T = kl^n$ then

$$\lg T = n \lg l + \lg k$$

and comparing with

$$Y = mX + c$$

shows that $\lg T$ is plotted vertically against $\lg l$ horizontally.

A table of values for $\lg T$ and $\lg l$ is drawn up as shown below.

T	1.0	1.3	1.5	1.8	2.0	2.3
$\lg T$	0	0.114	0.176	0.255	0.301	0.362
l	0.25	0.42	0.56	0.81	1.0	1.32
$\lg l$	−0.602	−0.377	−0.252	−0.092	0	0.121

A graph of $\lg T$ against $\lg l$ is shown in Fig. 15.5 and the law $T = kl^n$ is true since a straight line results.

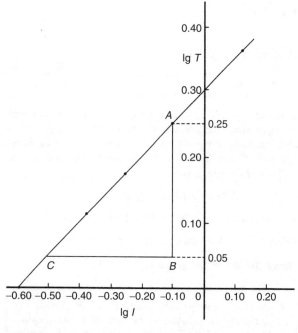

Fig. 15.5

From the graph, gradient of straight line,

$$n = \frac{AB}{BC} = \frac{0.25 - 0.05}{-0.10 - (-0.50)} = \frac{0.20}{0.40} = \frac{1}{2}$$

Vertical axis intercept, $\lg k = 0.30$
Hence $k = $ antilog $0.30 (= 10^{0.30}) = \mathbf{2.0}$

Hence the law of the graph is $T = 2.0 \, l^{1/2}$ or $T = 2.0\sqrt{l}$

When length $l = 0.75$ m then $T = 2.0\sqrt{0.75} = \mathbf{1.73\,s}$

Problem 6. Quantities x and y are believed to be related by a law of the form $y = ab^x$, where a and b are constants. Values of x and corresponding values of y are:

x	0	0.6	1.2	1.8	2.4	3.0
y	5.0	9.67	18.7	36.1	69.8	135.0

Verify the law and determine the approximate values of a and b. Hence determine (a) the value of y when x is 2.1 and (b) the value of x when y is 100

From para (ii), if $y = ab^x$ then

$$\lg y = (\lg b)x + \lg a$$

and comparing with

$$Y = mX + c$$

shows that $\lg y$ is plotted vertically and x horizontally.

Another table is drawn up as shown below.

x	0	0.6	1.2	1.8	2.4	3.0
y	5.0	9.67	18.7	36.1	69.8	135.0
$\lg y$	0.70	0.99	1.27	1.56	1.84	2.13

A graph of $\lg y$ against x is shown in Fig. 15.6 and since a straight line results, the law $y = ab^x$ is verified.

Gradient of straight line,

$$\lg b = \frac{AB}{BC} = \frac{2.13 - 1.17}{3.0 - 1.0} = \frac{0.96}{2.0} = 0.48$$

Hence $b = $ antilog $0.48 \ (= 10^{0.48}) = \mathbf{3.0}$, correct to 2 significant figures.

Vertical axis intercept, $\lg a = 0.70$,

from which $\boldsymbol{a} = $ antilog $0.70 \ (= 10^{0.70})$

$$= \mathbf{5.0}, \text{ correct to 2 significant figures}$$

Hence the law of the graph is $y = 5.0(3.0)^x$

(a) When $x = 2.1$, $y = 5.0(3.0)^{2.1} = \mathbf{50.2}$

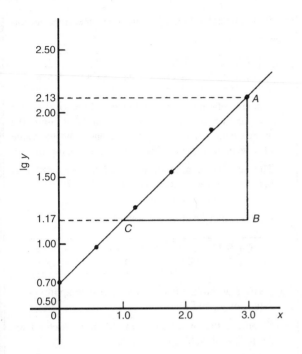

Fig. 15.6

(b) When $y = 100$, $100 = 5.0(3.0)^x$,

from which $100/5.0 = (3.0)^x$,

i.e. $\qquad\qquad 20 = (3.0)^x$

Taking logarithms of both sides gives

$$\lg 20 = \lg(3.0)^x = x \lg 3.0$$

Hence $x = \dfrac{\lg 20}{\lg 3.0} = \dfrac{1.3010}{0.4771} = \mathbf{2.73}$

Problem 7. The current i mA flowing in a capacitor which is being discharged varies with time t ms as shown below.

i mA	203	61.14	22.49	6.13	2.49	0.615
t ms	100	160	210	275	320	390

Show that these results are related by a law of the form $i = Ie^{t/T}$, where I and T are constants. Determine the approximate values of I and T.

Taking Napierian logarithms of both sides of $i = Ie^{t/T}$ gives

$$\ln i = \ln(Ie^{t/T}) = \ln I + \ln e^{t/T}$$

i.e. $\ln i = \ln I + \dfrac{t}{T}$ (since $\ln e = 1$)

or $\ln i = \left(\dfrac{1}{T}\right) t + \ln I$

which compares with $y = mx + c$, showing that $\ln i$ is plotted vertically against t horizontally. (For methods of evaluating Napierian logarithms see Chapter 14.) Another table of values is drawn up as shown below.

t	100	160	210	275	320	390
i	203	61.14	22.49	6.13	2.49	0.615
$\ln i$	5.31	4.11	3.11	1.81	0.91	−0.49

A graph of $\ln i$ against t is shown in Fig. 15.7 and since a straight line results the law $i = Ie^{t/T}$ is verified.

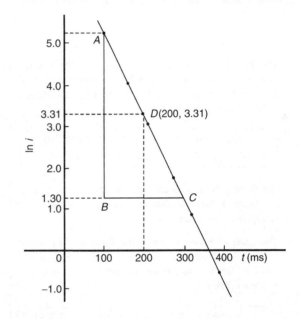

Fig. 15.7

Gradient of straight line,

$$\frac{1}{T} = \frac{AB}{BC} = \frac{5.30 - 1.30}{100 - 300} = \frac{4.0}{-200} = -0.02$$

Hence $T = \dfrac{1}{-0.02} = \mathbf{-50}$

Selecting any point on the graph, say point D, where $t = 200$ and $\ln i = 3.31$, and substituting into $\ln i = \left(\dfrac{1}{T}\right) t + \ln I$ gives

$$3.31 = -\frac{1}{50}(200) + \ln I$$

from which $\ln I = 3.31 + 4.0 = 7.31$

and $I = $ antilog 7.31 ($= e^{7.31}$) $= 1495$ or **1500** correct to 3 significant figures.

Hence the law of the graph is $i = 1500e^{-t/50}$

Now try the following exercise

Exercise 55   Further problems on reducing non-linear laws to linear form (Answers on page 259)

In Problems 1 to 3, x and y are two related variables and all other letters denote constants. For the stated laws to be verified it is necessary to plot graphs of the variables in a modified form. State for each (a) what should be plotted on the vertical axis, (b) what should be plotted on the horizontal axis, (c) the gradient and (d) the vertical axis intercept.

1. $y = ba^x$ 2. $y = kx^l$ 3. $\dfrac{y}{m} = e^{nx}$

4. The luminosity I of a lamp varies with the applied voltage V and the relationship between I and V is thought to be $I = kV^n$. Experimental results obtained are:

I candelas	1.92	4.32	9.72	15.87	23.52	30.72
V volts	40	60	90	115	140	160

Verify that the law is true and determine the law of the graph. Determine also the luminosity when 75 V is applied across the lamp.

5. The head of pressure h and the flow velocity v are measured and are believed to be connected by the law $v = ah^b$, where a and b are constants. The results are as shown below.

h	10.6	13.4	17.2	24.6	29.3
v	9.77	11.0	12.44	14.88	16.24

Verify that the law is true and determine values of a and b

6. Experimental values of x and y are measured as follows.

x	0.4	0.9	1.2	2.3	3.8
y	8.35	13.47	17.94	51.32	215.20

The law relating x and y is believed to be of the form $y = ab^x$, where a and b are constants. Determine the approximate values of a and b. Hence find the value of y when x is 2.0 and the value of x when y is 100

7. The activity of a mixture of radioactive isotope is believed to vary according to the law $R = R_0 t^{-c}$, where R_0 and c are constants.

Experimental results are shown below.

R	9.72	2.65	1.15	0.47	0.32	0.23
t	2	5	9	17	22	28

Verify that the law is true and determine approximate values of R_0 and c.

8. Determine the law of the form $y = ae^{kx}$ which relates the following values.

y	0.0306	0.285	0.841	5.21	173.2	1181
x	−4.0	5.3	9.8	17.4	32.0	40.0

9. The tension T in a belt passing round a pulley wheel and in contact with the pulley over an angle of θ radians is given by $T = T_0 e^{\mu\theta}$, where T_0 and μ are constants. Experimental results obtained are:

T newtons	47.9	52.8	60.3	70.1	80.9
θ radians	1.12	1.48	1.97	2.53	3.06

Determine approximate values of T_0 and μ. Hence find the tension when θ is 2.25 radians and the value of θ when the tension is 50.0 newtons.

16

Geometry and triangles

16.1 Angular measurement

Geometry is a part of mathematics in which the properties of points, lines, surfaces and solids are investigated.

An **angle** is the amount of rotation between two straight lines.

Angles may be measured in either **degrees** or **radians** (see Section 21.2).

1 revolution = 360 degrees, thus 1 degree = $\frac{1}{360}$ th of one revolution. Also 1 minute = $\frac{1}{60}$ th of a degree and

1 second = $\frac{1}{60}$ th of a minute. 1 minute is written as 1′ and 1 second is written as 1″

Thus **1° = 60′ and 1′ = 60″**

Problem 1. Add 14°53′ and 37°19′

$$14°53'$$
$$37°19'$$
$$\overline{52°12'}$$
$$\overline{1°}$$

53′ + 19′ = 72′. Since 60′ = 1°, 72′ = 1°12′. Thus the 12′ is placed in the minutes column and 1° is carried in the degrees column.

Then 14° + 37° + 1° (carried) = 52°

Thus **14°53′ + 37°19′ = 52°12′**

Problem 2. Subtract 15°47′ from 28°13′

27°
2̸8̸° 13′
15°47′
$$\overline{12°26'}$$

13′ − 47′ cannot be done. Hence 1° or 60′ is 'borrowed' from the degrees column, which leaves 27° in that column. Now (60′ + 13′) − 47′ = 26′, which is placed in the minutes column. 27° − 15° = 12°, which is placed in the degrees column.

Thus **28°13′ − 15°47′ = 12°26′**

Problem 3. Determine (a) 13°42′51″ + 48°22′17″ (b) 37°12′8″ − 21°17′25″

(a)
$$13°42'51''$$
$$48°22'17''$$
Adding: **62° 5′ 8″**

(b)
$$1° \; 1'$$
$$36°11'$$
$$3̸7̸° \; 1̸2̸' \, 8''$$
$$21°17'25''$$
Subtracting: **15°54′43″**

Problem 4. Convert (a) 24°42′ (b) 78°15′26″ to degrees and decimals of a degree.

(a) Since 1 minute = $\frac{1}{60}$ th of a degree,

$$42' = \left(\frac{42}{60}\right)^° = 0.70°$$

Hence **24°42′ = 24.70°**

(b) Since 1 second = $\dfrac{1}{60}$ th of a minute,

$$26'' = \left(\dfrac{26}{60}\right)' = 0.4333'$$

Hence $78°15'26'' = 78°15.4\dot{3}'$

$$15.4333' = \left(\dfrac{15.4\dot{3}}{60}\right)^\circ = 0.2572°,$$

correct to 4 decimal places.

Hence **78°15'26'' = 78.26°**, correct to 4 significant places.

Problem 5. Convert 45.371° into degrees, minutes and seconds.

Since $1° = 60'$, $0.371° = (0.371 \times 60)' = 22.26'$

Since: $1' = 60''$, $0.26' = (0.26 \times 60)'' = 15.6'' = 16''$ to the nearest second.

Hence   **45.371° = 45°22'16''**

Now try the following exercise

Exercise 56   Further problems on angular measurement (Answers on page 259)

1. Add together the following angles:

   (a) 32°19' and 49°52'

   (b) 29°42', 56°37' and 63°54'

   (c) 21°33'27'' and 78°42'36''

   (d) 48°11'19'', 31°41'27'' and 9°9'37''

2. Determine

   (a) $17° - 9°49'$ (b) $43°37' - 15°49'$

   (c) $78°29'41'' - 59°41'52''$ (d) $114° - 47°52'37''$

3. Convert the following angles to degrees and decimals of a degree, correct to 3 decimal places:

   (a) 15°11'   (b) 29°53'   (c) 49°42'17''   (d) 135°7'19''

4. Convert the following angles into degrees, minutes and seconds:

   (a) 25.4°   (b) 36.48°   (c) 55.724°   (d) 231.025°

16.2   Types and properties of angles

(a) (i) Any angle between 0° and 90° is called an **acute angle**.

   (ii) An angle equal to 90° is called a **right angle**.

(iii) Any angle between 90° and 180° is called an **obtuse angle**.

(iv) Any angle greater than 180° and less than 360° is called a **reflex angle**.

(b) (i) An angle of 180° lies on a straight line.

   (ii) If two angles add up to 90° they are called **complementary angles**.

   (iii) If two angles add up to 180° they are called **supplementary angles**.

   (iv) **Parallel lines** are straight lines which are in the same plane and never meet. (Such lines are denoted by arrows, as in Fig. 16.1).

   (v) A straight line which crosses two parallel lines is called a **transversal** (see MN in Fig. 16.1).

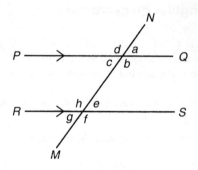

Fig. 16.1

(c) With reference to Fig. 16.1:

   (i) $a = c$, $b = d$, $e = g$ and $f = h$. Such pairs of angles are called **vertically opposite angles**.

   (ii) $a = e$, $b = f$, $c = g$ and $d = h$. Such pairs of angles are called **corresponding angles**.

   (iii) $c = e$ and $b = h$. Such pairs of angles are called **alternate angles**.

   (iv) $b + e = 180°$ and $c + h = 180°$. Such pairs of angles are called **interior angles**.

Problem 6. State the general name given to the following angles:
(a) 159°   (b) 63°   (c) 90°   (d) 227°

(a) 159° lies between 90° and 180° and is therefore called an **obtuse angle**.

(b) 63° lies between 0° and 90° and is therefore called an **acute angle**.

(c) 90° is called a **right angle**.

(d) 227° is greater than 180° and less than 360° and is therefore called a **reflex angle**.

Problem 7. Find the angles complementary to
(a) 41°   (b) 58°39′

(a) The complement of 41° is (90° − 41°), i.e. **49°**
(b) The complement of 58°39′ is (90° − 58°39′), i.e. **31°21′**

Problem 8. Find the angles supplementary to
(a) 27°   (b) 111°11′

(a) The supplement of 27° is (180° − 27°), i.e. **153°**
(b) The supplement of 111°11′ is (180°−111°11′), i.e. **68°49′**

Problem 9. Two straight lines *AB* and *CD* intersect at 0. If ∠*AOC* is 43°, find ∠*AOD*, ∠*DOB* and ∠*BOC*

From Fig. 16.2, ∠*AOD* is supplementary to ∠*AOC*. Hence ∠**AOD** = 180° − 43° = **137°**. When two straight lines intersect the vertically opposite angles are equal. Hence ∠**DOB** = **43°** and ∠**BOC** = **137°**

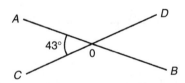

Fig. 16.2

Problem 10. Determine angle β in Fig. 16.3.

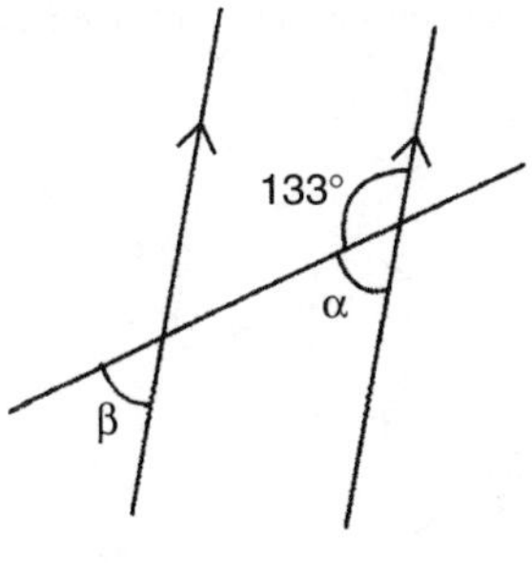

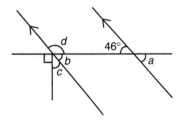

Fig. 16.3

α = 180° − 133° = 47° (i.e. supplementary angles).
α = β = **47°** (corresponding angles between parallel lines).

Problem 11. Determine the value of angle θ in Fig. 16.4.

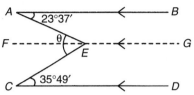

Fig. 16.4

Let a straight line *FG* be drawn through *E* such that *FG* is parallel to *AB* and *CD*. ∠*BAE* = ∠*AEF* (alternate angles between parallel lines *AB* and *FG*), hence ∠*AEF* = 23°37′. ∠*ECD* = ∠*FEC* (alternate angles between parallel lines *FG* and *CD*), hence ∠*FEC* = 35°49′

Angle θ = ∠*AEF* + ∠*FEC* = 23°37′ + 35°49′ = **59°26′**

Problem 12. Determine angles *c* and *d* in Fig. 16.5.

Fig. 16.5

b = 46° (corresponding angles between parallel lines).
Also *b* + *c* + 90° = 180° (angles on a straight line).
Hence 46° + *c* + 90° = 180°, from which **c = 44°**.
b and *d* are supplementary, hence **d** = 180° − 46° = **134°**.
Alternatively, 90° + *c* = *d* (vertically opposite angles).

Now try the following exercise

Exercise 57   Further problems on types and properties of angles (Answers on page 259)

1. State the general name given to the
   (a) 63°   (b) 147°   (c) 250°

2. Determine the angles complementary to the following:
   (a) 69°   (b) 27°37′   (c) 41°3′43″

3. Determine the angles supplementary to
   (a) 78°   (b) 15°   (c) 169°41′11″

4. With reference to Fig. 16.6, what is the name given to the line XY. Give examples of each of the following:
   (a) vertically opposite angles.

   (b) supplementary angles.

   (c) corresponding angles.

   (d) alternate angles.

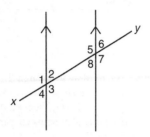

Fig. 16.6

5. In Fig. 16.7, find angle α.

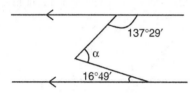

Fig. 16.7

6. In Fig. 16.8, find angles a, b and c.

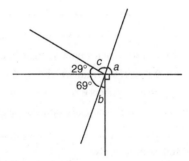

Fig. 16.8

7. Find angle β in Fig. 16.9.

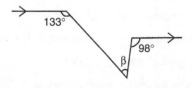

Fig. 16.9

16.3   Properties of triangles

A triangle is a figure enclosed by three straight lines. The sum of the three angles of a triangle is equal to 180°. Types of triangles:

(i) An **acute-angled triangle** is one in which all the angles are acute, i.e. all the angles are less than 90°.

(ii) A **right-angled triangle** is one which contains a right angle.

(iii) An **obtuse-angled triangle** is one which contains an obtuse angle, i.e. one angle which lies between 90° and 180°.

(iv) An **equilateral triangle** is one in which all the sides and all the angles are equal (i.e. each 60°).

(v) An **isosceles triangle** is one in which two angles and two sides are equal.

(vi) A **scalene triangle** is one with unequal angles and therefore unequal sides.

With reference to Fig. 16.10:

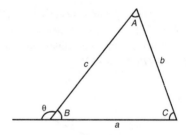

Fig. 16.10

(i) Angles A, B and C are called **interior angles** of the triangle.

(ii) Angle θ is called an **exterior angle** of the triangle and is equal to the sum of the two opposite interior angles, i.e. $\theta = A + C$

(iii) $a + b + c$ is called the **perimeter** of the triangle.

Problem 13. Name the types of triangles shown in Fig. 16.11.

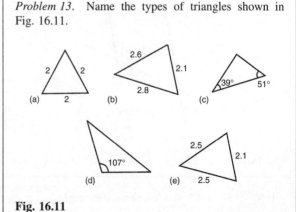

Fig. 16.11

(a) Equilateral triangle.

(b) Acute-angled scalene triangle.

(c) Right-angled triangle.

(d) Obtuse-angled scalene triangle.

(e) Isosceles triangle.

Problem 14. Determine the value of θ and α in Fig. 16.12.

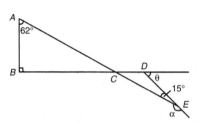

Fig. 16.12

In triangle ABC, $\angle A + \angle B + \angle C = 180°$ (angles in a triangle add up to $180°$), hence $\angle C = 180° - 90° - 62° = 28°$. Thus $\angle DCE = 28°$ (vertically opposite angles).

$\theta = \angle DCE + \angle DEC$ (exterior angle of a triangle is equal to the sum of the two opposite interior angles). Hence $\angle\theta = 28° + 15° = \mathbf{43°}$

$\angle\alpha$ and $\angle DEC$ are supplementary, thus

$$\alpha = 180° - 15° = \mathbf{165°}$$

Problem 15. ABC is an isosceles triangle in which the unequal angle BAC is $56°$. AB is extended to D as shown in Fig. 16.13. Determine the angle DBC.

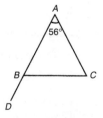

Fig. 16.13

Since the three interior angles of a triangle add up to $180°$ then $56° + \angle B + \angle C = 180°$, i.e. $\angle B + \angle C = 180° - 56° = 124°$.

Triangle ABC is isosceles hence $\angle B = \angle C = \dfrac{124°}{2} = 62°$.

$\angle DBC = \angle A + \angle C$ (exterior angle equals sum of two interior opposite angles), i.e. $\angle\mathbf{DBC} = 56° + 62° = \mathbf{118°}$ [Alternatively, $\angle DBC + \angle ABC = 180°$ (i.e. supplementary angles)].

Problem 16. Find angles a, b, c, d and e in Fig. 16.14.

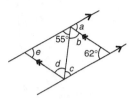

Fig. 16.14

$a = \mathbf{62°}$ and $c = \mathbf{55°}$ (alternate angles between parallel lines) $55° + b + 62° = 180°$ (angles in a triangle add up to $180°$), hence $\mathbf{b} = 180° - 55° - 62° = \mathbf{63°}$

$b = d = \mathbf{63°}$ (alternate angles between parallel lines).

$e + 55° + 63° = 180°$ (angles in a triangle add up to $180°$), hence $\mathbf{e} = 180° - 55° - 63° = \mathbf{62°}$

[Check: $e = a = 62°$ (corresponding angles between parallel lines)].

Now try the following exercise

Exercise 58 Further problems on properties of triangles (Answers on page 259)

1. In Fig. 16.15, (i) and (ii), find angles w, x, y and z. What is the name given to the types of triangle shown in (i) and (ii)?

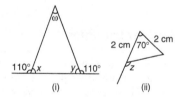

Fig. 16.15

2. Find the values of angles a to g in Fig. 16.16 (i) and (ii).

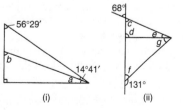

Fig. 16.16

3. Find the unknown angles *a* to *k* in Fig. 16.17.

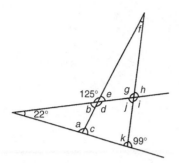

Fig. 16.17

4. Triangle *ABC* has a right angle at *B* and ∠*BAC* is 34°. *BC* is produced to *D*. If the bisectors of ∠*ABC* and ∠*ACD* meet at *E*, determine ∠*BEC*

5. If in Fig. 16.18, triangle *BCD* is equilateral, find the interior angles of triangle *ABE*.

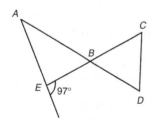

Fig. 16.18

16.4 Congruent triangles

Two triangles are said to be **congruent** if they are equal in all respects, i.e. three angles and three sides in one triangle are equal to three angles and three sides in the other triangle. Two triangles are congruent if:

(i) the three sides of one are equal to the three sides of the other (SSS),

(ii) they have two sides of the one equal to two sides of the other, and if the angles included by these sides are equal (SAS),

(iii) two angles of the one are equal to two angles of the other and any side of the first is equal to the corresponding side of the other (ASA), or

(iv) their hypotenuses are equal and if one other side of one is equal to the corresponding side of the other (RHS).

Problem 17. State which of the pairs of triangles shown in Fig. 16.19 are congruent and name their sequence.

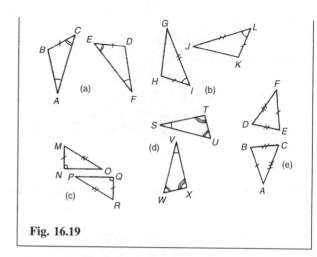

Fig. 16.19

(a) Congruent *ABC*, *FDE* (Angle, side, angle, i.e. ASA).

(b) Congruent *GIH*, *JLK* (Side, angle, side, i.e. SAS).

(c) Congruent *MNO*, *RQP* (Right-angle, hypotenuse, side, i.e. RHS).

(d) Not necessarily congruent. It is not indicated that any side coincides.

(e) Congruent *ABC*, *FED* (Side, side, side, i.e. SSS).

Problem 18. In Fig. 16.20, triangle *PQR* is isosceles with *Z* the mid-point of *PQ*. Prove that triangle *PXZ* and *QYZ* are congruent, and that triangles *RXZ* and *RYZ* are congruent. Determine the values of angles *RPZ* and *RXZ*.

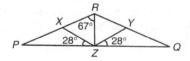

Fig. 16.20

Since triangle *PQR* is isosceles *PR = RQ* and thus

$$\angle QPR = \angle RQP$$

∠*RXZ* = ∠*QPR* + 28° and ∠*RYZ* = ∠*RQP* + 28° (exterior angles of a triangle equal the sum of the two interior opposite angles). Hence ∠*RXZ* = ∠*RYZ*.

∠*PXZ* = 180° − ∠*RXZ* and ∠*QYZ* = 180° − ∠*RYZ*. Thus ∠*PXZ* = ∠*QYZ*.

Triangles *PXZ* and *QYZ* are congruent since

$$\angle XPZ = \angle YQZ, PZ = ZQ \text{ and } \angle XZP = \angle YZQ \text{ (ASA)}$$

Hence *XZ = YZ*

Triangles *PRZ* and *QRZ* are congruent since *PR = RQ*, ∠*RPZ* = ∠*RQZ* and *PZ = ZQ* (SAS). Hence ∠*RZX* = ∠*RZY*

Triangles RXZ and RYZ are congruent since $\angle RXZ = \angle RYZ$, $XZ = YZ$ and $\angle RZX = \angle RZY$ (ASA). $\angle QRZ = 67°$ and thus $\angle PRQ = 67° + 67° = 134°$. Hence

$$\angle \mathbf{RPZ} = \angle \mathbf{RQZ} = \frac{180° - 134°}{2} = \mathbf{23°}$$

$\angle \mathbf{RXZ} = 23° + 28° = \mathbf{51°}$ (external angle of a triangle equals the sum of the two interior opposite angles).

___

Now try the following exercise

Exercise 59   Further problems on congruent triangles (Answers on page 259)

1. State which of the pairs of triangles in Fig. 16.21 are congruent and name their sequence.

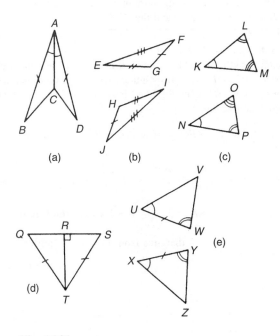

Fig. 16.21

2. In a triangle ABC, $AB = BC$ and D and E are points on AB and BC, respectively, such that $AD = CE$. Show that triangles AEB and CDB are congruent.

16.5   Similar triangles

Two triangles are said to be **similar** if the angles of one triangle are equal to the angles of the other triangle. With reference to Fig. 16.22: Triangles ABC and PQR are similar and the corresponding sides are in proportion to each other,

i.e.

$$\frac{\mathbf{p}}{\mathbf{a}} = \frac{\mathbf{q}}{\mathbf{b}} = \frac{\mathbf{r}}{\mathbf{c}}$$

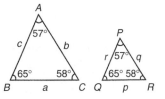

Fig. 16.22

Problem 19. In Fig. 16.23, find the length of side a.

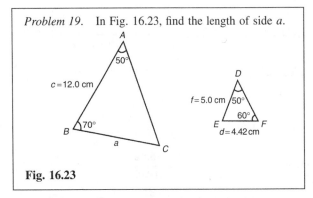

Fig. 16.23

In triangle ABC, $50° + 70° + \angle C = 180°$, from which $\angle C = 60°$

In triangle DEF, $\angle E = 180° - 50° - 60° = 70°$. Hence triangles ABC and DEF are similar, since their angles are the same. Since corresponding sides are in proportion to each other then:

$$\frac{a}{d} = \frac{c}{f} \quad \text{i.e.} \quad \frac{a}{4.42} = \frac{12.0}{5.0}$$

Hence $\quad \mathbf{a} = \dfrac{12.0}{5.0}(4.42) = \mathbf{10.61\,cm}$

Problem 20. In Fig. 16.24, find the dimensions marked r and p

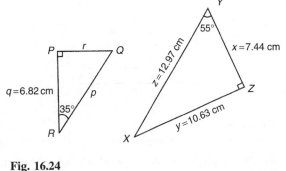

Fig. 16.24

In triangle PQR, $\angle Q = 180° - 90° - 35° = 55°$

In triangle XYZ, $\angle X = 180° - 90° - 55° = 35°$

Hence triangles PQR and ZYX are similar since their angles are the same. The triangles may be redrawn as shown in Fig. 16.25.

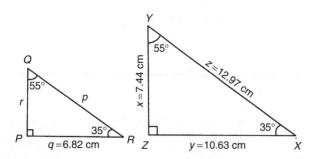

Fig. 16.25

By proportion: $\dfrac{p}{z} = \dfrac{r}{x} = \dfrac{q}{y}$

Hence $\dfrac{p}{z} = \dfrac{r}{7.44} = \dfrac{6.82}{10.63}$

from which, $r = 7.44\left(\dfrac{6.82}{10.63}\right) = \mathbf{4.77\,cm}$

By proportion: $\dfrac{p}{z} = \dfrac{q}{y}$ i.e. $\dfrac{p}{12.97} = \dfrac{6.82}{10.63}$

Hence $p = 12.97\left(\dfrac{6.82}{10.63}\right) = \mathbf{8.32\,cm}$

Problem 21. In Fig. 16.26, show that triangles CBD and CAE are similar and hence find the length of CD and BD.

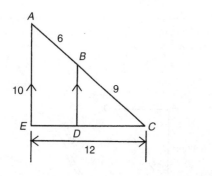

Fig. 16.26

Since BD is parallel to AE then $\angle CBD = \angle CAE$ and $\angle CDB = \angle CEA$ (corresponding angles between parallel

lines). Also $\angle C$ is common to triangles CBD and CAE. Since the angles in triangle CBD are the same as in triangle CAE the triangles are similar. Hence, by proportion:

$$\frac{CB}{CA} = \frac{CD}{CE}\left(= \frac{BD}{AE}\right)$$

i.e. $\dfrac{9}{6+9} = \dfrac{CD}{12}$, from which $\mathbf{CD} = 12\left(\dfrac{9}{15}\right)$

$$= \mathbf{7.2\,cm}$$

Also, $\dfrac{9}{15} = \dfrac{BD}{10}$, from which $\mathbf{BD} = 10\left(\dfrac{9}{15}\right)$

$$= \mathbf{6\,cm}$$

Problem 22. A rectangular shed 2 m wide and 3 m high stands against a perpendicular building of height 5.5 m. A ladder is used to gain access to the roof of the building. Determine the minimum distance between the bottom of the ladder and the shed.

A side view is shown in Fig. 16.27, where AF is the minimum length of ladder. Since BD and CF are parallel, $\angle ADB = \angle DFE$ (corresponding angles between parallel lines). Hence triangles BAD and EDF are similar since their angles are the same.

$$AB = AC - BC = AC - DE = 5.5 - 3 = 2.5\,\text{m}.$$

By proportion: $\dfrac{AB}{DE} = \dfrac{BD}{EF}$ i.e. $\dfrac{2.5}{3} = \dfrac{2}{EF}$

Hence $EF = 2\left(\dfrac{3}{2.5}\right) = \mathbf{2.4\ m = minimum}$

distance from bottom of ladder

to the shed

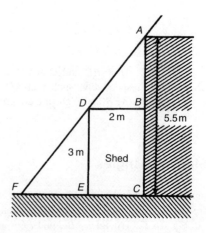

Fig. 16.27

Now try the following exercise

Exercise 60   Further problems on similar triangles
(Answers on page 259)

1. In Fig. 16.28, find the lengths x and y.

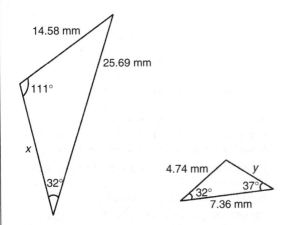

Fig. 16.28

2. *PQR* is an equilateral triangle of side 4 cm. When *PQ* and *PR* are produced to *S* and *T*, respectively, *ST* is found to be parallel with *QR*. If *PS* is 9 cm, find the length of *ST*. *X* is a point on *ST* between *S* and *T* such that the line *PX* is the bisector of ∠*SPT*. Find the length of *PX*.

3. In Fig. 16.29, find (a) the length of *BC* when *AB* = 6 cm, *DE* = 8 cm and *DC* = 3 cm, (b) the length of *DE* when *EC* = 2 cm, *AC* = 5 cm and *AB* = 10 cm.

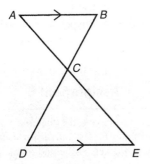

Fig. 16.29

4. In Fig. 16.30, *AF* = 8 m, *AB* = 5 m and *BC* = 3 m. Find the length of *BD*.

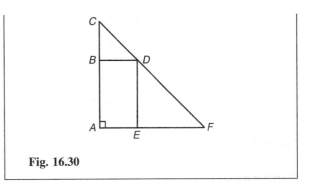

Fig. 16.30

16.6   Construction of triangles

To construct any triangle the following drawing instruments are needed:
(i) ruler and/or straight edge, (ii) compass, (iii) protractor, (iv) pencil. For actual constructions, see Problems 23 to 26 which follow.

Problem 23. Construct a triangle whose sides are 6 cm, 5 cm and 3 cm.

With reference to Fig. 16.31:

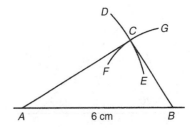

Fig. 16.31

(i) Draw a straight line of any length, and with a pair of compasses, mark out 6 cm length and label it *AB*.
(ii) Set compass to 5 cm and with centre at *A* describe arc *DE*.
(iii) Set compass to 3 cm and with centre at *B* describe arc *FG*.
(iv) The intersection of the two curves at *C* is the vertex of the required triangle. Join *AC* and *BC* by straight lines.

It may be proved by measurement that the ratio of the angles of a triangle is not equal to the ratio of the sides (i.e. in this problem, the angle opposite the 3 cm side is not equal to half the angle opposite the 6 cm side).

Problem 24. Construct a triangle *ABC* such that $a = 6$ cm, $b = 3$ cm and ∠*C* = 60°

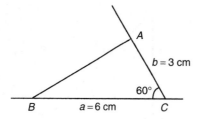

Fig. 16.32

With reference to Fig. 16.32:

(i) Draw a line BC, 6 cm long.

(ii) Using a protractor centred at C make an angle of 60° to BC.

(iii) From C measure a length of 3 cm and label A.

(iv) Join B to A by a straight line.

Problem 25. Construct a triangle PQR given that $QR = 5$ cm, $\angle Q = 70°$ and $\angle R = 44°$

With reference to Fig. 16.33:

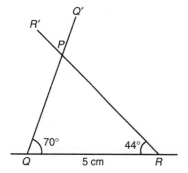

Fig. 16.33

(i) Draw a straight line 5 cm long and label it QR.

(ii) Use a protractor centred at Q and make an angle of 70°. Draw QQ'

(iii) Use a protractor centred at R and make an angle of 44°. Draw RR'.

(iv) The intersection of QQ' and RR' forms the vertex P of the triangle.

Problem 26. Construct a triangle XYZ given that $XY = 5$ cm, the hypotenuse $YZ = 6.5$ cm and $\angle X = 90°$

With reference to Fig. 16.34:

(i) Draw a straight line 5 cm long and label it XY.

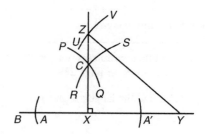

Fig. 16.34

(ii) Produce XY any distance to B. With compass centred at X make an arc at A and A'. (The length XA and XA' is arbitrary.) With compass centred at A draw the arc PQ. With the same compass setting and centred at A', draw the arc RS. Join the intersection of the arcs, C, to X, and a right angle to XY is produced at X. (Alternatively, a protractor can be used to construct a 90° angle).

(iii) The hypotenuse is always opposite the right angle. Thus YZ is opposite $\angle X$. Using a compass centred at Y and set to 6.5 cm, describe the arc UV.

(iv) The intersection of the arc UV with XC produced, forms the vertex Z of the required triangle. Join YZ by a straight line.

Now try the following exercise

Exercise 61　Further problems on the construction of triangles (Answers on page 259)

In problems 1 to 5, construct the triangles ABC for the given sides/angles.

1. $a = 8$ cm, $b = 6$ cm and $c = 5$ cm
2. $a = 40$ mm, $b = 60$ mm and $C = 60°$
3. $a = 6$ cm, $C = 45°$ and $B = 75°$
4. $c = 4$ cm, $A = 130°$ and $C = 15°$
5. $a = 90$ mm, $B = 90°$, hypotenuse $= 105$ mm

Assignment 8

This assignment covers the material contained in chapters 15 and 16. The marks for each question are shown in brackets at the end of each question.

1. In the following equations, x and y are two related variables and k and t are constants. For the stated equations to be verified it is necessary to plot graphs of the variables in modified form. State for each (a) what should be plotted on the horizontal axis, (b) what should

be plotted on the vertical axis, (c) the gradient, and (d) the vertical axis intercept.

(i) $y - \dfrac{k}{x} = t$ (ii) $\dfrac{y}{k} = x^t$ (8)

2. The following experimental values of x and y are believed to be related by the law $y = ax^2 + b$, where a and b are constants. By plotting a suitable graph verify this law and find the approximate values of a and b.

x	2.5	4.2	6.0	8.4	9.8	11.4
y	15.4	32.5	60.2	111.8	150.1	200.9

(9)

3. Determine the law of the form $y = ae^{kx}$ which relates the following values:

y	0.0306	0.285	0.841	5.21	173.2	1181
x	−4.0	5.3	9.8	17.4	32.0	40.0

(9)

4. Evaluate: $29°17' + 75°51' - 47°49'$ (3)

5. Convert $47.319°$ to degrees, minutes and seconds (2)

6. State the angle   (a) supplementary to $49°$

               (b) complementary to $49°$ (2)

7. In Fig. A8.1, determine angles x, y and z (3)

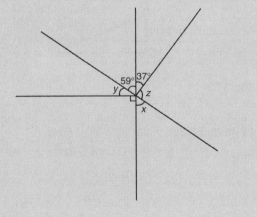

Fig. A8.1

8. In Fig. A8.2, determine angles a to e (5)

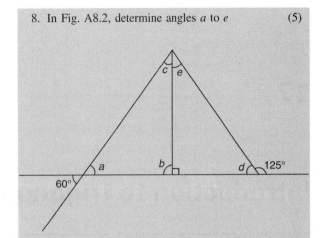

Fig. A8.2

9. In Fig. A8.3, determine the length of AC (4)

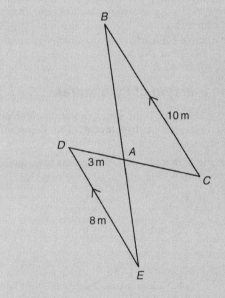

Fig. A8.3

10. Construct a triangle PQR given $PQ = 5\,\text{cm}$, $\angle QPR = 120°$ and $\angle PRQ = 35°$ (5)

17

Introduction to trigonometry

17.1 Trigonometry

Trigonometry is the branch of mathematics which deals with the measurement of sides and angles of triangles, and their relationship with each other. There are many applications in engineering where a knowledge of trigonometry is needed.

17.2 The theorem of Pythagoras

With reference to Fig. 17.1, the side opposite the right angle (i.e. side b) is called the **hypotenuse**. The **theorem of Pythagoras** states:

'In any right-angled triangle, the square on the hypotenuse is equal to the sum of the squares on the other two sides.'

Hence $\quad b^2 = a^2 + c^2$

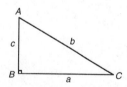

Fig. 17.1

Problem 1. In Fig. 17.2, find the length of BC

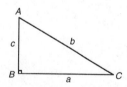

Fig. 17.2

By Pythagoras' theorem, $\quad a^2 = b^2 + c^2$

i.e. $\qquad a^2 = 4^2 + 3^2 = 16 + 9 = 25$

Hence $\qquad a = \sqrt{25} = \pm 5 (-5$ has no meaning in this context and is thus ignored)

Thus $\qquad \boldsymbol{bc = 5\,cm}$

Problem 2. In Fig. 17.3, find the length of EF

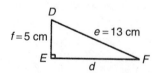

Fig. 17.3

By Pythagoras' theorem: $\quad e^2 = d^2 + f^2$

Hence $\qquad 13^2 = d^2 + 5^2$

$\qquad 169 = d^2 + 25$

$\qquad d^2 = 169 - 25 = 144$

Thus $\qquad d = \sqrt{144} = 12\,\text{cm}$

i.e. $\qquad \boldsymbol{EF = 12\,cm}$

Problem 3. Two aircraft leave an airfield at the same time. One travels due north at an average speed of 300 km/h and the other due west at an average speed of 220 km/h. Calculate their distance apart after 4 hours.

After 4 hours, the first aircraft has travelled $4 \times 300 = 1200\,\text{km}$, due north, and the second aircraft has travelled

$4 \times 220 = 880$ km due west, as shown in Fig. 17.4. Distance apart after 4 hours $= BC$

From Pythagoras' theorem:

$$BC^2 = 1200^2 + 880^2$$
$$= 1\,440\,000 + 774\,400$$

and $\quad BC = \sqrt{(2\,214\,400)}$

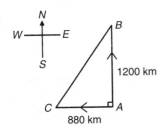

Fig. 17.4

Hence distance apart after 4 hours = 1488 km.

Now try the following exercise

Exercise 62 Further problems on the theorem of Pythagoras (Answers on page 260)

1. In a triangle ABC, $\angle B$ is a right angle, $AB = 6.92$ cm and $BC = 8.78$ cm. Find the length of the hypotenuse.

2. In a triangle CDE, $D = 90°$, $CD = 14.83$ mm and $CE = 28.31$ mm. Determine the length of DE.

3. Show that if a triangle has sides of 8, 15 and 17 cm it is right angled.

4. Triangle PQR is isosceles, Q being a right angle. If the hypotenuse is 38.47 cm find (a) the lengths of sides PQ and QR, and (b) the value of $\angle QPR$.

5. A man cycles 24 km due south and then 20 km due east. Another man, starting at the same time as the first man, cycles 32 km due east and then 7 km due south. Find the distance between the two men.

6. A ladder 3.5 m long is placed against a perpendicular wall with its foot 1.0 m from the wall. How far up the wall (to the nearest centimetre) does the ladder reach? If the foot of the ladder is now moved 30 cm further away from the wall, how far does the top of the ladder fall?

7. Two ships leave a port at the same time. One travels due west at 18.4 km/h and the other due south at 27.6 km/h. Calculate how far apart the two ships are after 4 hours.

17.3 Trigonometric ratios of acute angles

(a) With reference to the right-angled triangle shown in Fig. 17.5:

(i) sine $\theta = \dfrac{\text{opposite side}}{\text{hypotenuse}}$, i.e. $\textbf{sin}\,\boldsymbol{\theta} = \dfrac{b}{c}$

(ii) cosine $\theta = \dfrac{\text{adjacent side}}{\text{hypotenuse}}$, i.e. $\textbf{cos}\,\boldsymbol{\theta} = \dfrac{a}{c}$

(iii) tangent $\theta = \dfrac{\text{opposite side}}{\text{adjacent side}}$, i.e. $\textbf{tan}\,\boldsymbol{\theta} = \dfrac{b}{a}$

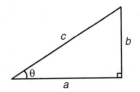

Fig. 17.5

Problem 4. From Fig. 17.6, find $\sin D$, $\cos D$ and $\tan F$

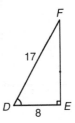

Fig. 17.6

By Pythagoras' theorem, $17^2 = 8^2 + EF^2$

from which, $\quad EF = \sqrt{17^2 - 8^2} = 15$

$$\sin D = \frac{EF}{DF} = \frac{15}{17} \quad \text{or} \quad \textbf{0.8824}$$

$$\cos D = \frac{DE}{DF} = \frac{8}{17} \quad \text{or} \quad \textbf{0.4706}$$

$$\tan F = \frac{DE}{EF} = \frac{8}{15} \quad \text{or} \quad \textbf{0.5333}$$

Problem 5. Determine the values of $\sin\theta$, $\cos\theta$ and $\tan\theta$ for the right-angled triangle ABC shown in Fig. 17.7.

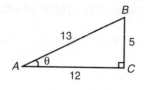

Fig. 17.7

By definition: $\sin\theta = \dfrac{\text{opposite side}}{\text{hypotenuse}} = \dfrac{5}{13} = \mathbf{0.3846}$

$\cos\theta = \dfrac{\text{adjacent side}}{\text{hypotenuse}} = \dfrac{12}{13} = \mathbf{0.9231}$

$\tan\theta = \dfrac{\text{opposite side}}{\text{adjacent side}} = \dfrac{5}{12} = \mathbf{0.4167}$

Problem 6.   If $\cos X = \dfrac{9}{41}$ determine the value of $\sin X$ and $\tan X$.

Figure 17.8 shows a right-angled triangle XYZ.

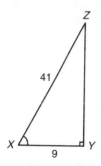

Fig. 17.8

Since $\cos X = \dfrac{9}{41}$, then $XY = 9$ units and $XZ = 41$ units.

Using Pythagoras' theorem: $41^2 = 9^2 + YZ^2$ from which $YZ = \sqrt{41^2 - 9^2} = 40$ units.
Thus

$$\mathbf{\sin X} = \frac{40}{41} \quad \text{and} \quad \mathbf{\tan X} = \frac{40}{9} = 4\frac{4}{9}$$

Problem 7.   Point A lies at co-ordinate (2,3) and point B at (8,7). Determine (a) the distance AB, (b) the gradient of the straight line AB, and (c) the angle AB makes with the horizontal.

(a) Points A and B are shown in Fig. 17.9(a).
In Fig. 17.9(b), the horizontal and vertical lines AC and BC are constructed.

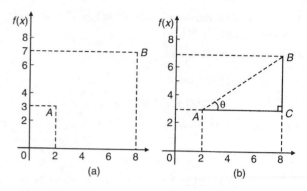

Fig. 17.9

Since ABC is a right-angled triangle, and $AC = (8-2) = 6$ and $BC = (7-3) = 4$, then by Pythagoras' theorem

$$AB^2 = AC^2 + BC^2 = 6^2 + 4^2$$

and $\mathbf{AB} = \sqrt{6^2 + 4^2} = \sqrt{52} = \mathbf{7.211}$, correct to 3 decimal places.

(b) The gradient of AB is given by $\tan\theta$,

i.e. $\mathbf{gradient} = \tan\theta = \dfrac{BC}{AC} = \dfrac{4}{6} = \dfrac{2}{3}$

(c) **The angle AB makes with the horizontal** is given by $\tan^{-1}\dfrac{2}{3} = \mathbf{33.69°}$

Now try the following exercise

Exercise 63   Further problems on trigonometric ratios of acute angles (Answers on page 260)

1. Sketch a triangle XYZ such that $\angle Y = 90°$, $XY = 9\,\text{cm}$ and $YZ = 40\,\text{cm}$. Determine $\sin Z$, $\cos Z$, $\tan X$ and $\cos X$.

2. In triangle ABC shown in Fig. 17.10, find $\sin A$, $\cos A$, $\tan A$, $\sin B$, $\cos B$ and $\tan B$.

Fig. 17.10

3. If $\cos A = \dfrac{15}{17}$ find $\sin A$ and $\tan A$, in fraction form.

4. If $\tan X = \dfrac{15}{112}$, find $\sin X$ and $\cos X$, in fraction form.

5. For the right-angled triangle shown in Fig. 17.11, find:
   (a) $\sin\alpha$ (b) $\cos\theta$ (c) $\tan\theta$

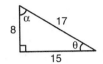

Fig. 17.11

6. If $\tan\theta = \dfrac{7}{24}$, find $\sin\theta$ and $\cos\theta$ in fraction form.

7. Point P lies at co-ordinate $(-3,1)$ and point Q at $(5,-4)$. Determine

   (a) the distance PQ,

   (b) the gradient of the straight line PQ and,

   (c) the angle PQ makes with the horizontal.

17.4   Solution of right-angled triangles

To 'solve a right-angled triangle' means 'to find the unknown sides and angles'. This is achieved by using (i) the theorem of Pythagoras, and/or (ii) trigonometric ratios. This is demonstrated in the following problems.

Problem 8.   Sketch a right-angled triangle ABC such that $B = 90°$, $AB = 5$ cm and $BC = 12$ cm. Determine the length of AC and hence evaluate $\sin A$, $\cos C$ and $\tan A$.

Triangle ABC is shown in Fig. 17.12.

By Pythagoras' theorem, $AC = \sqrt{5^2 + 12^2} = 13$

By definition: $\sin A = \dfrac{\text{opposite side}}{\text{hypotenuse}} = \dfrac{12}{13}$ or   **0.9231**

$\cos C = \dfrac{\text{adjacent side}}{\text{hypotenuse}} = \dfrac{12}{13}$ or   **0.9231**

and $\tan A = \dfrac{\text{opposite side}}{\text{adjacent side}} = \dfrac{12}{5}$ or   **2.400**

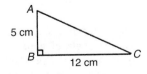

Fig. 17.12

Problem 9.   In triangle PQR shown in Fig. 17.13, find the lengths of PQ and PR.

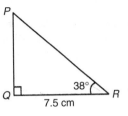

Fig. 17.13

$$\tan 38° = \frac{PQ}{QR} = \frac{PQ}{7.5}$$

hence $PQ = 7.5\tan 38° = 7.5(0.7813) = \mathbf{5.860\,cm}$

$$\cos 38° = \frac{QR}{PR} = \frac{7.5}{PR}$$

hence $PR = \dfrac{7.5}{\cos 38°} = \dfrac{7.5}{0.7880} = \mathbf{9.518\,cm}$

[Check: Using Pythagoras' theorem

$$(7.5)^2 + (5.860)^2 = 90.59 = (9.518)^2]$$

Problem 10.   Solve the triangle ABC shown in Fig. 17.14.

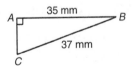

Fig. 17.14

To 'solve triangle ABC' means 'to find the length AC and angles B and C'

$$\sin C = \frac{35}{37} = 0.94595$$

hence $C = \sin^{-1} 0.94595 = 71.08°$ or $\mathbf{71°5'}$

$B = 180° - 90° - 71°5' = \mathbf{18°55'}$ (since angles in a triangle add up to $180°$)

$$\sin B = \frac{AC}{37}$$

hence $AC = 37\sin 18°55' = 37(0.3242) = \mathbf{12.0\,mm}$

or, using Pythagoras' theorem, $37^2 = 35^2 + AC^2$,

from which, $AC = \sqrt{(37^2 - 35^2)} = \mathbf{12.0\,mm}$.

> **Problem 11.** Solve triangle XYZ given $\angle X = 90°$, $\angle Y = 23°17'$ and $YZ = 20.0$ mm. Determine also its area.

It is always advisable to make a reasonably accurate sketch so as to visualize the expected magnitudes of unknown sides and angles. Such a sketch is shown in Fig. 17.15.

$$\angle Z = 180° - 90° - 23°17' = \mathbf{66°43'}$$

$$\sin 23°17' = \frac{XZ}{20.0} \quad \text{hence} \quad XZ = 20.0 \sin 23°17'$$

$$= 20.0(0.3953) = \mathbf{7.906\,mm}$$

$$\cos 23°17' = \frac{XY}{20.0} \quad \text{hence} \quad XY = 20.0 \cos 23°17'$$

$$= 20.0(0.9186) = \mathbf{18.37\,mm}$$

Fig. 17.15

(Check: Using Pythagoras' theorem

$$(18.37)^2 + (7.906)^2 = 400.0 = (20.0)^2)$$

Area of triangle $XYZ = \frac{1}{2}$ (base)(perpendicular height)

$$= \frac{1}{2}(XY)(XZ) = \frac{1}{2}(18.37)(7.906)$$

$$= \mathbf{72.62\,mm^2}$$

Now try the following exercise

> **Exercise 64 Further problems on the solution of right-angled triangles (Answers on page 260)**
>
> 1. Solve triangle ABC in Fig. 17.16(i).
>
> 2. Solve triangle DEF in Fig. 17.16(ii).

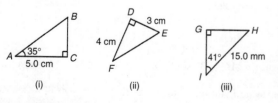

(i)   (ii)   (iii)

Fig. 17.16

3. Solve triangle GHI in Fig. 17.16(iii).

4. Solve the triangle JKL in Fig. 17.17(i) and find its area.

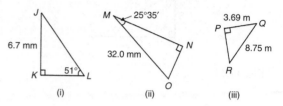

(i)   (ii)   (iii)

Fig. 17.17

5. Solve the triangle MNO in Fig. 17.17(ii) and find its area.

6. Solve the triangle PQR in Fig. 17.17(iii) and find its area.

7. A ladder rests against the top of the perpendicular wall of a building and makes an angle of $73°$ with the ground. If the foot of the ladder is 2 m from the wall, calculate the height of the building.

17.5 Angles of elevation and depression

(a) If, in Fig. 17.18, BC represents horizontal ground and AB a vertical flagpole, then the **angle of elevation** of the top of the flagpole, A, from the point C is the angle that the imaginary straight line AC must be raised (or elevated) from the horizontal CB, i.e. angle θ.

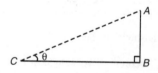

Fig. 17.18

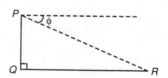

Fig. 17.19

(b) If, in Fig. 17.19, PQ represents a vertical cliff and R a ship at sea, then the **angle of depression** of the ship from point P is the angle through which the imaginary straight line PR must be lowered (or depressed) from the horizontal to the ship, i.e. angle ϕ.

(Note, $\angle PRQ$ is also ϕ – alternate angles between parallel lines.)

Problem 12. An electricity pylon stands on horizontal ground. At a point 80 m from the base of the pylon, the angle of elevation of the top of the pylon is 23°. Calculate the height of the pylon to the nearest metre.

Figure 17.20 shows the pylon AB and the angle of elevation of A from point C is 23°

$$\tan 23° = \frac{AB}{BC} = \frac{AB}{80}$$

Hence height of pylon $AB = 80 \tan 23° = 80(0.4245)$

$$= 33.96 \, \text{m}$$

$$= \textbf{34 m to the nearest metre}$$

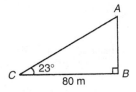

Fig. 17.20

Problem 13. A surveyor measures the angle of elevation of the top of a perpendicular building as 19°. He moves 120 m nearer the building and finds the angle of elevation is now 47°. Determine the height of the building.

The building PQ and the angles of elevation are shown in Fig. 17.21.

In triangle PQS, $\tan 19° = \dfrac{h}{x + 120}$

hence $h = \tan 19°(x + 120)$, i.e. $h = 0.3443(x + 120)$ (1)

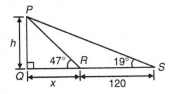

Fig. 17.21

In triangle PQR, $\tan 47° = \dfrac{h}{x}$

hence $h = \tan 47°(x)$, i.e. $h = 1.0724x$ (2)

Equating equations (1) and (2) gives:

$$0.3443(x + 120) = 1.0724x$$

$$0.3443x + (0.3443)(120) = 1.0724x$$

$$(0.3443)(120) = (1.0724 - 0.3443)x$$

$$41.316 = 0.7281x$$

$$x = \frac{41.316}{0.7281} = 56.74 \, \text{m}$$

From equation (2), **height of building,**

$$h = 1.0724x = 1.0724(56.74) = \textbf{60.85 m}$$

Problem 14. The angle of depression of a ship viewed at a particular instant from the top of a 75 m vertical cliff is 30°. Find the distance of the ship from the base of the cliff at this instant. The ship is sailing away from the cliff at constant speed and 1 minute later its angle of depression from the top of the cliff is 20°. Determine the speed of the ship in km/h.

Figure 17.22 shows the cliff AB, the initial position of the ship at C and the final position at D. Since the angle of depression is initially 30° then $\angle ACB = 30°$ (alternate angles between parallel lines).

$$\tan 30° = \frac{AB}{BC} = \frac{75}{BC} \quad \text{hence} \quad BC = \frac{75}{\tan 30°} = \frac{75}{0.5774}$$

$$= 129.9 \, \text{m} = \textbf{initial position of ship from base}$$
$$\textbf{of cliff}$$

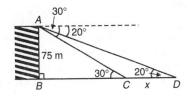

Fig. 17.22

In triangle ABD, $\tan 20° = \dfrac{AB}{BD} = \dfrac{75}{BC + CD} = \dfrac{75}{129.9 + x}$

Hence $129.9 + x = \dfrac{75}{\tan 20°} = \dfrac{75}{0.3640} = 206.0 \, \text{m}$

from which $x = 206.0 - 129.9 = 76.1 \, \text{m}$

Thus the ship sails 76.1 m in 1 minute, i.e. 60 s, hence speed of ship

$$= \frac{\text{distance}}{\text{time}} = \frac{76.1}{60} \text{m/s}$$

$$= \frac{76.1 \times 60 \times 60}{60 \times 1000} \text{km/h} = \textbf{4.57 km/h}$$

Now try the following exercise

Exercise 65   Further problems on angles of elevation and depression (Answers on page 260)

1. A vertical tower stands on level ground. At a point 105 m from the foot of the tower the angle of elevation of the top is 19°. Find the height of the tower.

2. If the angle of elevation of the top of a vertical 30 m high aerial is 32°, how far is it to the aerial?

3. From the top of a vertical cliff 90.0 m high the angle of depression of a boat is 19°50′. Determine the distance of the boat from the cliff.

4. From the top of a vertical cliff 80.0 m high the angles of depression of two buoys lying due west of the cliff are 23° and 15°, respectively. How far are the buoys apart?

5. From a point on horizontal ground a surveyor measures the angle of elevation of the top of a flagpole as 18°40′. He moves 50 m nearer to the flagpole and measures the angle of elevation as 26°22′. Determine the height of the flagpole.

6. A flagpole stands on the edge of the top of a building. At a point 200 m from the building the angles of elevation of the top and bottom of the pole are 32° and 30° respectively. Calculate the height of the flagpole.

7. From a ship at sea, the angles of elevation of the top and bottom of a vertical lighthouse standing on the edge of a vertical cliff are 31° and 26°, respectively. If the lighthouse is 25.0 m high, calculate the height of the cliff.

8. From a window 4.2 m above horizontal ground the angle of depression of the foot of a building across the road is 24° and the angle of elevation of the top of the building is 34°. Determine, correct to the nearest centimetre, the width of the road and the height of the building.

9. The elevation of a tower from two points, one due east of the tower and the other due west of it are 20° and 24°, respectively, and the two points of observation are 300 m apart. Find the height of the tower to the nearest metre.

17.6   Evaluating trigonometric ratios of any angles

Four-figure tables are available which gives sines, cosines, and tangents, for angles between 0° and 90°. However, the easiest method of evaluating trigonometric functions of any angle is by using a **calculator.**

The following values, correct to 4 decimal places, may be checked:

sine 18° = 0.3090	cosine 56° = 0.5592
tangent 29° = 0.5543	
sine 172° = 0.1392	cosine 115° = −0.4226
tangent 178° = −0.0349	
sine 241.63° = −0.8799	cosine 331.78° = 0.8811
tangent 296.42° = −2.0127	

To evaluate, say, sine 42°23′ using a calculator means finding $\text{sine } 42\dfrac{23°}{60}$ since there are 60 minutes in 1 degree.

$$\frac{23}{60} = 0.383\dot{3}, \quad \text{thus} \quad 42°23′ = 42.383\dot{3}°$$

Thus sine 42°23′ = sine 42.3833° = 0.6741, correct to 4 decimal places.

Similarly, cosine 72°38′ = $\text{cosine } 72\dfrac{38°}{60}$ = 0.2985, correct to 4 decimal places.

Problem 15.   Evaluate correct to 4 decimal places:
(a) sine 11°   (b) sine 121.68°   (c) sine 259°10′

(a) sine 11° = **0.1908**

(b) sine 121.68° = **0.8510**

(c) sine 259°10′ = $\text{sine } 259\dfrac{10°}{60}$ = **−0.9822**

Problem 16.   Evaluate, correct to 4 decimal places:

(a) cosine 23°   (b) cosine 159.32°

(c) cosine 321°41′

(a) cosine 23° = **0.9205**

(b) cosine 159.32° = **−0.9356**

(c) cosine 321°41′ = $\text{cosine } 321\dfrac{41°}{60}$ = **0.7846**

Problem 17.   Evaluate, correct to 4 significant figures:

(a) tangent 276°   (b) tangent 131.29°

(c) tangent 76°58′

(a) tangent 276° = **−9.514**

(b) tangent 131.29° = **−1.139**

(c) tangent 76°58′ = $\tan 76\dfrac{58°}{60}$ = **4.320**

Problem 18. Evaluate, correct to 4 significant figures:

(a) sin 1.481   (b) cos(3π/5)   (c) tan 2.93

(a) sin 1.481 means the sine of 1.481 radians. Hence a calculator needs to be on the radian function.

Hence sin 1.481 = **0.9960**

(b) cos(3π/5) = cos 1.884955 . . . = **−0.3090**

(c) tan 2.93 = **−0.2148**

Problem 19. Determine the acute angles:

(a) sin⁻¹ 0.7321   (b) cos⁻¹ 0.4174

(c) tan⁻¹ 1.4695

(a) $\sin^{-1}\theta$ is an abbreviation for 'the angle whose sine is equal to θ'. 0.7321 is entered into a calculator and then the inverse sine (or $\sin^{-1}$) key is pressed. Hence $\sin^{-1}0.7321 = 47.06273\ldots°$.

Subtracting 47 leaves 0.06273 . . .° and multiplying this by 60 gives 4′ to the nearest minute.

Hence $\sin^{-1}0.7321 =$ **47.06°** or **47°4′**

Alternatively, in radians, $\sin^{-1}0.7321 =$ **0.821 radians**.

(b) $\cos^{-1}0.4174 =$ **65.33°** or **65°20′** or **1.140 radians**.

(c) $\tan^{-1}1.4695 =$ **55.76°** or **55°46′** or **0.973 radians**.

Problem 20. Evaluate the following expression, correct to 4 significant figures:

$$\frac{4.2\tan 49°26' - 3.7\sin 66°1'}{7.1\cos 29°34'}$$

By calculator:

$$\tan 49°26' = \tan\left(49\frac{26}{60}\right)° = 1.1681,$$

$$\sin 66°1' = 0.9137 \quad\text{and}\quad \cos 29°34' = 0.8698$$

Hence

$$\begin{aligned}&= \frac{4.2\tan 49°26' - 3.7\sin 66°1'}{7.1\cos 29°34'}\\ &= \frac{(4.2\times 1.1681)-(3.7\times 0.9137)}{(7.1\times 0.8698)}\\ &= \frac{4.9060-3.3807}{6.1756} = \frac{1.5253}{6.1756}\\ &= 0.2470 = \textbf{0.247}, \quad\text{correct to 3 significant figures.}\end{aligned}$$

Problem 21. Evaluate correct to 4 decimal places:

(a) sin(−112°)   (b) cosine (−93°16′)

(c) tangent (−217.29°)

(a) Positive angles are considered by convention to be anticlockwise and negative angles as clockwise. From Fig. 17.23, −112° is actually the same as +248° (i.e. 360° − 112°).

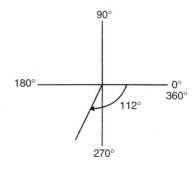

Fig. 17.23

Hence, by calculator, sin(−112°) = sin 248° = **−0.9272**

(b) $\cos(-93°16') = \cos\left(-93\frac{16}{60}\right)° =$ **−0.0570**

(c) tangent (−217.29°) = **−0.7615** (which is the same as tan(360° − 217.29°), i.e. tan 141.71°)

Now try the following exercise

Exercise 66   Further problems on evaluating trigonometric ratios of any angle (Answers on page 260)

In Problems 1 to 4, evaluate correct to 4 decimal places:

1. (a) sine 27°   (b) sine 172.41°   (c) sine 302°52′
2. (a) cosine 124°   (b) cosine 21.46°
   (c) cosine 284°10′
3. (a) tangent 145°   (b) tangent 310.59°
   (c) tangent 49°16′
4. (a) sine $\frac{2\pi}{3}$ (b) cos 1.681   (c) tan 3.672

In Problems 5 to 7, determine the acute angle in degrees (correct to 2 decimal places), degrees and minutes, and in radians (correct to 3 decimal places).

5. sin⁻¹ 0.2341   6. cos⁻¹ 0.8271   7. tan⁻¹ 0.8106

In problems 8 to 10, evaluate correct to 4 significant figures.

8. $4\cos 56°19' - 3\sin 21°57'$

9. $\dfrac{11.5\tan 49°11' - \sin 90°}{3\cos 45°}$

10. $\dfrac{5\sin 86°3'}{3\tan 14°29' - 2\cos 31°9'}$

11. Determine the acute angle, in degrees and minutes, correct to the nearest minute, given by

$$\sin^{-1}\left(\dfrac{4.32\sin 42°16'}{7.86}\right)$$

12. Evaluate $\dfrac{(\sin 34°27')(\cos 69°2')}{(2\tan 53°39')}$ correct to 4 significant figures

13. Evaluate correct to 4 decimal places:

(a) sine $(-125°)$ (b) $\tan(-241°)$

(c) $\cos(-49°15')$

18

Trigonometric waveforms

18.1 Graphs of trigonometric functions

By drawing up tables of values from 0° to 360°, graphs of $y = \sin A$, $y = \cos A$ and $y = \tan A$ may be plotted. Values obtained with a calculator (correct to 3 decimal places – which is more than sufficient for plotting graphs), using 30° intervals, are shown below, with the respective graphs shown in Fig. 18.1.

(a) $y = \sin A$

A	0	30°	60°	90°	120°	150°	180°
sin A	0	0.500	0.866	1.000	0.866	0.500	0

A	210°	240°	270°	300°	330°	360°
sin A	−0.500	−0.866	−1.000	−0.866	−0.500	0

(b) $y = \cos A$

A	0	30°	60°	90°	120°	150°	180°
cos A	1.000	0.866	0.500	0	−0.500	−0.866	−1.000

A	210°	240°	270°	300°	330°	360°
cos A	−0.866	−0.500	0	0.500	0.866	1.000

(c) $y = \tan A$

A	0	30°	60°	90°	120°	150°	180°
tan A	0	0.577	1.732	∞	−1.732	−0.577	0

A	210°	240°	270°	300°	330°	360°
tan A	0.577	1.732	∞	−1.732	−0.577	0

From Fig. 18.1 it is seen that:

(i) Sine and cosine graphs oscillate between peak values of ±1

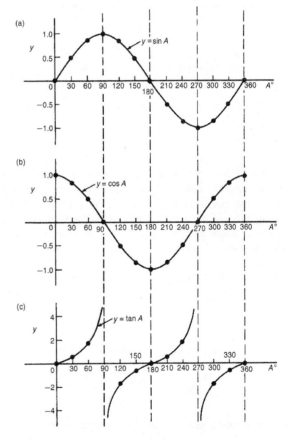

Fig. 18.1

(ii) The cosine curve is the same shape as the sine curve but displaced by 90°.

(iii) The sine and cosine curves are continuous and they repeat at intervals of 360°; the tangent curve appears to be discontinuous and repeats at intervals of 180°.

18.2   Angles of any magnitude

Figure 18.2 shows rectangular axes XX' and YY' intersecting at origin 0. As with graphical work, measurements made to the right and above 0 are positive, while those to the left and downwards are negative. Let $0A$ be free to rotate about 0. By convention, when $0A$ moves anticlockwise angular measurement is considered positive, and vice versa.

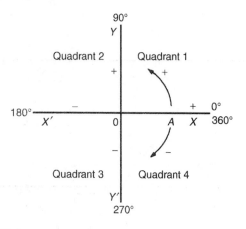

Fig. 18.2

Let $0A$ be rotated anticlockwise so that θ_1 is any angle in the first quadrant and let perpendicular AB be constructed to form the right-angled triangle $0AB$ in Fig. 18.3. Since all three sides of the triangle are positive, the trigonometric ratios sine, cosine and tangent will all be positive in the first quadrant. (Note: $0A$ is always positive since it is the radius of a circle).

Let $0A$ be further rotated so that θ_2 is any angle in the second quadrant and let AC be constructed to form the

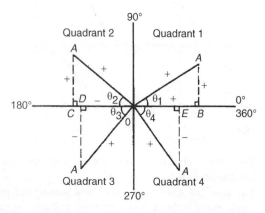

Fig. 18.3

right-angled triangle $0AC$. Then

$$\sin\theta_2 = \frac{+}{+} = + \qquad \cos\theta_2 = \frac{-}{+} = -$$

$$\tan\theta_2 = \frac{+}{-} = -$$

Let $0A$ be further rotated so that θ_3 is any angle in the third quadrant and let AD be constructed to form the right-angled triangle $0AD$. Then

$$\sin\theta_3 = \frac{-}{+} = - \qquad \cos\theta_3 = \frac{-}{+} = -$$

$$\tan\theta_3 = \frac{-}{-} = +$$

Let $0A$ be further rotated so that θ_4 is any angle in the fourth quadrant and let AE be constructed to form the right-angled triangle $0AE$. Then

$$\sin\theta_4 = \frac{-}{+} = - \qquad \cos\theta_4 = \frac{+}{+} = +$$

$$\tan\theta_4 = \frac{-}{+} = -$$

The above results are summarized in Fig. 18.4. The letters underlined spell the word CAST when starting in the fourth quadrant and moving in an anticlockwise direction.

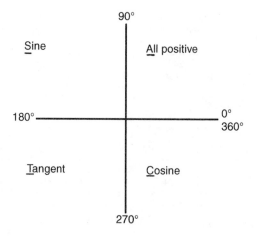

Fig. 18.4

In the first quadrant of Fig. 18.1 all of the curves have positive values; in the second only sine is positive; in the third only tangent is positive; in the fourth only cosine is positive – exactly as summarized in Fig. 18.4. A knowledge of angles of any magnitude is needed when finding, for example, all the angles between 0° and 360° whose sine is, say, 0.3261. If 0.3261 is entered into a calculator and then the inverse sine key pressed (or sin-key) the answer

19.03° appears. However, there is a second angle between 0° and 360° which the calculator does not give. Sine is also positive in the second quadrant [either from CAST or from Fig. 18.1(a)]. The other angle is shown in Fig. 18.5 as angle θ where $\theta = 180° - 19.03° = 160.97°$. Thus 19.03° **and** 160.97° are the angles between 0° and 360° whose sine is 0.3261 (check that sin 160.97° = 0.3261 on your calculator).

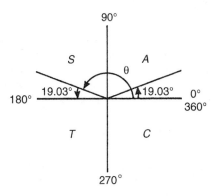

Fig. 18.5

Be careful! Your calculator only gives you one of these answers. The second answer needs to be deduced from a knowledge of angles of any magnitude, as shown in the following worked problems.

Problem 1. Determine all the angles between 0° and 360° whose sine is −0.4638

The angles whose sine is −0.4638 occurs in the third and fourth quadrants since sine is negative in these quadrants – see Fig. 18.6.

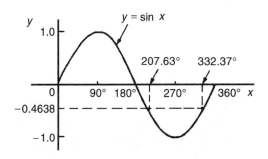

Fig. 18.6

From Fig. 18.7, $\theta = \sin^{-1} 0.4638 = 27.63°$. Measured from 0°, the two angles between 0° and 360° whose sine is −0.4638 are 180° + 27.63°, i.e. **207.63°** and 360° − 27.63°, i.e. **332.37°**

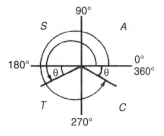

Fig. 18.7

(Note that a calculator only gives one answer, i.e. −27.632588°)

Problem 2. Determine all the angles between 0° and 360° whose tangent is 1.7629

A tangent is positive in the first and third quadrants – see Fig. 18.8.

From Fig. 18.9, $\theta = \tan^{-1} 1.7629 = 60.44°$

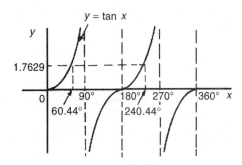

Fig. 18.8

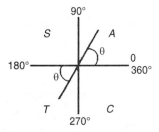

Fig. 18.9

Measured from 0°, the two angles between 0° and 360° whose tangent is 1.7629 are **60.44°** and 180° + 60.44°, i.e. **240.44°**

Problem 3. Solve the equation $\cos^{-1}(-0.2348) = \alpha$ for angles of α between 0° and 360°.

Cosine is positive in the first and fourth quadrants and thus negative in the second and third quadrants – from Fig. 18.5 or from Fig. 18.1(b).

In Fig. 18.10, angle $\theta = \cos^{-1}(0.2348) = 76.42°$

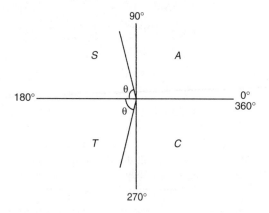

Fig. 18.10

Measured from 0°, the two angles whose cosine is -0.2348 are $\alpha = 180° - 76.42°$ i.e. **103.58°** and $\alpha = 180° + 76.42°$, i.e. **256.42°**

Now try the following exercise

Exercise 67   Further problems on angles of any magnitude (Answers on page 260)

1. Determine all of the angles between 0° and 360° whose sine is:

   (a) 0.6792   (b) -0.1483

2. Solve the following equations for values of x between 0° and 360°:

   (a) $x = \cos^{-1} 0.8739$ (b) $x = \cos^{-1}(-0.5572)$

3. Find the angles between 0° to 360° whose tangent is:

   (a) 0.9728   (b) -2.3418

18.3   The production of a sine and cosine wave

In Fig. 18.11, let OR be a vector 1 unit long and free to rotate anticlockwise about O. In one revolution a circle is produced and is shown with 15° sectors. Each radius arm has a vertical and a horizontal component. For example, at 30°, the vertical component is TS and the horizontal component is OS.

From trigonometric ratios,

$$\sin 30° = \frac{TS}{TO} = \frac{TS}{1}, \quad \text{i.e.} \quad TS = \sin 30°$$

and $\cos 30° = \dfrac{OS}{TO} = \dfrac{OS}{1}$,   i.e. $OS = \cos 30°$

The vertical component TS may be projected across to $T'S'$, which is the corresponding value of 30° on the graph of y against angle $x°$. If all such vertical components as TS are projected on to the graph, then a **sine wave** is produced as shown in Fig. 18.11.

If all horizontal components such as OS are projected on to a graph of y against angle $x°$, then a **cosine wave** is produced. It is easier to visualize these projections by redrawing the circle with the radius arm OR initially in a vertical position as shown in Fig. 18.12.

From Figs. 18.11 and 18.12 it is seen that a cosine curve is of the same form as the sine curve but is displaced by 90° (or $\pi/2$ radians).

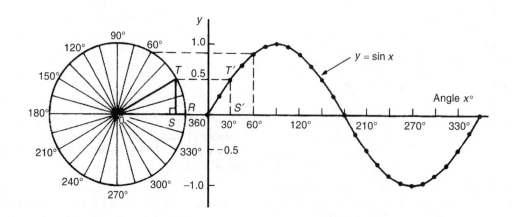

Fig. 18.11

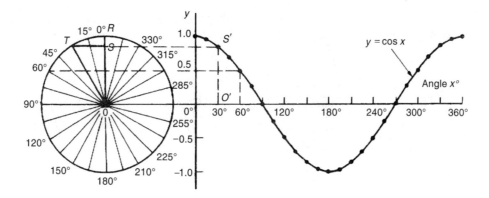

Fig. 18.12

18.4 Sine and cosine curves

Graphs of sine and cosine waveforms

(i) A graph of $y = \sin A$ is shown by the broken line in Fig. 18.13 and is obtained by drawing up a table of values as in Section 18.1. A similar table may be produced for $y = \sin 2A$.

$A°$	0	30	45	60	90	120
$2A$	0	60	90	120	180	240
$\sin 2A$	0	0.866	1.0	0.866	0	−0.866

$A°$	135	150	180	210	225	240
$2A$	270	300	360	420	450	480
$\sin 2A$	−1.0	−0.866	0	0.866	1.0	0.866

$A°$	270	300	315	330	360
$2A$	540	600	630	660	720
$\sin 2A$	0	−0.866	−1.0	−0.866	0

A graph of $y = \sin 2A$ is shown in Fig. 18.13.

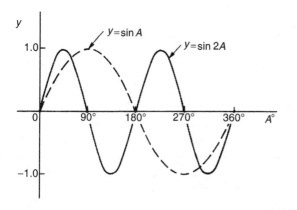

Fig. 18.13

(ii) A graph of $y = \sin \frac{1}{2}A$ is shown in Fig. 18.14 using the following table of values.

$A°$	0	30	60	90	120	150	180
$\frac{1}{2}A$	0	15	30	45	60	75	90
$\sin \frac{1}{2}A$	0	0.259	0.500	0.707	0.866	0.966	1.00

$A°$	210	240	270	300	330	360
$\frac{1}{2}A$	105	120	135	150	165	180
$\sin \frac{1}{2}A$	0.966	0.866	0.707	0.500	0.259	0

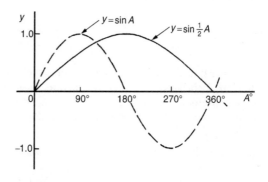

Fig. 18.14

(iii) A graph of $y = \cos A$ is shown by the broken line in Fig. 18.15 and is obtained by drawing up a table of values. A similar table may be produced for $y = \cos 2A$ with the result as shown.

(iv) A graph of $y = \cos \frac{1}{2}A$ is shown in Fig. 18.16 which may be produced by drawing up a table of values, similar to above.

Periodic time and period

(i) Each of the graphs shown in Figs. 18.13 to 18.16 will repeat themselves as angle A increases and are thus called **periodic functions**.

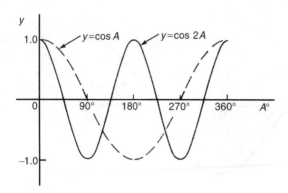

Fig. 18.15

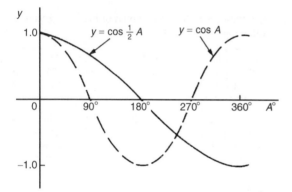

Fig. 18.16

(ii) $y = \sin A$ and $y = \cos A$ repeat themselves every 360° (or 2π radians); thus 360° is called the **period** of these waveforms. $y = \sin 2A$ and $y = \cos 2A$ repeat themselves every 180° (or π radians); thus 180° is the period of these waveforms.

(iii) In general, if $y = \sin pA$ or $y = \cos pA$ (where p is a constant) then the period of the waveform is $360°/p$ (or $2\pi/p$ rad). Hence if $y = \sin 3A$ then the period is 360/3, i.e. 120°, and if $y = \cos 4A$ then the period is 360/4, i.e. 90°

Amplitude

Amplitude is the name given to the maximum or peak value of a sine wave. Each of the graphs shown in Figs. 18.13 to 18.16 has an amplitude of +1 (i.e. they oscillate between +1 and −1). However, if $y = 4 \sin A$, each of the values in the table is multiplied by 4 and the maximum value, and thus amplitude, is 4. Similarly, if $y = 5 \cos 2A$, the amplitude is 5 and the period is 360°/2, i.e. 180°

Problem 4.   Sketch $y = \sin 3A$ between $A = 0°$ and $A = 360°$

Amplitude = 1 and period = $360°/3 = 120°$.

A sketch of $y = \sin 3A$ is shown in Fig. 18.17.

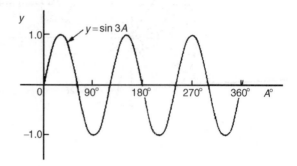

Fig. 18.17

Problem 5.   Sketch $y = 3 \sin 2A$ from $A = 0$ to $A = 2\pi$ radians

---

Amplitude = 3 and period = $2\pi/2 = \pi$ rads (or 180°)

A sketch of $y = 3 \sin 2A$ is shown in Fig. 18.18.

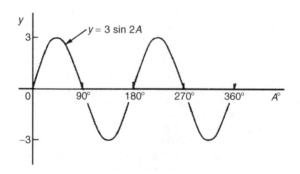

Fig. 18.18

Problem 6.   Sketch $y = 4 \cos 2x$ from $x = 0°$ to $x = 360°$

Amplitude = 4 and period = $360°/2 = 180°$.

A sketch of $y = 4 \cos 2x$ is shown in Fig. 18.19.

Problem 7.   Sketch $y = 2 \sin \frac{3}{5}A$ over one cycle.

Amplitude = 2; period = $\dfrac{360°}{\frac{3}{5}} = \dfrac{360° \times 5}{3} = 600°$.

A sketch of $y = 2 \sin \frac{3}{5}A$ is shown in Fig. 18.20.

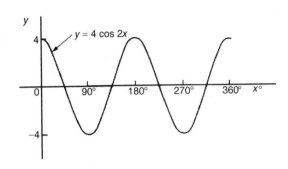

Fig. 18.19

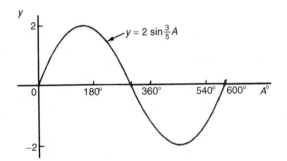

Fig. 18.20

Lagging and leading angles

(i) A sine or cosine curve may not always start at 0°. To show this a periodic function is represented by $y = \sin(A \pm \alpha)$ or $y = \cos(A \pm \alpha)$ where α is a phase displacement compared with $y = \sin A$ or $y = \cos A$.

(ii) By drawing up a table of values, a graph of $y = \sin(A - 60°)$ may be plotted as shown in Fig. 18.21. If $y = \sin A$ is assumed to start at 0° then $y = \sin(A - 60°)$ starts 60° later (i.e. has a zero value 60° later). Thus $y = \sin(A - 60°)$ is said to **lag** $y = \sin A$ by 60°

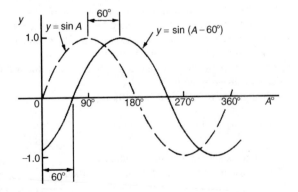

Fig. 18.21

(iii) By drawing up a table of values, a graph of $y = \cos(A + 45°)$ may be plotted as shown in Fig. 18.22. If $y = \cos A$ is assumed to start at 0° then $y = \cos(A + 45°)$ starts 45° earlier (i.e. has a maximum value 45° earlier).Thus $y = \cos(A + 45°)$ is said to **lead** $y = \cos A$ by 45°

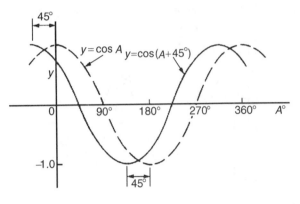

Fig. 18.22

(iv) Generally, a graph of $y = \sin(A - \alpha)$ lags $y = \sin A$ by angle α, and a graph of $y = \sin(A + \alpha)$ leads $y = \sin A$ by angle α

(v) A cosine curve is the same shape as a sine curve but starts 90° earlier, i.e. leads by 90°. Hence

$$\cos A = \sin(A + 90°)$$

Problem 8. Sketch $y = 5\sin(A + 30°)$ from $A = 0°$ to $A = 360°$

Amplitude $= 5$ and period $= 360°/1 = 360°$.

$5\sin(A + 30°)$ leads $5\sin A$ by 30° (i.e. starts 30° earlier).

A sketch of $y = 5\sin(A + 30°)$ is shown in Fig. 18.23.

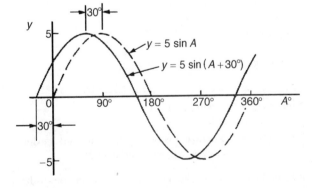

Fig. 18.23

Problem 9. Sketch $y = 7\sin(2A - \pi/3)$ over one cycle.

Amplitude $= 7$ and period $= 2\pi/2 = \pi$ radians.

In general, $y = \sin(pt - \alpha)$ lags $y = \sin pt$ by α/p, hence $7\sin(2A - \pi/3)$ lags $7\sin 2A$ by $(\pi/3)/2$, i.e. $\pi/6$ rad or $30°$.

A sketch of $y = 7\sin(2A - \pi/3)$ is shown in Fig. 18.24.

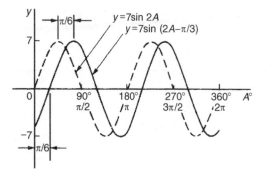

Fig. 18.24

Problem 10. Sketch $y = 2\cos(\omega t - 3\pi/10)$ over one cycle.

Amplitude $= 2$ and period $= 2\pi/\omega$ rad.

$2\cos(\omega t - 3\pi/10)$ lags $2\cos \omega t$ by $3\pi/10\omega$ seconds.

A sketch of $y = 2\cos(\omega t - 3\pi/10)$ is shown in Fig. 18.25.

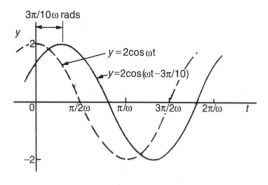

Fig. 18.25

Now try the following exercise

Exercise 68   Further problems on sine and cosine curves (Answers on page 260)

In Problems 1 to 7 state the amplitude and period of the waveform and sketch the curve between $0°$ and $360°$.

1. $y = \cos 3A$
2. $y = 2\sin\dfrac{5x}{2}$
3. $y = 3\sin 4t$
4. $y = 3\cos\dfrac{\theta}{2}$
5. $y = \dfrac{7}{2}\sin\dfrac{3x}{8}$
6. $y = 6\sin(t - 45°)$
7. $y = 4\cos(2\theta + 30°)$

18.5 Sinusoidal form $A\sin(\omega t \pm \alpha)$

In Fig. 18.26, let OR represent a vector that is free to rotate anti-clockwise about O at a velocity of ω rad/s. A rotating vector is called a **phasor**. After a time t seconds OR will have turned through an angle ωt radians (shown as angle TOR in Fig. 18.26). If ST is constructed perpendicular to OR, then $\sin \omega t = ST/OT$, i.e. $ST = OT\sin\omega t$.

If all such vertical components are projected on to a graph of y against ωt, a sine wave results of amplitude OR (as shown in Section 18.3).

If phasor OR makes one revolution (i.e. 2π radians) in T seconds, then the angular velocity, $\omega = 2\pi/T$ rad/s,

from which, $\boxed{T = 2\pi/\omega \text{ seconds}}$

T is known as the **periodic time**.

The number of complete cycles occurring per second is called the **frequency, f**

$$\text{Frequency} = \frac{\text{number of cycles}}{\text{second}} = \frac{1}{T} = \frac{\omega}{2\pi} \text{ Hz}$$

i.e. $\boxed{f = \dfrac{\omega}{2\pi} \text{ Hz}}$

Hence **angular velocity,** $\boxed{\omega = 2\pi f \text{ rad/s}}$

Amplitude is the name given to the maximum or peak value of a sine wave, as explained in Section 18.4. The amplitude of the sine wave shown in Fig. 18.26 has an amplitude of 1.

A sine or cosine wave may not always start at $0°$. To show this a periodic function is represented by $y = \sin(\omega t \pm \alpha)$ or $y = \cos(\omega t \pm \alpha)$, where α is a phase displacement compared with $y = \sin A$ or $y = \cos A$. A graph of $y = \sin(\omega t - \alpha)$ **lags** $y = \sin \omega t$ by angle α, and a graph of $y = \sin(\omega t + \alpha)$ **leads** $y = \sin \omega t$ by angle α.

The angle ωt is measured in **radians**

i.e. $\left(\omega\dfrac{\text{rad}}{\text{s}}\right)(t \text{ s}) = \omega t$ radians

hence angle α should also be in radians.

The relationship between degrees and radians is:

$360° = 2\pi$ radians   or $\boxed{180° = \pi \text{ radians}}$

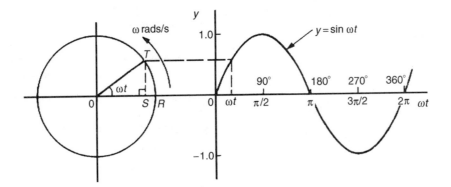

Fig. 18.26

Hence $1 \text{ rad} = \dfrac{180}{\pi} = 57.30°$ and, for example,

$$71° = 71 \times \dfrac{\pi}{180} = 1.239 \text{ rad}$$

Given a general sinusoidal function $y = A \sin(\omega t \pm \alpha)$, then

(i) A = amplitude

(ii) ω = angular velocity = $2\pi f$ rad/s

(iii) $\dfrac{2\pi}{\omega}$ = periodic time T seconds

(iv) $\dfrac{\omega}{2\pi}$ = frequency, f hertz

(v) α = angle of lead or lag (compared with $y = A \sin \omega t$)

Problem 11. An alternating current is given by $i = 30 \sin(100\pi t + 0.27)$ amperes. Find the amplitude, periodic time, frequency and phase angle (in degrees and minutes).

$i = 30 \sin(100\pi t + 0.27)$A, hence **amplitude = 30 A**

Angular velocity $\omega = 100\pi$, hence

periodic time, $T = \dfrac{2\pi}{\omega} = \dfrac{2\pi}{100\pi} = \dfrac{1}{50} = 0.02$ s or 20 ms

Frequency, $f = \dfrac{1}{T} = \dfrac{1}{0.02} = 50$ Hz

Phase angle, $\alpha = 0.27 \text{ rad} = \left(0.27 \times \dfrac{180}{\pi}\right)°$

$$= \textbf{15.47° or 15°28′ leading}$$

$$i = 30 \sin(100\pi t)$$

Problem 12. An oscillating mechanism has a maximum displacement of 2.5 m and a frequency of 60 Hz. At time $t = 0$ the displacement is 90 cm. Express the displacement in the general form $A \sin(\omega t \pm \alpha)$.

Amplitude = maximum displacement = 2.5 m

Angular velocity, $\omega = 2\pi f = 2\pi(60) = 120\pi$ rad/s

Hence displacement = $2.5 \sin(120\pi t + \alpha)$ m

When $t = 0$, displacement = 90 cm = 0.90 m

Hence $0.90 = 2.5 \sin(0 + \alpha)$ i.e. $\sin\alpha = \dfrac{0.90}{2.5} = 0.36$

Hence $\alpha = \sin^{-1} 0.36 = 21.10° = 21°6′ = 0.368$ rad

Thus **displacement = $2.5 \sin(120\pi t + 0.368)$ m**

Problem 13. The instantaneous value of voltage in an a.c. circuit at any time t seconds is given by $v = 340 \sin(50\pi t - 0.541)$ volts. Determine:

(a) the amplitude, periodic time, frequency and phase angle (in degrees)

(b) the value of the voltage when $t = 0$

(c) the value of the voltage when $t = 10$ ms

(d) the time when the voltage first reaches 200 V, and

(e) the time when the voltage is a maximum

Sketch one cycle of the waveform.

(a) **Amplitude = 340 V**

Angular velocity, $\omega = 50\pi$

Hence **periodic time, $T = \dfrac{2\pi}{\omega} = \dfrac{2\pi}{50\pi} = \dfrac{1}{25}$**

$$= \textbf{0.04 s} \quad \text{or} \quad \textbf{40 ms}$$

Frequency $f = \dfrac{1}{T} = \dfrac{1}{0.04} = 25$ Hz

Phase angle = $0.541 \text{ rad} = \left(0.541 \times \dfrac{180}{\pi}\right)$

$$= \textbf{31° lagging } v = 340 \sin(50\pi t)$$

(b) **When** $t = 0$, $v = 340 \sin(0 - 0.541)$

$$= 340 \sin(-31°) = -175.1 \text{ V}$$

(c) **When** $t = 10$ ms then $v = 340 \sin\left(50\pi \dfrac{10}{10^3} - 0.541\right)$

$$= 340 \sin(1.0298)$$

$$= 340 \sin 59° = 291.4 \text{ volts}$$

(d) When $v = 200$ volts then $200 = 340 \sin(50\pi t - 0.541)$

$$\frac{200}{340} = \sin(50\pi t - 0.541)$$

Hence $(50\pi t - 0.541) = \sin^{-1} \dfrac{200}{340}$

$$= 36.03° \quad \text{or}$$

$$0.6288 \text{ rad}$$

$$50\pi t = 0.6288 + 0.541$$

$$= 1.1698$$

Hence when $v = 200$ V,

time, $t = \dfrac{1.1698}{50\pi} = \textbf{7.447 ms}$

(e) When the voltage is a maximum, $v = 340$ V

Hence $340 = 340 \sin(50\pi t - 0.541)$

$$1 = \sin(50\pi t - 0.541)$$

$$50\pi t - 0.541 = \sin^{-1} 1 = 90° \quad \text{or} \quad 1.5708 \text{ rad}$$

$$50\pi t = 1.5708 + 0.541 = 2.1118$$

Hence time, $t = \dfrac{2.1118}{50\pi} = \textbf{13.44 m}$

A sketch of $v = 340 \sin(50\pi t - 0.541)$ volts is shown in Fig. 18.27

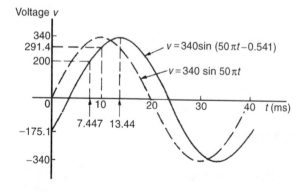

Voltage v

Fig. 18.27

Now try the following exercise

Exercise 69   Further problems on the sinusoidal form $A \sin(\omega t \pm \alpha)$ (Answers on page 260)

In Problems 1 to 3 find the amplitude, periodic time, frequency and phase angle (stating whether it is leading or lagging $\sin \omega t$) of the alternating quantities given.

1. $i = 40 \sin(50\pi t + 0.29)$ mA

2. $y = 75 \sin(40t - 0.54)$ cm

3. $v = 300 \sin(200\pi t - 0.412)$ V

4. A sinusoidal voltage has a maximum value of 120 V and a frequency of 50 Hz. At time $t = 0$, the voltage is (a) zero, and (b) 50 V.

   Express the instantaneous voltage v in the form $v = A \sin(\omega t \pm \alpha)$

5. An alternating current has a periodic time of 25 ms and a maximum value of 20 A. When time $t = 0$, current $i = -10$ amperes. Express the current i in the form $i = A \sin(\omega t \pm \alpha)$

6. An oscillating mechanism has a maximum displacement of 3.2 m and a frequency of 50 Hz. At time $t = 0$ the displacement is 150 cm. Express the displacement in the general form $A \sin(\omega t \pm \alpha)$

7. The current in an a.c. circuit at any time t seconds is given by:

$$i = 5 \sin(100\pi t - 0.432) \text{ amperes}$$

   Determine (a) the amplitude, periodic time, frequency and phase angle (in degrees) (b) the value of current at $t = 0$ (c) the value of current at $t = 8$ ms (d) the time when the current is first a maximum (e) the time when the current first reaches 3A. Sketch one cycle of the waveform showing relevant points.

Assignment 9

This assignment covers the material in Chapters 17 and 18. The marks for each question are shown in brackets at the end of each question.

1. Figure A9.1 shows a plan view of a kite design. Calculate the lengths of the dimensions shown as a and b   (4)

2. In Fig. A9.1, evaluate (a) angle θ (b) angle α (6)

3. Determine the area of the plan view of a kite shown in Fig. A9.1.   (4)

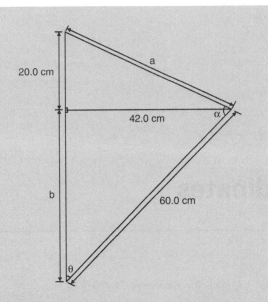

Fig. A9.1

4. If the angle of elevation of the top of a 25 m perpendicular building from point A is measured as 27°, determine the distance to the building. Calculate also the angle of elevation at a point B, 20 m closer to the building than point A. (5)

5. Evaluate, each correct to 4 significant figures:

   (a) $\sin 231.78°$ (b) $\cos 151°16'$ (c) $\tan \dfrac{3\pi}{8}$ (3)

6. (a) Determine the acute angle $\cos^{-1} 0.4117$ (i) in degrees and minutes and (ii) in radians (correct to 2 decimal places).

   (b) Calculate the other angle (in degrees) between 0° and 360° which satisfies $\cos^{-1} 0.4117$ (4)

7. Sketch the following curves labelling relevant points:

   (a) $y = 4\cos(\theta + 45°)$ (b) $y = 5\sin(2t - 60°)$ (8)

8. The current in an alternating current circuit at any time t seconds is given by:

$$i = 120 \sin(100\pi t + 0.274) \text{ amperes}$$

Determine

   (a) the amplitude, periodic time, frequency and phase angle (with reference to $120 \sin 100\pi t$)

   (b) the value of current when $t = 0$

   (c) the value of current when $t = 6$ ms

Sketch one cycle of the oscillation.   (16)

19

Cartesian and polar co-ordinates

19.1 Introduction

There are two ways in which the position of a point in a plane can be represented. These are

(a) by **Cartesian co-ordinates**, i.e. (x, y), and

(b) by **polar co-ordinates**, i.e. (r, θ), where r is a 'radius' from a fixed point and θ is an angle from a fixed point.

19.2 Changing from Cartesian into polar co-ordinates

In Fig. 19.1, if lengths x and y are known, then the length of r can be obtained from Pythagoras' theorem (see Chapter 17) since OPQ is a right-angled triangle.

Hence

$$r^2 = (x^2 + y^2)$$

from which, $\boxed{r = \sqrt{x^2 + y^2}}$

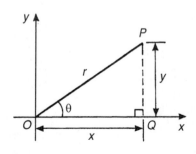

Fig. 19.1

From trigonometric ratios (see Chapter 17),

$$\tan\theta = \frac{y}{x}$$

from which $\boxed{\theta = \tan^{-1}\frac{y}{x}}$

$r = \sqrt{x^2 + y^2}$ and $\theta = \tan^{-1}\frac{y}{x}$ are the two formulae we need to change from Cartesian to polar co-ordinates. The angle θ, which may be expressed in degrees or radians, must **always** be measured from the positive x-axis, i.e. measured from the line OQ in Fig. 19.1. It is suggested that when changing from Cartesian to polar co-ordinates a diagram should always be sketched.

Problem 1. Change the Cartesian co-ordinates $(3, 4)$ into polar co-ordinates.

A diagram representing the point $(3, 4)$ is shown in Fig. 19.2.

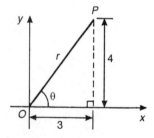

Fig. 19.2

From Pythagoras' theorem, $r = \sqrt{3^2 + 4^2} = 5$ (note that -5 has no meaning in this context).

By trigonometric ratios, $\theta = \tan^{-1}\frac{4}{3} = 53.13°$ or 0.927 rad.

[note that $53.13° = 53.13 \times (\pi/180)$ rad $= 0.927$ rad.]

Hence (3, 4) in Cartesian co-ordinates corresponds to (5, 53.13°) or (5, 0.927 rad) in polar co-ordinates.

Problem 2. Express in polar co-ordinates the position $(-4, 3)$

A diagram representing the point using the Cartesian co-ordinates $(-4, 3)$ is shown in Fig. 19.3.

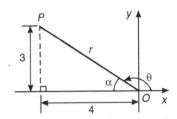

Fig. 19.3

From Pythagoras' theorem, $r = \sqrt{4^2 + 3^2} = 5$

By trigonometric ratios, $\alpha = \tan^{-1}\frac{3}{4} = 36.87°$ or 0.644 rad.

Hence $\theta = 180° - 36.87° = 143.13°$

or $\quad \theta = \pi - 0.644 = 2.498$ rad

Hence the position of point P in polar co-ordinate form is (5, 143.13°) or (5, 2.498 rad)

Problem 3. Express $(-5, -12)$ in polar co-ordinates.

A sketch showing the position $(-5, -12)$ is shown in Fig. 19.4.

$$r = \sqrt{5^2 + 12^2} = 13$$

and $\quad \alpha = \tan^{-1}\dfrac{12}{5} = 67.38°$ or $\quad 1.176$ rad

Hence $\theta = 180° + 67.38° = 247.38°$

or $\quad \theta = \pi + 1.176 = 4.318$ rad

Thus $(-5, -12)$ in Cartesian co-ordinates corresponds to (13, 247.38°) or (13, 4.318 rad) in polar co-ordinates.

Problem 4. Express $(2, -5)$ in polar co-ordinates.

A sketch showing the position $(2, -5)$ is shown in Fig. 19.5.

$$r = \sqrt{2^2 + 5^2} = \sqrt{29} = 5.385$$

correct to 3 decimal places

$\alpha = \tan^{-1}\frac{5}{2} = 68.20°$ or $\quad 1.190$ rad

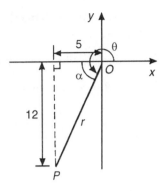

Fig. 19.4

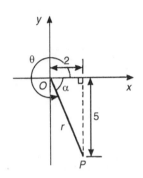

Fig. 19.5

Hence $\theta = 360° - 68.20° = 291.80°$

or $\quad \theta = 2\pi - 1.190 = 5.093$ rad

Thus $(2, -5)$ in Cartesian co-ordinates corresponds to (5.385, 291.80°) or (5.385, 5.093 rad) in polar co-ordinates.

Now try the following exercise

Exercise 70 Further problems on changing from Cartesian into polar co-ordinates (Answers on page 260)

In Problems 1 to 8, express the given Cartesian co-ordinates as polar co-ordinates, correct to 2 decimal places, in both degrees and in radians.

1. $(3, 5)$ 2. $(6.18, 2.35)$

3. $(-2, 4)$ 4. $(-5.4, 3.7)$

5. $(-7, -3)$ 6. $(-2.4, -3.6)$

7. $(5, -3)$ 8. $(9.6, -12.4)$

19.3 Changing from polar into Cartesian co-ordinates

From the right-angled triangle OPQ in Fig. 19.6.

$$\cos\theta = \frac{x}{r} \text{ and } \sin\theta = \frac{y}{r}, \text{ from trigonometric ratios}$$

Hence $\boxed{x = r\cos\theta}$ and $\boxed{y = r\sin\theta}$

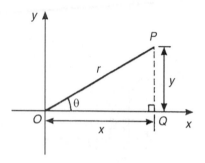

Fig. 19.6

If lengths r and angle θ are known then $x = r\cos\theta$ and $y = r\sin\theta$ are the two formulae we need to change from polar to Cartesian co-ordinates.

> **Problem 5.** Change $(4, 32°)$ into Cartesian co-ordinates.

A sketch showing the position $(4, 32°)$ is shown in Fig. 19.7.

Now $x = r\cos\theta = 4\cos 32° = 3.39$

and $y = r\sin\theta = 4\sin 32° = 2.12$

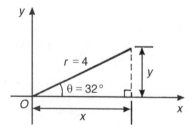

Fig. 19.7

Hence **$(4, 32°)$ in polar co-ordinates corresponds to $(3.39, 2.12)$ in Cartesian co-ordinates.**

> **Problem 6.** Express $(6, 137°)$ in Cartesian co-ordinates.

A sketch showing the position $(6, 137°)$ is shown in Fig. 19.8.

$$x = r\cos\theta = 6\cos 137° = -4.388$$

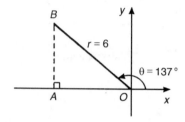

Fig. 19.8

which corresponds to length OA in Fig. 19.8.

$$y = r\sin\theta = 6\sin 137° = 4.092$$

which corresponds to length AB in Fig. 19.8.

Thus $(6, 137°)$ in polar co-ordinates corresponds to $(-4.388, 4.092)$ in Cartesian co-ordinates.

(Note that when changing from polar to Cartesian co-ordinates it is not quite so essential to draw a sketch. Use of $x = r\cos\theta$ and $y = r\sin\theta$ automatically produces the correct signs.)

> **Problem 7.** Express $(4.5, 5.16\,\text{rad})$ in Cartesian co-ordinates

A sketch showing the position $(4.5, 5.16\,\text{rad})$ is shown in Fig. 19.9.

$$x = r\cos\theta = 4.5\cos 5.16 = 1.948$$

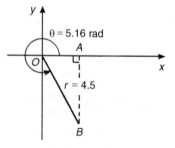

Fig. 19.9

which corresponds to length OA in Fig. 19.9.

$$y = r\sin\theta = 4.5\sin 5.16 = -4.057$$

which corresponds to length AB in Fig. 19.9.

Thus **(1.948, −4.057)** in Cartesian co-ordinates corresponds to **(4.5, 5.16 rad)** in polar co-ordinates.

19.4   Use of $R \rightarrow P$ and $P \rightarrow R$ functions on calculators

Another name for Cartesian co-ordinates is **rectangular** co-ordinates. Many scientific notation calculators possess $R \rightarrow P$ and $P \rightarrow R$ functions. The R is the first letter of the word rectangular and the P is the first letter of the word polar. Check the operation manual for your particular calculator to determine how to use these two functions. They make changing from Cartesian to polar co-ordinates, and vice-versa, so much quicker and easier.

Now try the following exercise

Exercise 71   **Further problems on changing polar into Cartesian co-ordinates (Answers on page 261)**

In Problems 1 to 8, express the given polar co-ordinates as Cartesian co-ordinates, correct to 3 decimal places.

1. $(5, 75°)$

2. $(4.4, 1.12 \text{ rad})$

3. $(7, 140°)$

4. $(3.6, 2.5 \text{ rad})$

5. $(10.8, 210°)$

6. $(4, 4 \text{ rad})$

7. $(1.5, 300°)$

8. $(6, 5.5 \text{ rad})$

20

Areas of plane figures

20.1 Mensuration

Mensuration is a branch of mathematics concerned with the determination of lengths, areas and volumes.

20.2 Properties of quadrilaterals

Polygon

A **polygon** is a closed plane figure bounded by straight lines. A polygon which has:

  (i) 3 sides is called a **triangle**
 (ii) 4 sides is called a **quadrilateral**
(iii) 5 sides is called a **pentagon**
 (iv) 6 sides is called a **hexagon**
  (v) 7 sides is called a **heptagon**
 (vi) 8 sides is called an **octagon**

There are five types of **quadrilateral**, these being:

  (i) rectangle
 (ii) square
(iii) parallelogram
 (iv) rhombus
  (v) trapezium

(The properties of these are given below).

If the opposite corners of any quadrilateral are joined by a straight line, two triangles are produced. Since the sum of the angles of a triangle is 180°, the sum of the angles of a quadrilateral is 360°.

In a **rectangle**, shown in Fig. 20.1:

  (i) all four angles are right angles,
 (ii) opposite sides are parallel and equal in length, and
(iii) diagonals AC and BD are equal in length and bisect one another.

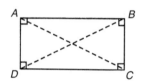

Fig. 20.1

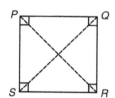

Fig. 20.2

In a **square**, shown in Fig. 20.2:

  (i) all four angles are right angles,
 (ii) opposite sides are parallel,
(iii) all four sides are equal in length, and
 (iv) diagonals PR and QS are equal in length and bisect one another at right angles.

In a **parallelogram**, shown in Fig. 20.3:

 (i) opposite angles are equal,
(ii) opposite sides are parallel and equal in length, and
(iii) diagonals WY and XZ bisect one another.

In a **rhombus**, shown in Fig. 20.4:

  (i) opposite angles are equal,
 (ii) opposite angles are bisected by a diagonal,
(iii) opposite sides are parallel,
 (iv) all four sides are equal in length, and
  (v) diagonals AC and BD bisect one another at right angles.

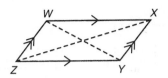

Fig. 20.3

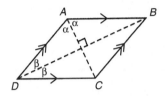

Fig. 20.4

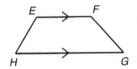

Fig. 20.5

In a **trapezium**, shown in Fig. 20.5:

(i) only one pair of sides is parallel

20.3 Worked problems on areas of plane figures

Table 20.1

(i) Square	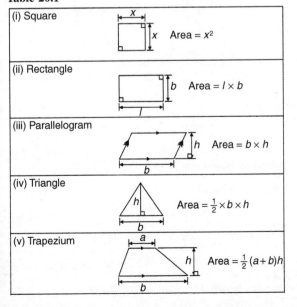
	Area = x^2
(ii) Rectangle	Area = $l \times b$
(iii) Parallelogram	Area = $b \times h$
(iv) Triangle	Area = $\frac{1}{2} \times b \times h$
(v) Trapezium	Area = $\frac{1}{2}(a+b)h$

Table 20.1 (*Continued*)

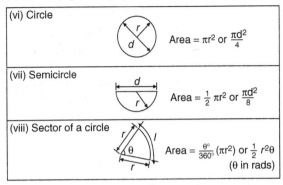

(vi) Circle	Area = πr^2 or $\frac{\pi d^2}{4}$
(vii) Semicircle	Area = $\frac{1}{2}\pi r^2$ or $\frac{\pi d^2}{8}$
(viii) Sector of a circle	Area = $\frac{\theta^\circ}{360^\circ}(\pi r^2)$ or $\frac{1}{2}r^2\theta$ (θ in rads)

Problem 1. State the types of quadrilateral shown in Fig. 20.6 and determine the angles marked a to l.

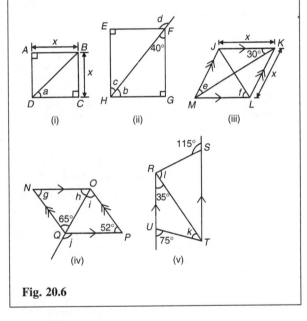

Fig. 20.6

(i) **ABCD is a square**

The diagonals of a square bisect each of the right angles, hence

$$a = \frac{90^\circ}{2} = 45^\circ$$

(ii) **EFGH is a rectangle**

In triangle FGH, $40^\circ + 90^\circ + b = 180^\circ$ (angles in a triangle add up to 180°) from which, $b = 50^\circ$. Also $c = 40^\circ$ (alternate angles between parallel lines EF and HG).

(Alternatively, b and c are complementary, i.e. add up to 90°)

$d = 90^\circ + c$ (external angle of a triangle equals the sum of the interior opposite angles), hence

$$d = 90^\circ + 40^\circ = 130^\circ$$

(iii) **JKLM is a rhombus**

The diagonals of a rhombus bisect the interior angles and opposite internal angles are equal.

Thus $\angle JKM = \angle MKL = \angle JMK = \angle LMK = 30°$, hence $e = \mathbf{30°}$

In triangle KLM, $30° + \angle KLM + 30° = 180°$ (angles in a triangle add up to 180°), hence $\angle KLM = 120°$.

The diagonal JL bisects $\angle KLM$, hence

$$f = \frac{120°}{2} = \mathbf{60°}$$

(iv) **NOPQ is a parallelogram**

$g = \mathbf{52°}$ (since opposite interior angles of a parallelogram are equal).

In triangle NOQ, $g + h + 65° = 180°$ (angles in a triangle add up to 180°), from which,

$$h = 180° - 65° - 52° = \mathbf{63°}$$

$i = \mathbf{65°}$ (alternate angles between parallel lines NQ and OP).

$j = 52° + i = 52° + 65° = \mathbf{117°}$ (external angle of a triangle equals the sum of the interior opposite angles).

(v) **RSTU is a trapezium**

$35° + k = 75°$ (external angle of a triangle equals the sum of the interior opposite angles), hence $k = \mathbf{40°}$

$\angle STR = 35°$ (alternate angles between parallel lines RU and ST).

$l + 35° = 115°$ (external angle of a triangle equals the sum of the interior opposite angles), hence

$$l = 115° - 35° = \mathbf{80°}$$

Problem 2. A rectangular tray is 820 mm long and 400 mm wide. Find its area in (a) mm², (b) cm², (c) m².

(a) Area = length × width = $820 \times 400 = \mathbf{328\,000\,mm^2}$

(b) $1\,cm^2 = 100\,mm^2$. Hence

$$328\,000\,mm^2 = \frac{328\,000}{100}\,cm^2 = \mathbf{3280\,cm^2}$$

(c) $1\,m^2 = 10\,000\,cm^2$. Hence

$$3280\,cm^2 = \frac{3280}{10\,000}\,m^2 = \mathbf{0.3280\,m^2}$$

Problem 3. Find (a) the cross-sectional area of the girder shown in Fig. 20.7(a) and (b) the area of the path shown in Fig. 20.7(b).

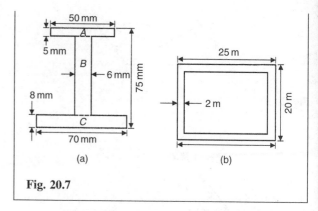

Fig. 20.7

(a) The girder may be divided into three separate rectangles as shown.

Area of rectangle $A = 50 \times 5 = 250\,mm^2$

Area of rectangle $B = (75 - 8 - 5) \times 6$

$$= 62 \times 6 = 372\,mm^2$$

Area of rectangle $C = 70 \times 8 = 560\,mm^2$

Total area of girder $= 250 + 372 + 560 = \mathbf{1182\,mm^2}$ or $\mathbf{11.82\,cm^2}$

(b) Area of path = area of large rectangle − area of small rectangle

$$= (25 \times 20) - (21 \times 16) = 500 - 336 = \mathbf{164\,m^2}$$

Problem 4. Find the area of the parallelogram shown in Fig. 20.8 (dimensions are in mm).

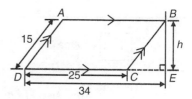

Fig. 20.8

Area of parallelogram = base × perpendicular height. The perpendicular height h is found using Pythagoras' theorem.

$$BC^2 = CE^2 + h^2$$

i.e. $15^2 = (34 - 25)^2 + h^2$

$$h^2 = 15^2 - 9^2 = 225 - 81 = 144$$

Hence, $h = \sqrt{144} = 12\,mm$ (−12 can be neglected).

Hence, area of $ABCD = 25 \times 12 = \mathbf{300\,mm^2}$

Problem 5. Figure 20.9 shows the gable end of a building. Determine the area of brickwork in the gable end.

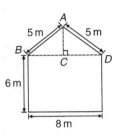

Fig. 20.9

The shape is that of a rectangle and a triangle.

Area of rectangle $= 6 \times 8 = 48\,\text{m}^2$

Area of triangle $= \frac{1}{2} \times \text{base} \times \text{height}$.

$CD = 4\,\text{m}$, $AD = 5\,\text{m}$, hence $AC = 3\,\text{m}$ (since it is a 3, 4, 5 triangle).

Hence, area of triangle $ABD = \frac{1}{2} \times 8 \times 3 = 12\,\text{m}^2$

Total area of brickwork $= 48 + 12 = \mathbf{60\,m^2}$

Problem 6. Determine the area of the shape shown in Fig. 20.10.

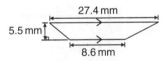

Fig. 20.10

The shape shown is a trapezium.

Area of trapezium $= \frac{1}{2}$(sum of parallel sides)(perpendicular distance between them)

$$= \tfrac{1}{2}(27.4 + 8.6)(5.5)$$

$$= \tfrac{1}{2} \times 36 \times 5.5 = \mathbf{99\,mm^2}$$

Problem 7. Find the areas of the circles having (a) a radius of 5 cm, (b) a diameter of 15 mm, (c) a circumference of 70 mm.

Area of a circle $= \pi r^2$ or $\dfrac{\pi d^2}{4}$

(a) Area $= \pi r^2 = \pi(5)^2 = 25\pi = \mathbf{78.54\,cm^2}$

(b) Area $= \dfrac{\pi d^2}{4} = \dfrac{\pi(15)^2}{4} = \dfrac{225\pi}{4} = \mathbf{176.7\,mm^2}$

(c) Circumference, $c = 2\pi r$, hence

$$r = \frac{c}{2\pi} = \frac{70}{2\pi} = \frac{35}{\pi}, \text{mm}$$

$$\text{Area of circle} = \pi r^2 = \pi\left(\frac{35}{\pi}\right)^2 = \frac{35^2}{\pi}$$

$$= \mathbf{389.9\,mm^2} \text{ or } \mathbf{3.899\,cm^2}$$

Problem 8. Calculate the areas of the following sectors of circles:

(a) having radius 6 cm with angle subtended at centre 50°
(b) having diameter 80 mm with angle subtended at centre 107°42′
(c) having radius 8 cm with angle subtended at centre 1.15 radians.

Area of sector of a circle $= \dfrac{\theta^2}{360}(\pi r^2)$ or $\dfrac{1}{2}r^2\theta$ (θ in radians).

(a) Area of sector

$$= \frac{50}{360}(\pi 6^2) = \frac{50 \times \pi \times 36}{360} = 5\pi = \mathbf{15.71\,cm^2}$$

(b) If diameter $= 80\,\text{mm}$, then radius, $r = 40\,\text{mm}$, and area of sector

$$= \frac{107°42'}{360}(\pi 40^2) = \frac{107\frac{42}{60}}{360}(\pi 40^2) = \frac{107.7}{360}(\pi 40^2)$$

$$= \mathbf{1504\,mm^2} \quad \text{or} \quad \mathbf{15.04\,cm^2}$$

(c) Area of sector $= \dfrac{1}{2}r^2\theta = \dfrac{1}{2} \times 8^2 \times 1.15 = \mathbf{36.8\,cm^2}$

Problem 9. A hollow shaft has an outside diameter of 5.45 cm and an inside diameter of 2.25 cm. Calculate the cross-sectional area of the shaft.

The cross-sectional area of the shaft is shown by the shaded part in Fig. 20.11 (often called an **annulus**).

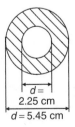

Fig. 20.11

Area of shaded part $=$ area of large circle $-$ area of small circle

$$= \frac{\pi D^2}{4} - \frac{\pi d^2}{4} = \frac{\pi}{4}(D^2 - d^2) = \frac{\pi}{4}(5.45^2 - 2.25^2)$$

$$= \mathbf{19.35\,cm^2}$$

Now try the following exercise

Exercise 72   Further problems on areas of plane figures (Answers on page 261)

1. A rectangular plate is 85 mm long and 42 mm wide. Find its area in square centimetres.

2. A rectangular field has an area of 1.2 hectares and a length of 150 m. Find (a) its width and (b) the length of a diagonal (1 hectare = 10 000 m^2).

3. Determine the area of each of the angle iron sections shown in Fig. 20.12.

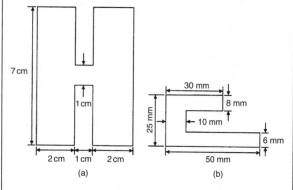

Fig. 20.12

4. A rectangular garden measures 40 m by 15 m. A 1 m flower border is made round the two shorter sides and one long side. A circular swimming pool of diameter 8 m is constructed in the middle of the garden. Find, correct to the nearest square metre, the area remaining.

5. The area of a trapezium is 13.5 cm^2 and the perpendicular distance between its parallel sides is 3 cm. If the length of one of the parallel sides is 5.6 cm, find the length of the other parallel side.

6. Find the angles p, q, r, s and t in Fig. 20.13(a) to (c).

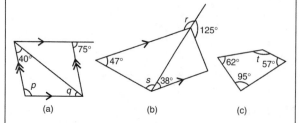

Fig. 20.13

7. Name the types of quadrilateral shown in Fig. 20.14(i) to (iv), and determine (a) the area, and (b) the perimeter of each.

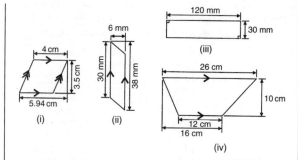

Fig. 20.14

8. Determine the area of circles having (a) a radius of 4 cm (b) a diameter of 30 mm (c) a circumference of 200 mm.

9. An annulus has an outside diameter of 60 mm and an inside diameter of 20 mm. Determine its area.

10. If the area of a circle is 320 mm^2, find (a) its diameter, and (b) its circumference.

11. Calculate the areas of the following sectors of circles:

    (a) radius 9 cm, angle subtended at centre 75°,
    (b) diameter 35 mm, angle subtended at centre 48°37′,
    (c) diameter 5 cm, angle subtended at centre 2.19 radians.

12. Determine the area of the template shown in Fig. 20.15.

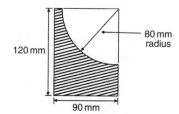

Fig. 20.15

13. An archway consists of a rectangular opening topped by a semi-circular arch as shown in Fig. 20.16. Determine the area of the opening if the width is 1 m and the greatest height is 2 m.

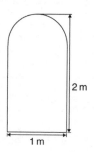

Fig. 20.16

14. Calculate the area of the steel plate shown in Fig. 20.17.

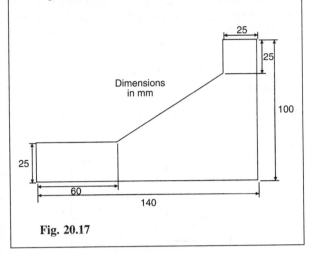

Fig. 20.17

20.4 Further worked problems on areas of plane figures

Problem 10. Calculate the area of a regular octagon, if each side is 5 cm and the width across the flats is 12 cm.

An octagon is an 8-sided polygon. If radii are drawn from the centre of the polygon to the vertices then 8 equal triangles are produced (see Fig. 20.18).

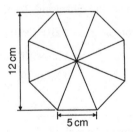

Fig. 20.18

Area of one triangle $= \dfrac{1}{2} \times$ base $\times$ height

$$= \frac{1}{2} \times 5 \times \frac{12}{2} = 15\,\text{cm}^2$$

Area of octagon $= 8 \times 15 = \mathbf{120\,cm^2}$

Problem 11. Determine the area of a regular hexagon which has sides 8 cm long.

A hexagon is a 6-sided polygon which may be divided into 6 equal triangles as shown in Fig. 20.19. The angle subtended at the centre of each triangle is $360°/6 = 60°$. The other two angles in the triangle add up to 120° and are equal to each other. Hence each of the triangles is equilateral with each angle 60° and each side 8 cm.

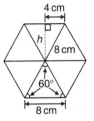

Fig. 20.19

Area of one triangle $= \frac{1}{2} \times$ base $\times$ height $= \frac{1}{2} \times 8 \times h$.

h is calculated using Pythagoras' theorem:

$$8^2 = h^2 + 4^2$$

from which $h = \sqrt{8^2 - 4^2} = 6.928\,\text{cm}$

Hence area of one triangle $= \frac{1}{2} \times 8 \times 6.928 = 27.71\,\text{cm}^2$

Area of hexagon $= 6 \times 27.71 = \mathbf{166.3\,cm^2}$

Problem 12. Figure 20.20 shows a plan of a floor of a building which is to be carpeted. Calculate the area of the floor in square metres. Calculate the cost, correct to the nearest pound, of carpeting the floor with carpet costing £16.80 per m², assuming 30% extra carpet is required due to wastage in fitting.

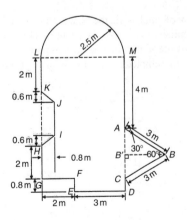

Fig. 20.20

Area of floor plan

= area of triangle ABC + area of semicircle

+ area of rectangle $CGLM$ + area of rectangle $CDEF$

− area of trapezium $HIJK$

Triangle ABC is equilateral since $AB = BC = 3$ m and hence angle $B'CB = 60°$

$\sin B'CB = BB'/3$, i.e. $BB' = 3\sin 60° = 2.598$ m

Area of triangle $ABC = \frac{1}{2}(AC)(BB') = \frac{1}{2}(3)(2.598)$

$$= 3.897\,\text{m}^2$$

Area of semicircle $= \frac{1}{2}\pi r^2 = \frac{1}{2}\pi(2.5)^2 = 9.817\,\text{m}^2$

Area of $CGLM = 5 \times 7 = 35\,\text{m}^2$

Area of $CDEF = 0.8 \times 3 = 2.4\,\text{m}^2$

Area of $HIJK = \frac{1}{2}(KH + IJ)(0.8)$

Since $MC = 7$ m then $LG = 7$ m, hence

$$JI = 7 - 5.2 = 1.8\,\text{m}$$

Hence area of $HIJK = \frac{1}{2}(3 + 1.8)(0.8) = 1.92\,\text{m}^2$

Total floor area $= 3.897 + 9.817 + 35 + 2.4 - 1.92 = 49.194\,\text{m}^2$

To allow for 30% wastage, amount of carpet required $= 1.3 \times 49.194 = 63.95\,\text{m}^2$

Cost of carpet at £16.80 per $\text{m}^2 = 63.95 \times 16.80 = £1074$, correct to the nearest pound.

Now try the following exercise

Exercise 73   Further problems on areas of plane figures (Answers on page 261)

1. Calculate the area of a regular octagon if each side is 20 mm and the width across the flats is 48.3 mm.
2. Determine the area of a regular hexagon which has sides 25 mm.
3. A plot of land is in the shape shown in Fig. 20.21. Determine (a) its area in hectares ($1\,\text{ha} = 10^4\,\text{m}^2$), and (b) the length of fencing required, to the nearest metre, to completely enclose the plot of land.

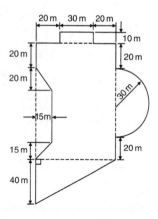

Fig. 20.21

20.5   Areas of similar shapes

The areas of similar shapes are proportional to the squares of corresponding linear dimensions. For example, Fig. 20.22 shows two squares, one of which has sides three times as long as the other.

$$\text{Area of Figure 20.22(a)} = (x)(x) = x^2$$
$$\text{Area of Figure 20.22(b)} = (3x)(3x) = 9x^2$$

Hence Figure 20.22(b) has an area $(3)^2$, i.e. 9 times the area of Figure 20.22(a).

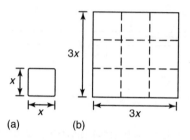

(a)   (b)

Fig. 20.22

Problem 13.   A rectangular garage is shown on a building plan having dimensions 10 mm by 20 mm. If the plan is drawn to a scale of 1 to 250, determine the true area of the garage in square metres.

Area of garage on the plan $= 10\,\text{mm} \times 20\,\text{mm} = 200\,\text{mm}^2$.

Since the areas of similar shapes are proportional to the squares of corresponding dimensions then:

$$\text{True area of garage} = 200 \times (250)^2$$
$$= 12.5 \times 10^6\,\text{mm}^2$$
$$= \frac{12.5 \times 10^6}{10^6}\,\text{m}^2 = \mathbf{12.5\,m^2}$$

Now try the following exercise

Exercise 74   Further problems on areas of similar shapes (Answers on page 261)

1. The area of a park on a map is $500\,\text{mm}^2$. If the scale of the map is 1 to 40 000 determine the true area of the park in hectares (1 hectare $= 10^4\,\text{m}^2$).
2. A model of a boiler is made having an overall height of 75 mm corresponding to an overall height of the actual boiler of 6 m. If the area of metal required for the model is $12\,500\,\text{mm}^2$ determine, in square metres, the area of metal required for the actual boiler.

Assignment 10

This assignment covers the material contained in Chapters 19 and 20. The marks for each question are shown in brackets at the end of each question.

1. Change the following Cartesian co-ordinates into polar co-ordinates, correct to 2 decimal places, in both degrees and in radians:

   (a) $(-2.3, 5.4)$ (b) $(7.6, -9.2)$ (10)

2. Change the following polar co-ordinates into Cartesian co-ordinates, correct to 3 decimal places:

   (a) $(6.5, 132°)$ (b) $(3, 3\,\text{rad})$ (6)

3. A rectangular park measures 150 m by 70 m. A 2 m flower border is constructed round the two longer sides and one short side. A circular fish pond of diameter 15 m is in the centre of the park and the remainder of the park is grass. Calculate, correct to the nearest square metre, the area of (a) the fish pond, (b) the flower borders, (c) the grass. (10)

4. A swimming pool is 55 m long and 10 m wide. The perpendicular depth at the deep end is 5 m and at the shallow end is 1.5 m, the slope from one end to the other being uniform. The inside of the pool needs two coats of a protective paint before it is filled with water. Determine how many litres of paint will be needed if 1 litre covers 10 m². (7)

5. Find the area of an equilateral triangle of side 20.0 cm. (4)

6. A steel template is of the shape shown in Fig. A10.1, the circular area being removed. Determine the area of the template, in square centimetres, correct to 1 decimal place. (7)

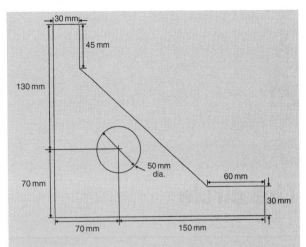

Fig. A10.1

7. The area of a plot of land on a map is 400 mm². If the scale of the map is 1 to 50 000, determine the true area of the land in hectares (1 hectare = 10^4 m²). (3)

8. Determine the shaded area in Fig. A10.2, correct to the nearest square centimetre. (3)

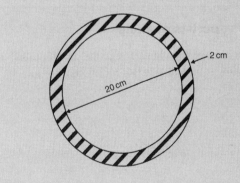

Fig. A10.2

21

The circle

21.1 Introduction

A **circle** is a plain figure enclosed by a curved line, every point on which is equidistant from a point within, called the **centre**.

21.2 Properties of circles

(i) The distance from the centre to the curve is called the **radius**, r, of the circle (see OP in Fig. 21.1).

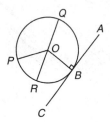

Fig. 21.1

(ii) The boundary of a circle is called the **circumference**, c.

(iii) Any straight line passing through the centre and touching the circumference at each end is called the **diameter**, d (see QR in Fig. 21.1). Thus $d = 2r$

(iv) The ratio $\dfrac{\text{circumference}}{\text{diameter}}$ = a constant for any circle.

This constant is denoted by the Greek letter π (pronounced 'pie'), where $\pi = 3.14159$, correct to 5 decimal places.

Hence $c/d = \pi$ or $c = \pi d$ or $c = 2\pi r$

(v) A **semicircle** is one half of the whole circle.

(vi) A **quadrant** is one quarter of a whole circle.

(vii) A **tangent** to a circle is a straight line which meets the circle in one point only and does not cut the circle when produced. AC in Fig. 21.1 is a tangent to the circle since it touches the curve at point B only. If radius OB is drawn, then angle ABO is a right angle.

(viii) A **sector** of a circle is the part of a circle between radii (for example, the portion OXY of Fig. 21.2 is a sector). If a sector is less than a semicircle it is called a **minor sector**, if greater than a semicircle it is called a **major sector**.

Fig. 21.2

(ix) A **chord** of a circle is any straight line which divides the circle into two parts and is terminated at each end by the circumference. ST, in Fig. 21.2 is a chord.

(x) A **segment** is the name given to the parts into which a circle is divided by a chord. If the segment is less than a semicircle it is called a **minor segment** (see shaded area in Fig. 21.2). If the segment is greater than a semicircle it is called a **major segment** (see the unshaded area in Fig. 21.2).

(xi) An **arc** is a portion of the circumference of a circle. The distance SRT in Fig. 21.2 is called a **minor arc** and the distance $SXYT$ is called a **major arc**.

(xii) The angle at the centre of a circle, subtended by an arc, is double the angle at the circumference subtended by the same arc. With reference to Fig. 21.3, **Angle *AOC* = 2 × angle *ABC*.**

Fig. 21.3

(xiii) The angle in a semicircle is a right angle (see angle *BQP* in Fig. 21.3).

Problem 1. Find the circumference of a circle of radius 12.0 cm.

Circumference,

$$c = 2 \times \pi \times \text{radius} = 2\pi r = 2\pi(12.0) = \textbf{75.40 cm}$$

Problem 2. If the diameter of a circle is 75 mm, find its circumference.

Circumference,

$$c = \pi \times \text{diameter} = \pi d = \pi(75) = \textbf{235.6 mm}$$

Problem 3. Determine the radius of a circle if its perimeter is 112 cm.

Perimeter = circumference, $c = 2\pi r$

Hence **radius** $r = \dfrac{c}{2\pi} = \dfrac{112}{2\pi} = \textbf{17.83 cm}$

Problem 4. In Fig. 21.4, *AB* is a tangent to the circle at *B*. If the circle radius is 40 mm and *AB* = 150 mm, calculate the length *AO*.

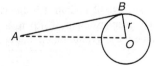

Fig. 21.4

A tangent to a circle is at right angles to a radius drawn from the point of contact, i.e. *ABO* = 90°. Hence, using

Pythagoras' theorem:

$$AO^2 = AB^2 + OB^2$$

from which, $AO = \sqrt{AB^2 + OB^2} = \sqrt{150^2 + 40^2}$

$$= \textbf{155.2 mm}$$

Now try the following exercise

Exercise 75 Further problems on properties of a circle (Answers on page 261)

1. Calculate the length of the circumference of a circle of radius 7.2 cm.
2. If the diameter of a circle is 82.6 mm, calculate the circumference of the circle.
3. Determine the radius of a circle whose circumference is 16.52 cm.
4. Find the diameter of a circle whose perimeter is 149.8 cm.

21.3 Arc length and area of a sector

One **radian** is defined as the angle subtended at the centre of a circle by an arc equal in length to the radius. With reference to Fig. 21.5, for arc length *s*,

$$\theta \text{ radians} = s/r \quad \text{or} \quad \textbf{arc length,} \quad \boxed{s = r\theta} \qquad (1)$$

where θ is in radians.

Fig. 21.5

When s = whole circumference ($= 2\pi r$) then

$$\theta = s/r = 2\pi r/r = 2\pi$$

i.e. 2π radians = 360° or $\boxed{\pi \textbf{ radians} = 180°}$

Thus 1 rad = 180°/π = 57.30°, correct to 2 decimal places. Since π rad = 180°, then $\pi/2 = 90°$, $\pi/3 = 60°$, $\pi/4 = 45°$, and so on.

$$\textbf{Area of a sector} = \frac{\theta}{360}(\pi r^2)$$
when θ is in degrees

$$= \frac{\theta}{2\pi}(\pi r^2) = \frac{1}{2}r^2\theta \qquad (2)$$
when θ is in radians

Problem 5. Convert to radians: (a) 125° (b) 69°47′

(a) Since $180° = \pi$ rad then $1° = \pi/180$ rad, therefore

$$125° = 125 \left(\frac{\pi}{180}\right)^c = \textbf{2.182 radians}$$

(Note that ^{c} means 'circular measure' and indicates radian measure.)

(b) $69°47' = 69\dfrac{47°}{60} = 69.783°$

$$69.783° = 69.783 \left(\frac{\pi}{180}\right)^c = \textbf{1.218 radians}$$

Problem 6. Convert to degrees and minutes:
(a) 0.749 radians (b) $3\pi/4$ radians

(a) Since π rad $= 180°$ then 1 rad $= 180°/\pi$, therefore

$$0.749 = 0.749 \left(\frac{180}{\pi}\right)^° = 42.915°$$

$0.915° = (0.915 \times 60)' = 55'$, correct to the nearest minute, hence

$$\textbf{0.749 radians} = \textbf{42°55′}$$

(b) Since 1 rad $= \left(\dfrac{180}{\pi}\right)^°$ then $\dfrac{3\pi}{4}$ rad $= \dfrac{3\pi}{4}\left(\dfrac{180}{\pi}\right)^°$

$$= \frac{3}{4}(180)° = \textbf{135°}$$

Problem 7. Express in radians, in terms of π (a) 150°
(b) 270° (c) 37.5°

Since $180° = \pi$ rad then $1° = 180/\pi$, hence

(a) $150° = 150 \left(\dfrac{\pi}{180}\right)$ rad $= \dfrac{5\pi}{6}$ **rad**

(b) $270° = 270 \left(\dfrac{\pi}{180}\right)$ rad $= \dfrac{3\pi}{2}$ **rad**

(c) $37.5° = 37.5 \left(\dfrac{\pi}{180}\right)$ rad $= \dfrac{75\pi}{360}$ rad $= \dfrac{5\pi}{24}$ **rad**

Problem 8. Find the length of arc of a circle of radius 5.5 cm when the angle subtended at the centre is 1.20 radians.

From equation (1), length of arc, $s = r\theta$, where θ is in radians, hence

$$s = (5.5)(1.20) = \textbf{6.60 cm}$$

Problem 9. Determine the diameter and circumference of a circle if an arc of length 4.75 cm subtends an angle of 0.91 radians.

Since $s = r\theta$ then $r = \dfrac{s}{\theta} = \dfrac{4.75}{0.91} = 5.22$ cm.

Diameter $= 2 \times$ radius $= 2 \times 5.22 = \textbf{10.44 cm}$.
Circumference, $c = \pi d = \pi(10.44) = \textbf{32.80 cm}$.

Problem 10. If an angle of 125° is subtended by an arc of a circle of radius 8.4 cm, find the length of (a) the minor arc, and (b) the major arc, correct to 3 significant figures.

Since $180° = \pi$ rad then $1° = \left(\dfrac{\pi}{180}\right)$ rad

and $125° = 125 \left(\dfrac{\pi}{180}\right)$ rad

Length of minor arc,

$$s = r\theta = (8.4)(125)\left(\frac{\pi}{180}\right) = \textbf{18.3 cm}$$

correct to 3 significant figures.

Length of major arc $=$ (circumference $-$ minor arc)
$= 2\pi(8.4) - 18.3 = \textbf{34.5 cm}$, correct to 3 significant figures.
(Alternatively, major arc $= r\theta = 8.4(360 - 125)(\pi/180)$
$= \textbf{34.5 cm}$.)

Problem 11. Determine the angle, in degrees and minutes, subtended at the centre of a circle of diameter 42 mm by an arc of length 36 mm. Calculate also the area of the minor sector formed.

Since length of arc, $s = r\theta$ then $\theta = s/r$

Radius, $r = \dfrac{\text{diameter}}{2} = \dfrac{42}{2} = 21$ mm

hence $\theta = \dfrac{s}{r} = \dfrac{36}{21} = 1.7143$ radians

1.7143 rad $= 1.7143 \times (180/\pi)° = 98.22° = \textbf{98°13′}$ = angle subtended at centre of circle.
From equation (2),

$$\textbf{area of sector} = \frac{1}{2}r^2\theta = \frac{1}{2}(21)^2(1.7143) = \textbf{378 mm}^2$$

Problem 12. A football stadium floodlight can spread its illumination over an angle of 45° to a distance of 55 m. Determine the maximum area that is floodlit.

Floodlit area $=$ area of sector $= \dfrac{1}{2}r^2\theta$

$$= \frac{1}{2}(55)^2 \left(45 \times \frac{\pi}{180}\right) \quad \text{from equation (2)}$$

$$= \textbf{1188 m}^2$$

Problem 13. An automatic garden spray produces a spray to a distance of 1.8 m and revolves through an angle α which may be varied. If the desired spray catchment area is to be 2.5 m², to what should angle α be set, correct to the nearest degree.

Area of sector $= \frac{1}{2}r^2\theta$, hence $2.5 = \frac{1}{2}(1.8)^2\alpha$

from which, $\alpha = \frac{2.5 \times 2}{1.8^2} = 1.5432$ radians

$1.5432\,\text{rad} = \left(1.5432 \times \frac{180}{\pi}\right)^{\circ} = 88.42^{\circ}$

Hence **angle $\alpha = 88^{\circ}$**, correct to the nearest degree.

Problem 14. The angle of a tapered groove is checked using a 20 mm diameter roller as shown in Fig. 21.6. If the roller lies 2.12 mm below the top of the groove, determine the value of angle θ.

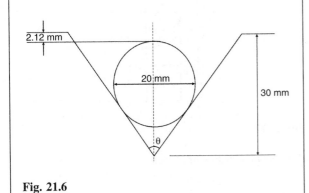

Fig. 21.6

In Fig. 21.7, triangle ABC is right-angled at C (see Section 21.2(vii), page 160).

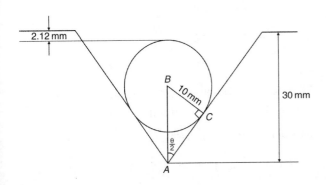

Fig. 21.7

Length $BC = 10$ mm (i.e. the radius of the circle), and $AB = 30 - 10 - 2.12 = 17.88$ mm from Fig. 21.6.

Hence $\sin\dfrac{\theta}{2} = \dfrac{10}{17.88}$ and $\dfrac{\theta}{2} = \sin^{-1}\left(\dfrac{10}{17.88}\right) = 34^{\circ}$

and **angle $\theta = 68^{\circ}$**

Now try the following exercise

Exercise 76 Further problems on arc length and area of a sector (Answers on page 261)

1. Convert to radians in terms of π:

   (a) 30° (b) 75° (c) 225°

2. Convert to radians:

   (a) 48° (b) 84°51′ (c) 232°15′

3. Convert to degrees:

   (a) $\dfrac{5\pi}{6}$ rad (b) $\dfrac{4\pi}{9}$ rad (c) $\dfrac{7\pi}{12}$ rad

4. Convert to degrees and minutes:

   (a) 0.0125 rad (b) 2.69 rad (c) 7.241 rad

5. Find the length of an arc of a circle of radius 8.32 cm when the angle subtended at the centre is 2.14 radians. Calculate also the area of the minor sector formed.

6. If the angle subtended at the centre of a circle of diameter 82 mm is 1.46 rad, find the lengths of the (a) minor arc (b) major arc.

7. A pendulum of length 1.5 m swings through an angle of 10° in a single swing. Find, in centimetres, the length of the arc traced by the pendulum bob.

8. Determine the length of the radius and circumference of a circle if an arc length of 32.6 cm subtends an angle of 3.76 radians.

9. Determine the angle of lap, in degrees and minutes, if 180 mm of a belt drive are in contact with a pulley of diameter 250 mm.

10. Determine the number of complete revolutions a motorcycle wheel will make in travelling 2 km, if the wheel's diameter is 85.1 cm.

11. The floodlights at a sports ground spread its illumination over an angle of 40° to a distance of 48 m. Determine (a) the angle in radians, and (b) the maximum area that is floodlit.

12. Determine (a) the shaded area in Fig. 21.8 (b) the percentage of the whole sector that the shaded area represents.

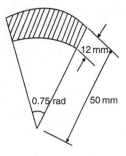

Fig. 21.8

13. Determine the length of steel strip required to make the clip shown in Fig. 21.9.

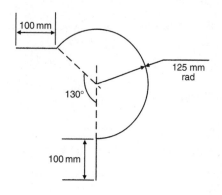

Fig. 21.9

14. A 50° tapered hole is checked with a 40 mm diameter ball as shown in Fig. 21.10. Determine the length shown as x.

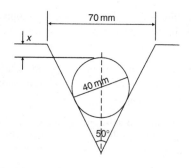

Fig. 21.10

21.4   The equation of a circle

The simplest equation of a circle, centre at the origin, radius r, is given by:

$$x^2 + y^2 = r^2$$

For example, Fig. 21.11 shows a circle $x^2 + y^2 = 9$.

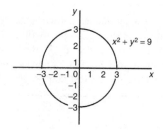

Fig. 21.11

More generally, the equation of a circle, centre (a, b), radius r, is given by:

$$(x - a)^2 + (y - b)^2 = r^2 \tag{1}$$

Figure 21.12 shows a circle $(x - 2)^2 + (y - 3)^2 = 4$

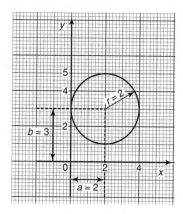

Fig. 21.12

The general equation of a circle is:

$$x^2 + y^2 + 2ex + 2fy + c = 0 \tag{2}$$

Multiplying out the bracketed terms in equation (1) gives:

$$x^2 - 2ax + a^2 + y^2 - 2by + b^2 = r^2$$

Comparing this with equation (2) gives:

$$2e = -2a, \quad \text{i.e.} \quad \boldsymbol{a = -\dfrac{2e}{2}}$$

and $2f = -2b$, i.e. $\boldsymbol{b = -\dfrac{2f}{2}}$

and $c = a^2 + b^2 - r^2$, i.e. $\boldsymbol{\sqrt{a^2 + b^2 - c}}$

Thus, for example, the equation

$$x^2 + y^2 - 4x - 6y + 9 = 0$$

represents a circle with centre

$$a = -\left(\frac{-4}{2}\right), \quad b = -\left(\frac{-6}{2}\right)$$

i.e., at (2, 3) and radius $r = \sqrt{2^2 + 3^2 - 9} = 2$

Hence $x^2 + y^2 - 4x - 6y + 9 = 0$ is the circle shown in Fig. 21.12 (which may be checked by multiplying out the brackets in the equation

$$(x - 2)^2 + (y - 3)^2 = 4$$

Problem 15. Determine (a) the radius, and (b) the co-ordinates of the centre of the circle given by the equation:

$$x^2 + y^2 + 8x - 2y + 8 = 0$$

$x^2 + y^2 + 8x - 2y + 8 = 0$ is of the form shown in equation (2),

where $a = -\left(\dfrac{8}{2}\right) = -4$, $b = -\left(\dfrac{-2}{2}\right) = 1$

and $r = \sqrt{(-4)^2 + 1^2 - 8} = \sqrt{9} = 3$

Hence $x^2 + y^2 + 8x - 2y + 8 - 0$ represents a circle **centre $(-4, 1)$** and **radius 3**, as shown in Fig. 21.13.

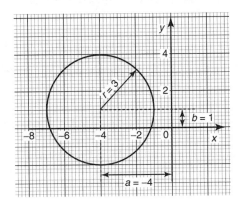

Fig. 21.13

Problem 16. Sketch the circle given by the equation:

$$x^2 + y^2 - 4x + 6y - 3 = 0$$

The equation of a circle, centre (a, b), radius r is given by:

$$(x - a)^2 + (y - b)^2 = r^2$$

The general equation of a circle is

$$x^2 + y^2 + 2ex + 2fy + c = 0$$

From above $a = -\dfrac{2e}{2}$, $b = -\dfrac{2f}{2}$

and $r = \sqrt{a^2 + b^2 - c}$

Hence if $x^2 + y^2 - 4x + 6y - 3 = 0$.

then $a = -\left(\dfrac{-4}{2}\right) = 2$, $b = -\left(\dfrac{6}{2}\right) = -3$

and $r = \sqrt{2^2 + (-3)^2 - (-3)} = \sqrt{16} = 4$

Thus **the circle has centre $(2, -3)$ and radius 4**, as shown in Fig. 21.14.

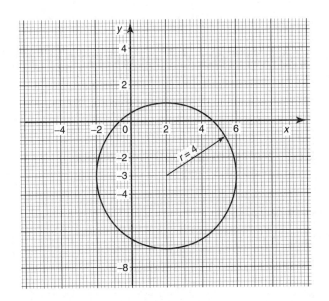

Fig. 21.14

Now try the following exercise

Exercise 77   Further problems on the equation of a circle (Answers on page 261)

1. Determine (a) the radius, and (b) the co-ordinates of the centre of the circle given by the equation

$$x^2 + y^2 + 8x - 2y + 8 = 0$$

2. Sketch the circle given by the equation

$$x^2 + y^2 - 6x + 4y - 3 = 0$$

3. Sketch the curve

$$x^2 + (y - 1)^2 - 25 = 0$$

4. Sketch the curve

$$x = 6\sqrt{\left[1 - \left(\dfrac{y}{6}\right)^2\right]}$$

22

Volumes of common solids

22.1 Volumes and surface areas of regular solids

A summary of volumes and surface areas of regular solids is shown in Table 22.1.

Table 22.1

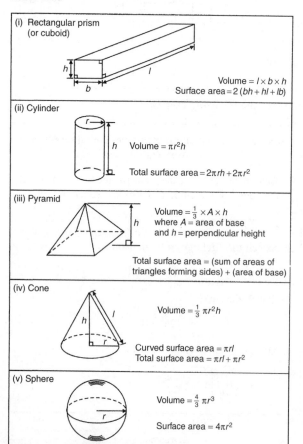

(i) Rectangular prism (or cuboid)

Volume $= l \times b \times h$
Surface area $= 2(bh + hl + lb)$

(ii) Cylinder

Volume $= \pi r^2 h$

Total surface area $= 2\pi rh + 2\pi r^2$

(iii) Pyramid

Volume $= \frac{1}{3} \times A \times h$
where $A =$ area of base
and $h =$ perpendicular height

Total surface area = (sum of areas of triangles forming sides) + (area of base)

(iv) Cone

Volume $= \frac{1}{3}\pi r^2 h$

Curved surface area $= \pi rl$
Total surface area $= \pi rl + \pi r^2$

(v) Sphere

Volume $= \frac{4}{3}\pi r^3$

Surface area $= 4\pi r^2$

22.2 Worked problems on volumes and surface areas of regular solids

Problem 1. A water tank is the shape of a rectangular prism having length 2 m, breadth 75 cm and height 50 cm. Determine the capacity of the tank in (a) m³ (b) cm³ (c) litres.

Volume of rectangular prism $= l \times b \times h$ (see Table 22.1)

(a) Volume of tank $= 2 \times 0.75 \times 0.5 = \mathbf{0.75\,m^3}$

(b) $1\,m^3 = 10^6\,cm^3$, hence $0.75\,m^3 = 0.75 \times 10^6\,cm^3$
$$= \mathbf{750\,000\,cm^3}$$

(c) $1\,litre = 1000\,cm^3$, hence

$$750\,000\,cm^3 = \frac{750\,000}{1000}\,litres = \mathbf{750\,litres}$$

Problem 2. Find the volume and total surface area of a cylinder of length 15 cm and diameter 8 cm.

Volume of cylinder $= \pi r^2 h$ (see Table 22.1)

Since diameter $= 8\,cm$, then radius $r = 4\,cm$

Hence volume $= \pi \times 4^2 \times 15 = \mathbf{754\,cm^3}$

Total surface area (i.e. including the two ends)

$$= 2\pi rh + 2\pi r^2$$
$$= (2 \times \pi \times 4 \times 15) + (2 \times \pi \times 4^2)$$
$$= \mathbf{477.5\,cm^2}$$

Problem 3. Determine the volume (in cm³) of the shape shown in Fig. 22.1.

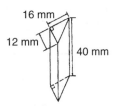

Fig. 22.1

The solid shown in Fig. 22.1 is a triangular prism. The volume V of any prism is given by: $V = Ah$, where A is the cross-sectional area and h is the perpendicular height.

Hence volume $= \frac{1}{2} \times 16 \times 12 \times 40$

$= 3840\,\text{mm}^3$

$= \mathbf{3.840\,cm^3}$ (since $1\,\text{cm}^3 = 1000\,\text{mm}^3$)

Problem 4. Calculate the volume and total surface area of the solid prism shown in Fig. 22.2.

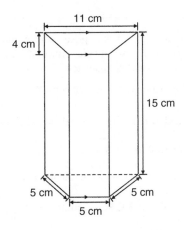

Fig. 22.2

The solid shown in Fig. 22.2 is a trapezoidal prism.

Volume $=$ cross-sectional area $\times$ height

$= \frac{1}{2}(11 + 5)4 \times 15 = 32 \times 15 = \mathbf{480\,cm^3}$

Surface area $=$ sum of two trapeziums $+$ 4 rectangles

$= (2 \times 32) + (5 \times 15) + (11 \times 15) + 2(5 \times 15)$

$= 64 + 75 + 165 + 150 = \mathbf{454\,cm^2}$

Problem 5. Determine the volume and the total surface area of the square pyramid shown in Fig. 22.3 if its perpendicular height is 12 cm.

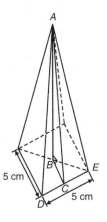

Fig. 22.3

Volume of pyramid $= \frac{1}{3}$(area of base) (perpendicular height)

$= \frac{1}{3}(5 \times 5) \times 12 = \mathbf{100\,cm^3}$

The total surface area consists of a square base and 4 equal triangles.

Area of triangle $ADE = \frac{1}{2} \times$ base $\times$ perpendicular height

$= \frac{1}{2} \times 5 \times AC$

The length AC may be calculated using Pythagoras' theorem on triangle ABC, where $AB = 12\,\text{cm}$, $BC = \frac{1}{2} \times 5 = 2.5\,\text{cm}$

$$AC = \sqrt{AB^2 + BC^2} = \sqrt{12^2 + 2.5^2} = 12.26\,\text{cm}$$

Hence area of triangle $ADE = \frac{1}{2} \times 5 \times 12.26 = 30.65\,\text{cm}^2$

Total surface area of pyramid $= (5 \times 5) + 4(30.65)$

$= \mathbf{147.6\,cm^2}$

Problem 6. Determine the volume and total surface area of a cone of radius 5 cm and perpendicular height 12 cm.

The cone is shown in Fig. 22.4.

Volume of cone $= \frac{1}{3}\pi r^2 h = \frac{1}{3} \times \pi \times 5^2 \times 12 = \mathbf{314.2\,cm^3}$

Total surface area $=$ curved surface area $+$ area of base

$$= \pi r l + \pi r^2$$

From Fig. 22.4, slant height l may be calculated using Pythagoras' theorem

$$l = \sqrt{12^2 + 5^2} = 13\,\text{cm}$$

Hence total surface area $= (\pi \times 5 \times 13) + (\pi \times 5^2)$

$= \mathbf{282.7\,cm^2}$

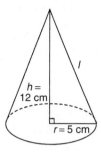

Fig. 22.4

> *Problem 7.* Find the volume and surface area of a sphere of diameter 8 cm.

Since diameter = 8 cm, then radius, $r = 4$ cm.

Volume of sphere = $\frac{4}{3}\pi r^3 = \frac{4}{3} \times \pi \times 4^3 = \mathbf{268.1\, cm^3}$

Surface area of sphere = $4\pi r^2 = 4 \times \pi \times 4^2 = \mathbf{201.1\, cm^2}$

Now try the following exercise

> **Exercise 78   Further problems on volumes and surface areas of regular solids (Answers on page 261)**
>
> 1. A rectangular block of metal has dimensions of 40 mm by 25 mm by 15 mm. Determine its volume. Find also its mass if the metal has a density of 9 g/cm³
>
> 2. Determine the maximum capacity, in litres, of a fish tank measuring 50 cm by 40 cm by 2.5 m (1 litre = 1000 cm³).
>
> 3. Determine how many cubic metres of concrete are required for a 120 m long path, 150 mm wide and 80 mm deep.
>
> 4. Calculate the volume of a metal tube whose outside diameter is 8 cm and whose inside diameter is 6 cm, if the length of the tube is 4 m.
>
> 5. The volume of a cylinder is 400 cm³. If its radius is 5.20 cm, find its height. Determine also its curved surface area.
>
> 6. If a cone has a diameter of 80 mm and a perpendicular height of 120 mm calculate its volume in cm³ and its curved surface area.
>
> 7. A cylinder is cast from a rectangular piece of alloy 5 cm by 7 cm by 12 cm. If the length of the cylinder is to be 60 cm, find its diameter.
>
> 8. Find the volume and the total surface area of a regular hexagonal bar of metal of length 3 m if each side of the hexagon is 6 cm.
>
> 9. A square pyramid has a perpendicular height of 4 cm. If a side of the base is 2.4 cm long find the volume and total surface area of the pyramid.
>
> 10. A sphere has a diameter of 6 cm. Determine its volume and surface area.
>
> 11. Find the total surface area of a hemisphere of diameter 50 mm.

22.3   Further worked problems on volumes and surface areas of regular solids

> *Problem 8.* A wooden section is shown in Fig. 22.5. Find (a) its volume (in m³), and (b) its total surface area.

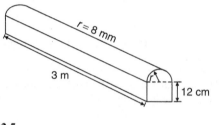

Fig. 22.5

(a) The section of wood is a prism whose end comprises a rectangle and a semicircle. Since the radius of the semicircle is 8 cm, the diameter is 16 cm.

Hence the rectangle has dimensions 12 cm by 16 cm.

Area of end = $(12 \times 16) + \frac{1}{2}\pi 8^2 = 292.5\, cm^2$

Volume of wooden section

= area of end × perpendicular height

= $292.5 \times 300 = 87750\, cm^3 = \dfrac{87750\, m^3}{10^6}$

= $\mathbf{0.08775\, m^3}$

(b) The total surface area comprises the two ends (each of area 292.5 cm²), three rectangles and a curved surface (which is half a cylinder), hence total surface area

= $(2 \times 292.5) + 2(12 \times 300) + (16 \times 300)$

$\quad + \frac{1}{2}(2\pi \times 8 \times 300)$

= $585 + 7200 + 4800 + 2400\pi$

= $\mathbf{20125\, cm^2}$ or $\mathbf{2.0125\, m^2}$

Problem 9. A pyramid has a rectangular base 3.60 cm by 5.40 cm. Determine the volume and total surface area of the pyramid if each of its sloping edges is 15.0 cm.

The pyramid is shown in Fig. 22.6. To calculate the volume of the pyramid the perpendicular height EF is required. Diagonal BD is calculated using Pythagoras' theorem,

i.e. $BD = \sqrt{[3.60^2 + 5.40^2]} = 6.490$ cm

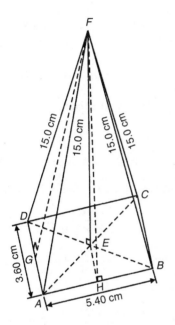

Fig. 22.6

Hence $EB = \frac{1}{2}BD = \frac{6.490}{2} = 3.245$ cm.

Using Pythagoras' theorem on triangle BEF gives

$$BF^2 = EB^2 + EF^2$$

from which, $EF = \sqrt{(BF^2 - EB^2)}$

$$= \sqrt{15.0^2 - 3.245^2} = 14.64 \text{ cm}.$$

Volume of pyramid

$$= \tfrac{1}{3}(\text{area of base})(\text{perpendicular height})$$

$$= \tfrac{1}{3}(3.60 \times 5.40)(14.64) = \textbf{94.87 cm}^3$$

Area of triangle ADF (which equals triangle BCF) $= \frac{1}{2}(AD)(FG)$, where G is the midpoint of AD. Using Pythagoras' theorem on triangle FGA gives

$$FG = \sqrt{[15.0^2 - 1.80^2]} = 14.89 \text{ cm}$$

Hence area of triangle $ADF = \frac{1}{2}(3.60)(14.89) = 26.80 \text{ cm}^2$

Similarly, if H is the mid-point of AB, then

$$FH = \sqrt{15.0^2 - 2.70^2} = 14.75 \text{ cm},$$

hence area of triangle ABF (which equals triangle CDF)

$$= \tfrac{1}{2}(5.40)(14.75) = 39.83 \text{ cm}^2$$

Total surface area of pyramid

$$= 2(26.80) + 2(39.83) + (3.60)(5.40)$$

$$= 53.60 + 79.66 + 19.44$$

$$= \textbf{152.7 cm}^2$$

Problem 10. Calculate the volume and total surface area of a hemisphere of diameter 5.0 cm.

Volume of hemisphere $= \dfrac{1}{2}(\text{volume of sphere})$

$$= \frac{2}{3}\pi r^3 = \frac{2}{3}\pi \left(\frac{5.0}{2}\right)^3 = \textbf{32.7 cm}^3$$

Total surface area

$$= \text{curved surface area} + \text{area of circle}$$

$$= \frac{1}{2}(\text{surface area of sphere}) + \pi r^2$$

$$= \frac{1}{2}(4\pi r^2) + \pi r^2$$

$$= 2\pi r^2 + \pi r^2 = 3\pi r^2 = 3\pi \left(\frac{5.0}{2}\right)^2 = \textbf{58.9 cm}^2$$

Problem 11. A rectangular piece of metal having dimensions 4 cm by 3 cm by 12 cm is melted down and recast into a pyramid having a rectangular base measuring 2.5 cm by 5 cm. Calculate the perpendicular height of the pyramid.

Volume of rectangular prism of metal $= 4 \times 3 \times 12 = 144 \text{ cm}^3$

Volume of pyramid

$$= \tfrac{1}{3}(\text{area of base})(\text{perpendicular height})$$

Assuming no waste of metal,

$$144 = \frac{1}{3}(2.5 \times 5)(\text{height})$$

i.e. perpendicular height $= \dfrac{144 \times 3}{2.5 \times 5} = \textbf{34.56 cm}$

Problem 12. A rivet consists of a cylindrical head, of diameter 1 cm and depth 2 mm, and a shaft of diameter 2 mm and length 1.5 cm. Determine the volume of metal in 2000 such rivets.

Radius of cylindrical head $= \dfrac{1}{2}$ cm $= 0.5$ cm and

height of cylindrical head $= 2$ mm $= 0.2$ cm

Hence, volume of cylindrical head

$$= \pi r^2 h = \pi (0.5)^2 (0.2) = 0.1571 \text{ cm}^3$$

Volume of cylindrical shaft

$$= \pi r^2 h = \pi \left(\dfrac{0.2}{2} \right)^2 (1.5) = 0.0471 \text{ cm}^3$$

Total volume of 1 rivet $= 0.1571 + 0.0471 = 0.2042 \text{ cm}^3$

Volume of metal in 2000 such rivets $= 2000 \times 0.2042$

$$= 408.4 \text{ cm}^3$$

Problem 13. A solid metal cylinder of radius 6 cm and height 15 cm is melted down and recast into a shape comprising a hemisphere surmounted by a cone. Assuming that 8% of the metal is wasted in the process, determine the height of the conical portion, if its diameter is to be 12 cm.

Volume of cylinder $= \pi r^2 h = \pi \times 6^2 \times 15 = 540\pi \text{ cm}^3$

If 8% of metal is lost then 92% of 540π gives the volume of the new shape (shown in Fig. 22.7).

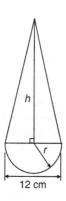

Fig. 22.7

Hence the volume of (hemisphere+cone) $= 0.92 \times 540\pi \text{ cm}^3$,

i.e. $\dfrac{1}{2} \left(\dfrac{4}{3} \pi r^3 \right) + \dfrac{1}{3} \pi r^2 h = 0.92 \times 540\pi$

Dividing throughout by π gives:

$$\dfrac{2}{3} r^3 + \dfrac{1}{3} r^2 h = 0.92 \times 540$$

Since the diameter of the new shape is to be 12 cm, then radius $r = 6$ cm,

hence $\dfrac{2}{3}(6)^3 + \dfrac{1}{3}(6)^2 h = 0.92 \times 540$

$$144 + 12h = 496.8$$

i.e. height of conical portion,

$$h = \dfrac{496.8 - 144}{12} = 29.4 \text{ cm}$$

Problem 14. A block of copper having a mass of 50 kg is drawn out to make 500 m of wire of uniform cross-section. Given that the density of copper is 8.91 g/cm³, calculate (a) the volume of copper, (b) the cross-sectional area of the wire, and (c) the diameter of the cross-section of the wire.

(a) A density of 8.91 g/cm³ means that 8.91 g of copper has a volume of 1 cm³, or 1 g of copper has a volume of $(1/8.91) \text{ cm}^3$

Hence 50 kg, i.e. 50 000 g, has a volume

$$\dfrac{50\,000}{8.91} \text{ cm}^3 = 5612 \text{ cm}^3$$

(b) Volume of wire

$$= \text{area of circular cross-section} \times \text{length of wire.}$$

Hence 5612 cm³ $= \text{area} \times (500 \times 100 \text{ cm})$,

from which, area $= \dfrac{5612}{500 \times 100} \text{ cm}^2 = 0.1122 \text{ cm}^2$

(c) Area of circle $= \pi r^2$ or $\dfrac{\pi d^2}{4}$, hence $0.1122 = \dfrac{\pi d^2}{4}$

from which

$$d = \sqrt{\left(\dfrac{4 \times 0.1122}{\pi} \right)} = 0.3780 \text{ cm}$$

i.e. **diameter of cross-section is 3.780 mm**.

Problem 15. A boiler consists of a cylindrical section of length 8 m and diameter 6 m, on one end of which is surmounted a hemispherical section of diameter 6 m, and on the other end a conical section of height 4 m and base diameter 6 m. Calculate the volume of the boiler and the total surface area.

The boiler is shown in Fig. 22.8.

Volume of hemisphere,

$$P = \tfrac{2}{3} \pi r^3 = \tfrac{2}{3} \times \pi \times 3^3 = 18\pi \text{ m}^3$$

Volume of cylinder,

$$Q = \pi r^2 h = \pi \times 3^2 \times 8 = 72\pi \text{ m}^3$$

Volume of cone,

$$R = \tfrac{1}{3} \pi r^2 h = \tfrac{1}{3} \times \pi \times 3^2 \times 4 = 12\pi \text{ m}^3$$

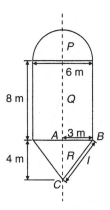

Fig. 22.8

Total volume of boiler $= 18\pi + 72\pi + 12\pi$

$$= 102\pi = \mathbf{320.4\,m^3}$$

Surface area of hemisphere,

$$P = \tfrac{1}{2}(4\pi r^2) = 2 \times \pi \times 3^2 = 18\pi\,m^2$$

Curved surface area of cylinder,

$$Q = 2\pi r h = 2 \times \pi \times 3 \times 8 = 48\pi\,m^2$$

The slant height of the cone, l, is obtained by Pythagoras' theorem on triangle ABC, i.e. $l = \sqrt{(4^2 + 3^2)} = 5$

Curved surface area of cone,

$$R = \pi r l = \pi \times 3 \times 5 = 15\pi\,m^2$$

Total surface area of boiler $= 18\pi + 48\pi + 15\pi$

$$= 81\pi = \mathbf{254.5\,m^2}$$

Now try the following exercise

Exercise 79 Further problems on volumes and surface areas of regular solids (Answers on page 261)

1. Determine the mass of a hemispherical copper container whose external and internal radii are 12 cm and 10 cm. Assuming that 1 cm³ of copper weighs 8.9 g.

2. If the volume of a sphere is 566 cm³, find its radius.

3. A metal plumb bob comprises a hemisphere surmounted by a cone. If the diameter of the hemisphere and cone are each 4 cm and the total length is 5 cm, find its total volume.

4. A marquee is in the form of a cylinder surmounted by a cone. The total height is 6 m and the cylindrical portion has a height of 3.5 m, with a diameter of 15 m. Calculate the surface area of material needed to make the marquee assuming 12% of the material is wasted in the process.

5. Determine (a) the volume and (b) the total surface area of the following solids:

   (i) a cone of radius 8.0 cm and perpendicular height 10 cm
   (ii) a sphere of diameter 7.0 cm
   (iii) a hemisphere of radius 3.0 cm
   (iv) a 2.5 cm by 2.5 cm square pyramid of perpendicular height 5.0 cm
   (v) a 4.0 cm by 6.0 cm rectangular pyramid of perpendicular height 12.0 cm
   (vi) a 4.2 cm by 4.2 cm square pyramid whose sloping edges are each 15.0 cm
   (vii) a pyramid having an octagonal base of side 5.0 cm and perpendicular height 20 cm.

6. The volume of a sphere is 325 cm³. Determine its diameter.

7. A metal sphere weighing 24 kg is melted down and recast into a solid cone of base radius 8.0 cm. If the density of the metal is 8000 kg/m³ determine (a) the diameter of the metal sphere and (b) the perpendicular height of the cone, assuming that 15% of the metal is lost in the process.

8. Find the volume of a regular hexagonal pyramid if the perpendicular height is 16.0 cm and the side of base is 3.0 cm.

9. A buoy consists of a hemisphere surmounted by a cone. The diameter of the cone and hemisphere is 2.5 m and the slant height of the cone is 4.0 m. Determine the volume and surface area of the buoy.

10. A petrol container is in the form of a central cylindrical portion 5.0 m long with a hemispherical section surmounted on each end. If the diameters of the hemisphere and cylinder are both 1.2 m determine the capacity of the tank in litres (1 litre = 1000 cm³).

11. Figure 22.9 shows a metal rod section. Determine its volume and total surface area.

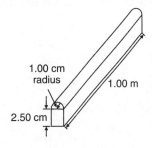

Fig. 22.9

22.4   Volumes and surface areas of frusta of pyramids and cones

The **frustum** of a pyramid or cone is the portion remaining when a part containing the vertex is cut off by a plane parallel to the base.

The **volume of a frustum of a pyramid or cone** is given by the volume of the whole pyramid or cone minus the volume of the small pyramid or cone cut off.

The **surface area of the sides of a frustum of a pyramid or cone** is given by the surface area of the whole pyramid or cone minus the surface area of the small pyramid or cone cut off. This gives the lateral surface area of the frustum. If the total surface area of the frustum is required then the surface area of the two parallel ends are added to the lateral surface area.

There is an alternative method for finding the volume and surface area of a **frustum of a cone**. With reference to Fig. 22.10:

$$\text{Volume} = \tfrac{1}{3}\pi h \left(R^2 + Rr + r^2\right)$$

$$\text{Curved surface area} = \pi l \left(R + r\right)$$

$$\text{Total surface area} = \pi l \left(R + r\right) + \pi r^2 + \pi R^2$$

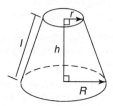

Fig. 22.10

Problem 16. Determine the volume of a frustum of a cone if the diameter of the ends are 6.0 cm and 4.0 cm and its perpendicular height is 3.6 cm.

Method 1

A section through the vertex of a complete cone is shown in Fig. 22.11.

Using similar triangles

$$\frac{AP}{DP} = \frac{DR}{BR}$$

Hence

$$\frac{AP}{2.0} = \frac{3.6}{1.0}$$

from which $AP = \dfrac{(2.0)(3.6)}{1.0} = 7.2\,\text{cm}$

The height of the large cone $= 3.6 + 7.2 = 10.8\,\text{cm}$.

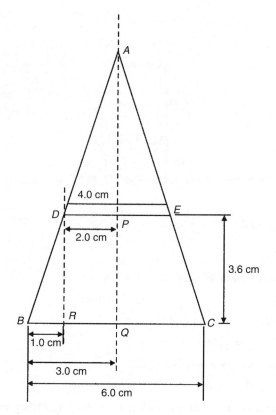

Fig. 22.11

Volume of frustum of cone

$$= \text{volume of large cone} - \text{volume of small cone cut off}$$

$$= \tfrac{1}{3}\pi(3.0)^2(10.8) - \tfrac{1}{3}\pi(2.0)^2(7.2)$$

$$= 101.79 - 30.16 = \mathbf{71.6\,cm^3}$$

Method 2

From above, volume of the frustum of a cone

$$= \tfrac{1}{3}\pi h(R^2 + Rr + r^2), \quad \text{where} \quad R = 3.0\,\text{cm},$$

$$r = 2.0\,\text{cm} \quad \text{and} \quad h = 3.6\,\text{cm}$$

Hence volume of frustum

$$= \tfrac{1}{3}\pi(3.6)\left[(3.0)^2 + (3.0)(2.0) + (2.0)^2\right]$$

$$= \tfrac{1}{3}\pi(3.6)(19.0) = \mathbf{71.6\,cm^3}$$

Problem 17. Find the total surface area of the frustum of the cone in Problem 16.

Method 1

Curved surface area of frustum = curved surface area of large cone − curved surface area of small cone cut off.

From Fig. 22.11, using Pythagoras' theorem:

$$AB^2 = AQ^2 + BQ^2, \quad \text{from which}$$

$$AB = \sqrt{[10.8^2 + 3.0^2]} = 11.21 \text{ cm}$$

and

$$AD^2 = AP^2 + DP^2, \quad \text{from which}$$

$$AD = \sqrt{[7.2^2 + 2.0^2]} = 7.47 \text{ cm}$$

Curved surface area of large cone

$$= \pi r l = \pi(BQ)(AB) = \pi(3.0)(11.21) = 105.65 \text{ cm}^2$$

and curved surface area of small cone

$$= \pi(DP)(AD) = \pi(2.0)(7.47) = 46.94 \text{ cm}^2$$

Hence, curved surface area of frustum $= 105.65 - 46.94$

$$= 58.71 \text{ cm}^2$$

Total surface area of frustum

$$= \text{curved surface area} + \text{area of two circular ends}$$

$$= 58.71 + \pi(2.0)^2 + \pi(3.0)^2$$

$$= 58.71 + 12.57 + 28.27 = \textbf{99.6 cm}^2$$

Method 2

From page 172, total surface area of frustum

$$= \pi l (R + r) + \pi r^2 + \pi R^2,$$

where $l = BD = 11.21 - 7.47 = 3.74 \text{ cm}$, $R = 3.0 \text{ cm}$ and $r = 2.0 \text{ cm}$.

Hence total surface area of frustum

$$= \pi(3.74)(3.0 + 2.0) + \pi(2.0)^2 + \pi(3.0)^2 = \textbf{99.6 cm}^2$$

Problem 18. A storage hopper is in the shape of a frustum of a pyramid. Determine its volume if the ends of the frustum are squares of sides 8.0 m and 4.6 m, respectively, and the perpendicular height between its ends is 3.6 m.

The frustum is shown shaded in Fig. 22.12(a) as part of a complete pyramid. A section perpendicular to the base through the vertex is shown in Fig. 22.12(b).

By similar triangles: $\dfrac{CG}{BG} = \dfrac{BH}{AH}$

Height $CG = BG\left(\dfrac{BH}{AH}\right) = \dfrac{(2.3)(3.6)}{1.7} = 4.87 \text{ m}$

Height of complete pyramid $= 3.6 + 4.87 = 8.47 \text{ m}$

Volume of large pyramid $= \dfrac{1}{3}(8.0)^2(8.47) = 180.69 \text{ m}^3$

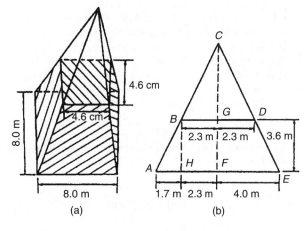

Fig. 22.12

Volume of small pyramid cut off

$$= \frac{1}{3}(4.6)^2(4.87) = 34.35 \text{ m}^3$$

Hence volume of storage hopper

$$= 180.69 - 34.35 = \textbf{146.3 m}^3$$

Problem 19. Determine the lateral surface area of the storage hopper in Problem 18.

The lateral surface area of the storage hopper consists of four equal trapeziums.

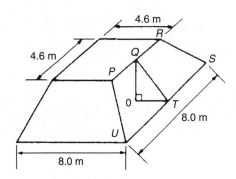

Fig. 22.13

From Fig. 22.13, area of trapezium $PRSU$

$$= \tfrac{1}{2}(PR + SU)(QT)$$

$OT = 1.7 \text{ m}$ (same as AH in Fig. 22.13(b) and $OQ = 3.6 \text{ m}$.

By Pythagoras' theorem,

$$QT = \sqrt{(OQ^2 + OT^2)} = \sqrt{[3.6^2 + 1.7^2]} = 3.98 \text{ m}$$

Area of trapezium $PRSU = \tfrac{1}{2}(4.6 + 8.0)(3.98) = 25.07 \text{ m}^2$

Lateral surface area of hopper $= 4(25.07) = \textbf{100.3 m}^2$

Problem 20. A lampshade is in the shape of a frustum of a cone. The vertical height of the shade is 25.0 cm and the diameters of the ends are 20.0 cm and 10.0 cm, respectively. Determine the area of the material needed to form the lampshade, correct to 3 significant figures.

The curved surface area of a frustum of a cone

$$= \pi l(R + r) \text{ from page 172.}$$

Since the diameters of the ends of the frustum are 20.0 cm and 10.0 cm, then from Fig. 22.14,

$$r = 5.0 \text{ cm}, \quad R = 10.0 \text{ cm}$$

and $\quad l = \sqrt{[25.0^2 + 5.0^2]} = 25.50 \text{ cm},$

from Pythagoras' theorem.

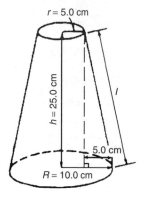

Fig. 22.14

Hence curved surface area $= \pi(25.50)(10.0 + 5.0) = 1201.7 \text{ cm}^2$, i.e. the area of material needed to form the lampshade is **1200 cm²**, correct to 3 significant figures.

Problem 21. A cooling tower is in the form of a cylinder surmounted by a frustum of a cone as shown in Fig. 22.15. Determine the volume of air space in the tower if 40% of the space is used for pipes and other structures.

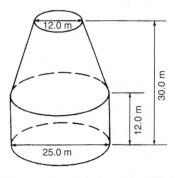

Fig. 22.15

Volume of cylindrical portion

$$= \pi r^2 h = \pi \left(\frac{25.0}{2}\right)^2 (12.0) = 5890 \text{ m}^3$$

Volume of frustum of cone

$$= \frac{1}{3}\pi h(R^2 + Rr + r^2)$$

where $h = 30.0 - 12.0 = 18.0 \text{ m}$, $R = 25.0/2 = 12.5 \text{ m}$ and $r = 12.0/2 = 6.0 \text{ m}$.

Hence volume of frustum of cone

$$= \frac{1}{3}\pi(18.0)\left[(12.5)^2 + (12.5)(6.0) + (6.0)^2\right] = 5038 \text{ m}^3$$

Total volume of cooling tower $= 5890 + 5038 = 10\,928 \text{ m}^3$.

If 40% of space is occupied then volume of air space

$$= 0.6 \times 10\,928 = \mathbf{6557 \text{ m}^3}$$

Now try the following exercise

Exercise 80   Further problems on volumes and surface areas of frustra of pyramids and cones (Answers on page 262)

1. The radii of the faces of a frustum of a cone are 2.0 cm and 4.0 cm and the thickness of the frustum is 5.0 cm. Determine its volume and total surface area.

2. A frustum of a pyramid has square ends, the squares having sides 9.0 cm and 5.0 cm, respectively. Calculate the volume and total surface area of the frustum if the perpendicular distance between its ends is 8.0 cm.

3. A cooling tower is in the form of a frustum of a cone. The base has a diameter of 32.0 m, the top has a diameter of 14.0 m and the vertical height is 24.0 m. Calculate the volume of the tower and the curved surface area.

4. A loudspeaker diaphragm is in the form of a frustum of a cone. If the end diameters are 28.0 cm and 6.00 cm and the vertical distance between the ends is 30.0 cm, find the area of material needed to cover the curved surface of the speaker.

5. A rectangular prism of metal having dimensions 4.3 cm by 7.2 cm by 12.4 cm is melted down and recast into a frustum of a square pyramid, 10% of the metal being lost in the process. If the ends of the frustum are squares of side 3 cm and 8 cm respectively, find the thickness of the frustum.

6. Determine the volume and total surface area of a bucket consisting of an inverted frustum of a cone, of slant height 36.0 cm and end diameters 55.0 cm and 35.0 cm.

7. A cylindrical tank of diameter 2.0 m and perpendicular height 3.0 m is to be replaced by a tank of the same capacity but in the form of a frustum of a cone. If the diameters of the ends of the frustum are 1.0 m and 2.0 m, respectively, determine the vertical height required.

22.5 Volumes of similar shapes

The volumes of similar bodies are proportional to the cubes of corresponding linear dimensions. For example, Fig. 22.16 shows two cubes, one of which has sides three times as long as those of the other.

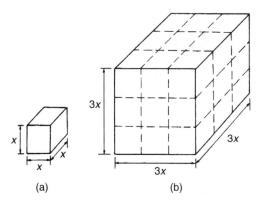

Fig. 22.16

Volume of Fig. 22.16(a) $= (x)(x)(x) = x^3$

Volume of Fig. 22.16(b) $= (3x)(3x)(3x) = 27x^3$

Hence Fig. 22.16(b) has a volume $(3)^3$, i.e. 27 times the volume of Fig. 22.16(a).

Problem 22. A car has a mass of 1000 kg. A model of the car is made to a scale of 1 to 50. Determine the mass of the model if the car and its model are made of the same material.

$$\frac{\text{Volume of model}}{\text{Volume of car}} = \left(\frac{1}{50}\right)^3$$

since the volume of similar bodies are proportional to the cube of corresponding dimensions.

Mass = density × volume, and since both car and model are made of the same material then:

$$\frac{\text{Mass of model}}{\text{Mass of car}} = \left(\frac{1}{50}\right)^3$$

Hence mass of model $= (\text{mass of car}) \left(\frac{1}{50}\right)^3$

$$= \frac{1000}{50^3} = \mathbf{0.008\,kg} \quad \text{or} \quad \mathbf{8\,g}$$

Now try the following exercise

Exercise 81   Further problems on volumes of similar shapes (Answers on page 262)

1. The diameter of two spherical bearings are in the ratio 2:5. What is the ratio of their volumes?
2. An engineering component has a mass of 400 g. If each of its dimensions are reduced by 30% determine its new mass.

Assignment 11

This assignment covers the material contained in Chapters 21 and 22. The marks for each question are shown in brackets at the end of each question.

1. Determine the diameter of a circle whose circumference is 178.4 cm. (2)
2. Convert (a) 125°47′ to radians
   (b) 1.724 radians to degrees and minutes (4)
3. Calculate the length of metal strip needed to make the clip shown in Fig. A11.1. (6)

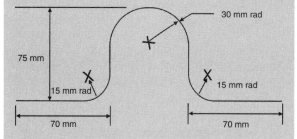

Fig. A11.1

4. A lorry has wheels of radius 50 cm. Calculate the number of complete revolutions a wheel makes (correct to the nearest revolution) when travelling 3 miles (assume 1 mile = 1.6 km). (5)
5. The equation of a circle is:

$$x^2 + y^2 + 12x - 4y + 4 = 0$$

Determine (a) the diameter of the circle, and (b) the co-ordinates of the centre of the circle. (5)

6. Determine the volume (in cubic metres) and the total surface area (in square metres) of a solid metal cone of base radius 0.5 m and perpendicular height 1.20 m. Give answers correct to 2 decimal places.   (5)

7. A rectangular storage container has dimensions 3.2 m by 90 cm by 60 cm.

   Determine its volume in (a) m^3 (b) cm^3 (4)

8. Calculate

   (a) the volume, and
   (b) the total surface area of a 10 cm by 15 cm rectangular pyramid of height 20 cm.   (7)

9. A water container is of the form of a central cylindrical part 3.0 m long and diameter 1.0 m, with a hemispherical section surmounted at each end as shown in Fig. A11.2. Determine the maximum capacity of the container, correct to the nearest litre. (1 litre = 1000 cm^3)   (5)

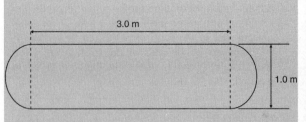

Fig. A11.2

10. Find the total surface area of a bucket consisting of an inverted frustum of a cone, of slant height 35.0 cm and end diameters 60.0 cm and 40.0 cm.   (4)

11. A boat has a mass of 20 000 kg. A model of the boat is made to a scale of 1 to 80. If the model is made of the same material as the boat, determine the mass of the model (in grams).   (3)

23

Irregular areas and volumes and mean values of waveforms

23.1 Areas of irregular figures

Areas of irregular plane surfaces may be approximately determined by using (a) a planimeter, (b) the trapezoidal rule, (c) the mid-ordinate rule, and (d) Simpson's rule. Such methods may be used, for example, by engineers estimating areas of indicator diagrams of steam engines, surveyors estimating areas of plots of land or naval architects estimating areas of water planes or transverse sections of ships.

(a) **A planimeter** is an instrument for directly measuring small areas bounded by an irregular curve.

(b) **Trapezoidal rule**

To determine the areas $PQRS$ in Fig. 23.1:

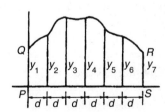

Fig. 23.1

(i) Divide base PS into any number of equal intervals, each of width d (the greater the number of intervals, the greater the accuracy).

(ii) Accurately measure ordinates y_1, y_2, y_3, etc.

(iii) Area $PQRS$

$$= d \left[\frac{y_1 + y_7}{2} + y_2 + y_3 + y_4 + y_5 + y_6 \right]$$

In general, the trapezoidal rule states:

$$\text{Area} = \left(\begin{array}{c} \text{width of} \\ \text{interval} \end{array} \right) \left[\frac{1}{2} \left(\begin{array}{c} \text{first} + \text{last} \\ \text{ordinate} \end{array} \right) + \left(\begin{array}{c} \text{sum of} \\ \text{remaining} \\ \text{ordinates} \end{array} \right) \right]$$

(c) **Mid-ordinate rule**

To determine the area $ABCD$ of Fig. 23.2:

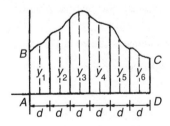

Fig. 23.2

(i) Divide base AD into any number of equal intervals, each of width d (the greater the number of intervals, the greater the accuracy).

(ii) Erect ordinates in the middle of each interval (shown by broken lines in Fig. 23.2).

(iii) Accurately measure ordinates y_1, y_2, y_3, etc.

(iv) Area $ABCD = d(y_1 + y_2 + y_3 + y_4 + y_5 + y_6)$.

In general, the mid-ordinate rule states:

Area = (width of interval)(sum of mid-ordinates)

(d) Simpson's rule

To determine the area *PQRS* of Fig. 23.1:

(i) Divide base *PS* into an **even** number of intervals, each of width *d* (the greater the number of intervals, the greater the accuracy).

(ii) Accurately measure ordinates y_1, y_2, y_3, etc.

(iii) Area *PQRS*

$$= \frac{d}{3}[(y_1 + y_7) + 4(y_2 + y_4 + y_6) + 2(y_3 + y_5)]$$

In general, Simpson's rule states:

> Area $= \dfrac{1}{3}$(width of interval) [(first + last ordinate)
>
> $+4$ (sum of even ordinates)
>
> $+2$ (sum of remaining odd ordinates)]

Problem 1. A car starts from rest and its speed is measured every second for 6 s:

Time t (s)	0	1	2	3	4	5	6
Speed v (m/s)	0	2.5	5.5	8.75	12.5	17.5	24.0

Determine the distance travelled in 6 seconds (i.e. the area under the v/t graph), by (a) the trapezoidal rule, (b) the mid-ordinate rule, and (c) Simpson's rule.

A graph of speed/time is shown in Fig. 23.3.

(a) Trapezoidal rule (see para. (b) above).

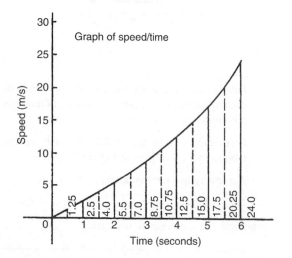

Fig. 23.3

The time base is divided into 6 strips each of width 1 s, and the length of the ordinates measured. Thus

$$\textbf{area} = (1)\left[\left(\frac{0 + 24.0}{2}\right) + 2.5 + 5.5 + 8.75\right.$$
$$\left. + 12.5 + 17.5\right] = \textbf{58.75 m}$$

(b) Mid-ordinate rule (see para. (c) above).

The time base is divided into 6 strips each of width 1 second. Mid-ordinates are erected as shown in Fig. 23.3 by the broken lines. The length of each mid-ordinate is measured. Thus

$$\textbf{area} = (1)[1.25 + 4.0 + 7.0 + 10.75 + 15.0 + 20.25]$$
$$= \textbf{58.25 m}$$

(c) Simpson's rule (see para. (d) above).

The time base is divided into 6 strips each of width 1 s, and the length of the ordinates measured. Thus

$$\textbf{area} = \tfrac{1}{3}(1)\,[(0 + 24.0) + 4(2.5 + 8.75 + 17.5)$$
$$+ 2(5.5 + 12.5)] = \textbf{58.33 m}$$

Problem 2. A river is 15 m wide. Soundings of the depth are made at equal intervals of 3 m across the river and are as shown below.

Depth (m)	0	2.2	3.3	4.5	4.2	2.4	0

Calculate the cross-sectional area of the flow of water at this point using Simpson's rule.

From para. (d) above,

$$\text{Area} = \tfrac{1}{3}(3)[(0 + 0) + 4(2.2 + 4.5 + 2.4) + 2(3.3 + 4.2)]$$
$$= (1)[0 + 36.4 + 15] = \textbf{51.4 m}^2$$

Now try the following exercise

> **Exercise 82** **Further problems on areas of irregular figures (Answers on page 262)**
>
> 1. Plot a graph of $y = 3x - x^2$ by completing a table of values of y from $x = 0$ to $x = 3$. Determine the area enclosed by the curve, the x-axis and ordinate $x = 0$ and $x = 3$ by (a) the trapezoidal rule, (b) the mid-ordinate rule and (c) by Simpson's rule.
>
> 2. Plot the graph of $y = 2x^2 + 3$ between $x = 0$ and $x = 4$. Estimate the area enclosed by the curve, the ordinates $x = 0$ and $x = 4$, and the x-axis by an approximate method.

3. The velocity of a car at one second intervals is given in the following table:

time t (s)	0	1	2	3	4	5	6
velocity v (m/s)	0	2.0	4.5	8.0	14.0	21.0	29.0

Determine the distance travelled in 6 seconds (i.e. the area under the v/t graph) using an approximate method.

4. The shape of a piece of land is shown in Fig. 23.4. To estimate the area of the land, a surveyor takes measurements at intervals of 50 m, perpendicular to the straight portion with the results shown (the dimensions being in metres). Estimate the area of the land in hectares (1 ha $= 10^4$ m^2).

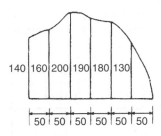

140 160 200 190 180 130

50 50 50 50 50 50

Fig. 23.4

5. The deck of a ship is 35 m long. At equal intervals of 5 m the width is given by the following table:

Width (m)	0	2.8	5.2	6.5	5.8	4.1	3.0	2.3

Estimate the area of the deck.

23.2 Volumes of irregular solids

If the cross-sectional areas A_1, A_2, A_3, .. of an irregular solid bounded by two parallel planes are known at equal intervals of width d (as shown in Fig. 23.5), then by Simpson's rule:

$$\textbf{Volume, V} = \frac{d}{3}[(A_1 + A_7) + 4(A_2 + A_4 + A_6)$$
$$+ 2(A_3 + A_5)]$$

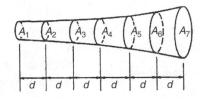

A_1 A_2 A_3 A_4 A_5 A_6 A_7

d d d d d d

Fig. 23.5

Problem 3. A tree trunk is 12 m in length and has a varying cross-section. The cross-sectional areas at intervals of 2 m measured from one end are:

0.52, 0.55, 0.59, 0.63, 0.72, 0.84, 0.97 m^2

Estimate the volume of the tree trunk.

A sketch of the tree trunk is similar to that shown in Fig. 23.5, where $d = 2$ m, $A_1 = 0.52$ m^2, $A_2 = 0.55$ m^2, and so on. Using Simpson's rule for volumes gives:

$$\text{Volume} = \tfrac{2}{3}[(0.52 + 0.97) + 4(0.55 + 0.63 + 0.84)$$
$$+ 2(0.59 + 0.72)]$$
$$= \tfrac{2}{3}[1.49 + 8.08 + 2.62] = \textbf{8.13 m}^3$$

Problem 4. The areas of seven horizontal cross-sections of a water reservoir at intervals of 10 m are:

210, 250, 320, 350, 290, 230, 170 m^2

Calculate the capacity of the reservoir in litres.

Using Simpson's rule for volumes gives:

$$\text{Volume} = \frac{10}{3}[(210 + 170) + 4(250 + 350 + 230)$$
$$+ 2(320 + 290)]$$
$$= \frac{10}{3}[380 + 3320 + 1220] = \textbf{16400 m}^3$$

$$16400 \text{ m} = 16400 \times 10^6 \text{ cm}^3$$

Since 1 litre $= 1000$ cm^3, capacity of reservoir

$$= \frac{16400 \times 10^6}{1000} \text{ litres}$$

$$= 16400000 = \textbf{1.64} \times \textbf{10}^7 \textbf{ litres}$$

Now try the following exercise

Exercise 83 Further problems on volumes of irregular solids (Answers on page 262)

1. The areas of equidistantly spaced sections of the underwater form of a small boat are as follows:

1.76, 2.78, 3.10, 3.12, 2.61, 1.24, 0.85 m^2

Determine the underwater volume if the sections are 3 m apart.

2. To estimate the amount of earth to be removed when constructing a cutting the cross-sectional area

at intervals of 8 m were estimated as follows:

$$0, \quad 2.8, \quad 3.7, \quad 4.5, \quad 4.1, \quad 2.6, \quad 0 \, \text{m}^3$$

Estimate the volume of earth to be excavated.

3. The circumference of a 12 m long log of timber of varying circular cross-section is measured at intervals of 2 m along its length and the results are:

Distance from one end (m)	0	2	4	6	8	10	12
Circumference (m)	2.80	3.25	3.94	4.32	5.16	5.82	6.36

Estimate the volume of the timber in cubic metres.

23.3   The mean or average value of a waveform

The mean or average value, y, of the waveform shown in Fig. 23.6 is given by:

$$y = \frac{\textbf{area under curve}}{\textbf{length of base, } b}$$

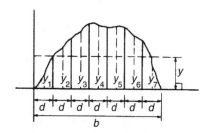

Fig. 23.6

If the mid-ordinate rule is used to find the area under the curve, then:

$$y = \frac{\text{sum of mid-ordinates}}{\text{number of mid-ordinates}}$$

$$\left(= \frac{y_1 + y_2 + y_3 + y_4 + y_5 + y_6 + y_7}{7} \right) \text{ for Fig. 23.6}$$

For a **sine wave**, the mean or average value:

(i) over one complete cycle is zero (see Fig. 23.7(a)),

(ii) over half a cycle is **0.637 × maximum value**, or **$2/\pi$ × maximum value**,

(iii) of a full-wave rectified waveform (see Fig. 23.7(b)) is **0.637 × maximum value**,

(iv) of a half-wave rectified waveform (see Fig. 23.7(c)) is **0.318 × maximum value**, or **$1/\pi$ × maximum value**.

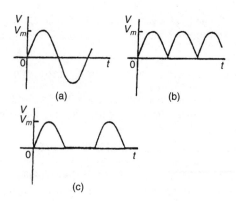

Fig. 23.7

Problem 5. Determine the average values over half a cycle of the periodic waveforms shown in Fig. 23.8.

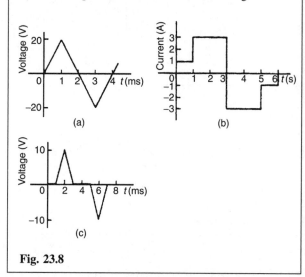

Fig. 23.8

(a) Area under triangular waveform (a) for a half cycle is given by:

$$\text{Area} = \frac{1}{2}(\text{base})(\text{perpendicular height})$$

$$= \frac{1}{2}(2 \times 10^{-3})(20) = 20 \times 10^{-3} \, \text{Vs}$$

$$\text{Average value of waveform} = \frac{\text{area under curve}}{\text{length of base}}$$

$$= \frac{20 \times 10^{-3} \, \text{Vs}}{2 \times 10^{-3} \, \text{s}} = \textbf{10 V}$$

(b) Area under waveform (b) for a half cycle

$$= (1 \times 1) + (3 \times 2) = 7 \, \text{As}$$

$$\text{Average value of waveform} = \frac{\text{area under curve}}{\text{length of base}}$$

$$= \frac{7 \, \text{As}}{3 \, \text{s}} = \textbf{2.33 A}$$

(c) A half cycle of the voltage waveform (c) is completed in 4 ms.

$$\text{Area under curve} = \frac{1}{2}\{(3-1)10^{-3}\}(10) = 10 \times 10^{-3}\,\text{Vs}$$

$$\text{Average value of waveform} = \frac{\text{area under curve}}{\text{length of base}}$$

$$= \frac{10 \times 10^{-3}\,\text{Vs}}{4 \times 10^{-3}\,\text{s}} = \mathbf{2.5\,V}$$

Problem 6. Determine the mean value of current over one complete cycle of the periodic waveforms shown in Fig. 23.9.

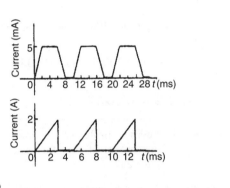

Fig. 23.9

(a) One cycle of the trapezoidal waveform (a) is completed in 10 ms (i.e. the periodic time is 10 ms).

Area under curve = area of trapezium

$$= \frac{1}{2}(\text{sum of parallel sides})(\text{perpendicular distance}$$

$$\text{between parallel sides})$$

$$= \frac{1}{2}\{(4+8) \times 10^{-3}\}(5 \times 10^{-3}) = 30 \times 10^{-6}\,\text{As}$$

$$\text{Mean value over one cycle} = \frac{\text{area under curve}}{\text{length of base}}$$

$$= \frac{30 \times 10^{-6}\,\text{As}}{10 \times 10^{-3}\,\text{s}} = \mathbf{3\,mA}$$

(b) One cycle of the sawtooth waveform (b) is completed in 5 ms.
Area under curve $= \frac{1}{2}(3 \times 10^{-3})(2) = 3 \times 10^{-3}\,\text{As}$

$$\text{Mean value over one cycle} = \frac{\text{area under curve}}{\text{length of base}}$$

$$= \frac{3 \times 10^{-3}\,\text{As}}{5 \times 10^{-3}\,\text{s}} = \mathbf{0.6\,A}$$

Problem 7. The power used in a manufacturing process during a 6 hour period is recorded at intervals of

1 hour as shown below.

Time (h)	0	1	2	3	4	5	6
Power (kW)	0	14	29	51	45	23	0

Plot a graph of power against time and, by using the mid-ordinate rule, determine (a) the area under the curve and (b) the average value of the power.

The graph of power/time is shown in Fig. 23.10.

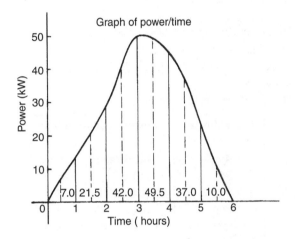

Fig. 23.10

(a) The time base is divided into 6 equal intervals, each of width 1 hour.
Mid-ordinates are erected (shown by broken lines in Fig. 23.10) and measured. The values are shown in Fig. 23.10.

Area under curve

$$= (\text{width of interval})(\text{sum of mid-ordinates})$$

$$= (1)\,[7.0 + 21.5 + 42.0 + 49.5 + 37.0 + 10.0]$$

$$= \mathbf{167\,kWh} \text{ (i.e. a measure of electrical energy)}$$

(b) Average value of waveform

$$= \frac{\text{area under curve}}{\text{length of base}} = \frac{167\,\text{kWh}}{6\,\text{h}} = \mathbf{27.83\,kW}$$

Alternatively, average value

$$= \frac{\text{Sum of mid-ordinates}}{\text{number of mid-ordinate}}$$

Problem 8. Figure 23.11 shows a sinusoidal output voltage of a full-wave rectifier. Determine, using the mid-ordinate rule with 6 intervals, the mean output voltage.

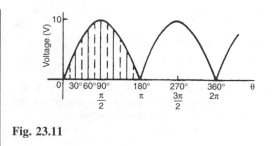

Fig. 23.11

One cycle of the output voltage is completed in π radians or 180°. The base is divided into 6 intervals, each of width 30°. The mid-ordinate of each interval will lie at 15°, 45°, 75°, etc.

At 15° the height of the mid-ordinate is $10 \sin 15° = 2.588$ V.

At 45° the height of the mid-ordinate is $10 \sin 45° = 7.071$ V, and so on.

The results are tabulated below:

Mid-ordinate	Height of mid-ordinate
15°	$10 \sin 15° = 2.588$ V
45°	$10 \sin 45° = 7.071$ V
75°	$10 \sin 75° = 9.659$ V
105°	$10 \sin 105° = 9.659$ V
135°	$10 \sin 135° = 7.071$ V
165°	$10 \sin 165° = 2.588$ V
	Sum of mid-ordinates $= 38.636$ V

Mean or average value of output voltage

$$= \frac{\text{Sum of mid-ordinates}}{\text{number of mid-ordinate}} = \frac{38.636}{6} = \mathbf{6.439\ V}$$

(With a larger number of intervals a more accurate answer may be obtained.)

For a sine wave the actual mean value is $0.637 \times$ maximum value, which in this problem gives 6.37 V.

Problem 9. An indicator diagram for a steam engine is shown in Fig. 23.12. The base line has been divided into 6 equally spaced intervals and the lengths of the 7 ordinates measured with the results shown in centimetres. Determine (a) the area of the indicator diagram using Simpson's rule, and (b) the mean pressure in the cylinder given that 1 cm represents 100 kPa.

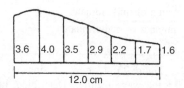

Fig. 23.12

(a) The width of each interval is $\dfrac{12.0}{6}$ cm. Using Simpson's rule,

$$\text{area} = \frac{1}{3}(2.0)[(3.6 + 1.6) + 4(4.0 + 2.9 + 1.7)$$
$$+ 2(3.5 + 2.2)]$$
$$= \frac{2}{3}[5.2 + 34.4 + 11.4] = \mathbf{34\ cm^2}$$

(b) Mean height of ordinates $= \dfrac{\text{area of diagram}}{\text{length of base}} = \dfrac{34}{12}$

$$= 2.83\ cm$$

Since 1 cm represents 100 kPa, the mean pressure in the cylinder

$$= 2.83\ cm \times 100\ kPa/cm = \mathbf{283\ kPa}$$

Now try the following exercise

Exercise 84 **Further problems on mean or average values of waveforms (Answers on page 262)**

1. Determine the mean value of the periodic waveforms shown in Fig. 23.13 over a half cycle.

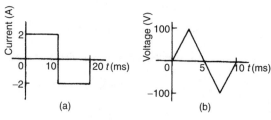

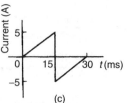

Fig. 23.13

2. Find the average value of the periodic waveforms shown in Fig. 23.14 over one complete cycle.

3. An alternating current has the following values at equal intervals of 5 ms.

Time (ms)	0	5	10	15	20	25	30
Current (A)	0	0.9	2.6	4.9	5.8	3.5	0

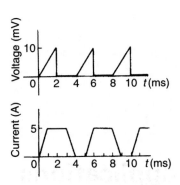

Fig. 23.14

Plot a graph of current against time and estimate the area under the curve over the 30 ms period using the mid-ordinate rule and determine its mean value.

4. Determine, using an approximate method, the average value of a sine wave of maximum value 50 V for (a) a half cycle and (b) a complete cycle.

5. An indicator diagram of a steam engine is 12 cm long. Seven evenly spaced ordinates, including the end ordinates, are measured as follows:

   5.90, 5.52, 4.22, 3.63, 3.32, 3.24, 3.16 cm

   Determine the area of the diagram and the mean pressure in the cylinder if 1 cm represents 90 kPa.

24

Triangles and some practical applications

24.1 Sine and cosine rules

To 'solve a triangle' means 'to find the values of unknown sides and angles'. If a triangle is **right angled**, trigonometric ratios and the theorem of Pythagoras may be used for its solution, as shown in Chapter 17. However, for a **non-right-angled triangle**, trigonometric ratios and Pythagoras' theorem **cannot** be used. Instead, two rules, called the **sine rule** and the **cosine rule**, are used.

Sine rule

With reference to triangle ABC of Fig. 24.1, the **sine rule** states:

$$\frac{a}{\sin A} = \frac{b}{\sin B} = \frac{c}{\sin C}$$

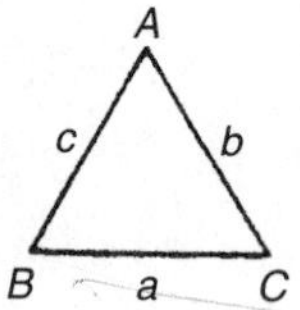

Fig. 24.1

The rule may be used only when:

(i) 1 side and any 2 angles are initially given, or

(ii) 2 sides and an angle (not the included angle) are initially given.

Cosine rule

With reference to triangle ABC of Fig. 24.1, the **cosine rule** states:

$$a^2 = b^2 + c^2 - 2bc \cos A$$

$$\text{or} \quad b^2 = a^2 + c^2 - 2ac \cos B$$

$$\text{or} \quad c^2 = a^2 + b^2 - 2ab \cos C$$

The rule may be used only when:

(i) 2 sides and the included angle are initially given, or

(ii) 3 sides are initially given.

24.2 Area of any triangle

The **area of any triangle** such as ABC of Fig. 24.1 is given by:

(i) $\dfrac{1}{2} \times$ **base** $\times$ **perpendicular height**, or

(ii) $\dfrac{1}{2}ab \sin C$ or $\dfrac{1}{2}ac \sin B$ or $\dfrac{1}{2}bc \sin A$, or

(iii) $\sqrt{[s(s-a)(s-b)(s-c)]}$, where

$$s = \frac{a+b+c}{2}$$

24.3 Worked problems on the solution of triangles and their areas

Problem 1. In a triangle XYZ, $\angle X = 51°$, $\angle Y = 67°$ and $YZ = 15.2$ cm. Solve the triangle and find its area.

Fig. 24.2

The triangle XYZ is shown in Fig. 24.2. Since the angles in a triangle add up to 180°, then $z = 180° - 51° - 67° = \mathbf{62°}$.

Applying the sine rule:

$$\frac{15.2}{\sin 51°} = \frac{y}{\sin 67°} = \frac{z}{\sin 62°}$$

Using $\dfrac{15.2}{\sin 51°} = \dfrac{y}{\sin 67°}$ and transposing gives:

$$y = \frac{15.2 \sin 67°}{\sin 51°} = \mathbf{18.00\,cm} = XZ$$

Using $\dfrac{15.2}{\sin 51°} = \dfrac{z}{\sin 62°}$ and transposing gives:

$$z = \frac{15.2 \sin 62°}{\sin 51°} = \mathbf{17.27\,cm} = XY$$

Area of triangle XYZ

$$= \frac{1}{2}xy \sin Z = \frac{1}{2}(15.2)(18.00) \sin 62° = \mathbf{120.8\,cm^2}$$

(or area $= \dfrac{1}{2}xz \sin Y = \dfrac{1}{2}(15.2)(17.27) \sin 67° = \mathbf{120.8\,cm^2}$)

It is always worth checking with triangle problems that the longest side is opposite the largest angle, and vice-versa. In this problem, Y is the largest angle and XZ is the longest of the three sides.

Problem 2. Solve the triangle ABC given $B = 78°51'$, $AC = 22.31\,mm$ and $AB = 17.92\,mm$. Find also its area.

Triangle ABC is shown in Fig. 24.3.

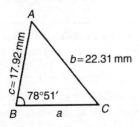

Fig. 24.3

Applying the sine rule:

$$\frac{22.31}{\sin 78°51'} = \frac{17.92}{\sin C}$$

from which, $\sin C = \dfrac{17.92 \sin 78°51'}{22.31} = 0.7881$

Hence $C = \sin^{-1} 0.7881 = 52°0'$ or $128°0'$ (see Chapters 17 and 18).

Since $B = 78°51'$, C cannot be $128°0'$, since $128°0' + 78°51'$ is greater than 180°.

Thus only $C = 52°0'$ is valid.

Angle $A = 180° - 78°51' - 52°0' = 49°9'$

Applying the sine rule:

$$\frac{a}{\sin 49°9'} = \frac{22.31}{\sin 78°51'}$$

from which, $a = \dfrac{22.31 \sin 49°9'}{\sin 78°51'} = 17.20\,mm$

Hence $A = \mathbf{49°9'}$, $C = \mathbf{52°0'}$ and $BC = \mathbf{17.20\,mm}$.

Area of triangle ABC

$$= \frac{1}{2}ac \sin B = \frac{1}{2}(17.20)(17.92) \sin 78°51'$$

$$= \mathbf{151.2\,mm^2}$$

Problem 3. Solve the triangle PQR and find its area given that $QR = 36.5\,mm$, $PR = 29.6\,mm$ and $\angle Q = 36°$

Triangle PQR is shown in Fig. 24.4.

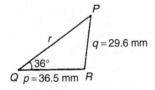

Fig. 24.4

Applying the sine rule:

$$\frac{29.6}{\sin 36°} = \frac{36.5}{\sin P}$$

from which, $\sin P = \dfrac{36.5 \sin 36°}{29.6} = 0.7248$

Hence $P = \sin^{-1} 0.7248 = 46°27'$ or $133°33'$

When $P = 46°27'$ and $Q = 36°$

then $R = 180° - 46°27' - 36° = 97°33'$

When $P = 133°33'$ and $Q = 36°$

then $R = 180° - 133°33' - 36° = 10°27'$

Thus, in this problem, there are **two** separate sets of results and both are feasible solutions. Such a situation is called the **ambiguous case**.

Case 1. $P = 46°27'$, $Q = 36°$, $R = 97°33'$, $p = 36.5\text{ mm}$ and $q = 29.6\text{ mm}$.

From the sine rule:

$$\frac{r}{\sin 97°33'} = \frac{29.6}{\sin 36°}$$

from which, $r = \dfrac{29.6 \sin 97°33'}{\sin 36°} = \textbf{49.92 mm}$

Area $= \dfrac{1}{2} pq \sin R = \dfrac{1}{2}(36.5)(29.6) \sin 97°33' = \textbf{535.5 mm}^2$.

Case 2. $P = 133°33'$, $Q = 36°$, $R = 10°27'$, $p = 36.5\text{ mm}$ and $q = 29.6\text{ mm}$.

From the sine rule:

$$\frac{r}{\sin 10°27'} = \frac{29.6}{\sin 36°}$$

from which, $r = \dfrac{29.6 \sin 10°27'}{\sin 36°} = \textbf{9.134 mm}$

Area $= \dfrac{1}{2} pq \sin R = \dfrac{1}{2}(36.5)(29.6) \sin 10°27' = \textbf{97.98 mm}^2$.

Triangle PQR for case 2 is shown in Fig. 24.5.

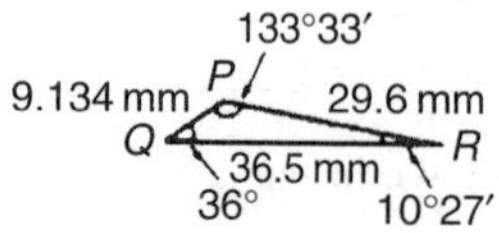

133°33'
9.134 mm P
Q 29.6 mm
36.5 mm R
36°   10°27'

Fig. 24.5

Now try the following exercise

Exercise 85   Further problems on the solution of triangles and their areas (Answers on page 262)

In Problems 1 and 2, use the sine rule to solve the triangles ABC and find their areas.

1. $A = 29°$, $B = 68°$, $b = 27\text{ mm}$.
2. $B = 71°26'$, $C = 56°32'$, $b = 8.60\text{ cm}$.

In Problems 3 and 4, use the sine rule to solve the triangles DEF and find their areas.

3. $d = 17\text{ cm}$, $f = 22\text{ cm}$, $F = 26°$
4. $d = 32.6\text{ mm}$, $e = 25.4\text{ mm}$, $D = 104°22'$

In Problems 5 and 6, use the sine rule to solve the triangles JKL and find their areas.

5. $j = 3.85\text{ cm}$, $k = 3.23\text{ cm}$, $K = 36°$
6. $k = 46\text{ mm}$, $l = 36\text{ mm}$, $L = 35°$

24.4   Further worked problems on the solution of triangles and their areas

Problem 4.   Solve triangle DEF and find its area given that $EF = 35.0\text{ mm}$, $DE = 25.0\text{ mm}$ and $\angle E = 64°$

Triangle DEF is shown in Fig. 24.6.

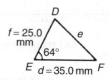

D
$f = 25.0$ mm
e
64°
E $d = 35.0$ mm F

Fig. 24.6

Applying the cosine rule:

$$e^2 = d^2 + f^2 - 2df \cos E$$

i.e. $e^2 = (35.0)^2 + (25.0)^2 - [2(35.0)(25.0) \cos 64°]$

$$= 1225 + 625 - 767.1 = 1083$$

from which, $\mathbf{e} = \sqrt{1083} = \textbf{32.91 mm}$

Applying the sine rule:

$$\frac{32.91}{\sin 64°} = \frac{25.0}{\sin F}$$

from which, $\sin F = \dfrac{25.0 \sin 64°}{32.91} = 0.6828$

Thus $\angle F = \sin^{-1} 0.6828 = 43°4'$ or $136°56'$

$F = 136°56'$ is not possible in this case since $136°56' + 64°$ is greater than $180°$. Thus only $F = \textbf{43°4'}$ is valid.

$$\angle D = 180° - 64° - 43°4' = \textbf{72°56'}$$

Area of triangle $DEF = \frac{1}{2} df \sin E$

$$= \frac{1}{2}(35.0)(25.0) \sin 64° = \textbf{393.2 mm}^2$$

Problem 5.   A triangle ABC has sides $a = 9.0\text{ cm}$, $b = 7.5\text{ cm}$ and $c = 6.5\text{ cm}$. Determine its three angles and its area.

Triangle ABC is shown in Fig. 24.7. It is usual first to calculate the largest angle to determine whether the triangle is acute or obtuse. In this case the largest angle is A (i.e. opposite the longest side).

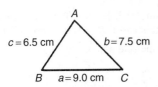

Fig. 24.7

Applying the cosine rule:

$$a^2 = b^2 + c^2 - 2bc \cos A$$

from which, $2bc \cos A = b^2 + c^2 - a^2$

and

$$\cos A = \frac{b^2 + c^2 - a^2}{2bc}$$

$$= \frac{7.5^2 + 6.5^2 - 9.0^2}{2(7.5)(6.5)} = 0.1795$$

Hence $A = \cos^{-1} 0.1795 = \mathbf{79°40'}$ (or $280°20'$, which is obviously impossible). The triangle is thus acute angled since $\cos A$ is positive. (If $\cos A$ had been negative, angle A would be obtuse, i.e. lie between $90°$ and $180°$).

Applying the sine rule:

$$\frac{9.0}{\sin 79°40'} = \frac{7.5}{\sin B}$$

from which, $\quad \sin B = \dfrac{7.5 \sin 79°40'}{9.0} = 0.8198$

Hence $B = \sin^{-1} 0.8198 = \mathbf{55°4'}$

and $\quad C = 180° - 79°40' - 55°4' = \mathbf{45°16'}$

Area $= \sqrt{[s(s-a)(s-b)(s-c)]}$, where

$$s = \frac{a+b+c}{2} = \frac{9.0 + 7.5 + 6.5}{2} = 11.5 \text{ cm}$$

Hence **area** $= \sqrt{[11.5(11.5 - 9.0)(11.5 - 7.5)(11.5 - 6.5)]}$

$$= \sqrt{[11.5(2.5)(4.0)(5.0)]} = \mathbf{23.98 \text{ cm}^2}$$

Alternatively, area $= \frac{1}{2} ab \sin C = \frac{1}{2}(9.0)(7.5) \sin 45°16'$

$$= \mathbf{23.98 \text{ cm}^2}$$

Problem 6. Solve triangle XYZ, shown in Fig. 24.8, and find its area given that $Y = 128°$, $XY = 7.2 \text{ cm}$ and $YZ = 4.5 \text{ cm}$.

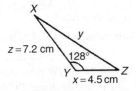

Fig. 24.8

Applying the cosine rule:

$$y^2 = x^2 + z^2 - 2xz \cos Y$$

$$= 4.5^2 + 7.2^2 - [2(4.5)(7.2) \cos 128°]$$

$$= 20.25 + 51.84 - [-39.89]$$

$$= 20.25 + 51.84 + 39.89 = 112.0$$

$$y = \sqrt{112.0} = \mathbf{10.58 \text{ cm}}$$

Applying the sine rule:

$$\frac{10.58}{\sin 128°} = \frac{7.2}{\sin Z}$$

from which, $\quad \sin Z = \dfrac{7.2 \sin 128°}{10.58} = 0.5363$

Hence $Z = \sin^{-1} 0.5363 = \mathbf{32°26'}$ (or $147°34'$ which, here, is impossible).

$X = 180° - 128° - 32°26' = \mathbf{19°34'}$

Area $= \dfrac{1}{2} xz \sin Y = \dfrac{1}{2}(4.5)(7.2) \sin 128° = \mathbf{12.77 \text{ cm}^2}$

Now try the following exercise

Exercise 86 **Further problems on the solution of triangles and their areas (Answers on page 262)**

In Problems 1 and 2, use the cosine and sine rules to solve the triangles PQR and find their areas.

1. $q = 12 \text{ cm}$, $r = 16 \text{ cm}$, $P = 54°$

2. $q = 3.25 \text{ m}$, $r = 4.42 \text{ m}$, $P = 105°$

In Problems 3 and 4, use the cosine and sine rules to solve the triangles XYZ and find their areas.

3. $x = 10.0 \text{ cm}$, $y = 8.0 \text{ cm}$, $z = 7.0 \text{ cm}$.

4. $x = 21 \text{ mm}$, $y = 34 \text{ mm}$, $z = 42 \text{ mm}$.

24.5 Practical situations involving trigonometry

There are a number of **practical situations** where the use of trigonometry is needed to find unknown sides and angles of triangles. This is demonstrated in the following worked problems.

Problem 7. A room 8.0 m wide has a span roof which slopes at $33°$ on one side and $40°$ on the other. Find the length of the roof slopes, correct to the nearest centimetre.

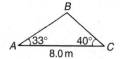

Fig. 24.9

A section of the roof is shown in Fig. 24.9.

Angle at ridge, $B = 180° - 33° - 40° = 107°$

From the sine rule:

$$\frac{8.0}{\sin 107°} = \frac{a}{\sin 33°}$$

from which, $a = \dfrac{8.0 \sin 33°}{\sin 107°} = 4.556\,\text{m}$

Also from the sine rule:

$$\frac{8.0}{\sin 107°} = \frac{c}{\sin 40°}$$

from which, $c = \dfrac{8.0 \sin 40°}{\sin 107°} = 5.377\,\text{m}$

Hence the roof slopes are 4.56 m and 5.38 m, correct to the nearest centimetre.

Problem 8. A man leaves a point walking at 6.5 km/h in a direction E 20° N (i.e. a bearing of 70°). A cyclist leaves the same point at the same time in a direction E 40° S (i.e. a bearing of 130°) travelling at a constant speed. Find the average speed of the cyclist if the walker and cyclist are 80 km apart after 5 hours.

After 5 hours the walker has travelled $5 \times 6.5 = 32.5\,\text{km}$ (shown as AB in Fig. 24.10). If AC is the distance the cyclist travels in 5 hours then $BC = 80\,\text{km}$.

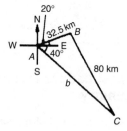

Fig. 24.10

Applying the sine rule:

$$\frac{80}{\sin 60°} = \frac{32.5}{\sin C}$$

from which, $\sin C = \dfrac{32.5 \sin 60°}{80} = 0.3518$

Hence $C = \sin^{-1} 0.3518 = 20°36'$ (or $159°24'$, which is impossible in this case). $B = 180° - 60° - 20°36' = 99°24'$.

Applying the sine rule again:

$$\frac{80}{\sin 60°} = \frac{b}{\sin 99°24'}$$

from which, $b = \dfrac{80 \sin 99°24'}{\sin 60°} = 91.14\,\text{km}$

Since the cyclist travels 91.14 km in 5 hours then

$$\textbf{average speed} = \frac{\text{distance}}{\text{time}} = \frac{91.14}{5} = \textbf{18.23 km/h}$$

Problem 9. Two voltage phasors are shown in Fig. 24.11. If $V_1 = 40$ V and $V_2 = 100$ V determine the value of their resultant (i.e. length OA) and the angle the resultant makes with V_1

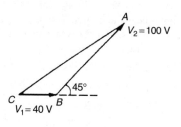

Fig. 24.11

Angle $OBA = 180° - 45° = 135°$

Applying the cosine rule:

$$
\begin{aligned}
OA^2 &= V_1^2 + V_2^2 - 2V_1V_2 \cos OBA \\
&= 40^2 + 100^2 - \{2(40)(100)\cos 135°\} \\
&= 1600 + 10\,000 - \{-5657\} \\
&= 1600 + 10\,000 + 5657 = 17\,257
\end{aligned}
$$

The resultant $OA = \sqrt{17\,257} = 131.4\,\text{V}$

Applying the sine rule:

$$\frac{131.4}{\sin 135°} = \frac{100}{\sin AOB}$$

from which, $\sin AOB = \dfrac{100 \sin 135°}{131.4} = 0.5381$

Hence angle $AOB = \sin^{-1} 0.5381 = 32°33'$ (or $147°27'$, which is impossible in this case).

Hence the resultant voltage is 131.4 volts at 32°33' to V_1

Problem 10. In Fig. 24.12, PR represents the inclined jib of a crane and is 10.0 m long. PQ is 4.0 m long. Determine the inclination of the jib to the vertical and the length of tie QR.

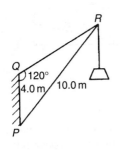

Fig. 24.12

Applying the sine rule:

$$\frac{PR}{\sin 120°} = \frac{PQ}{\sin R}$$

from which, $\quad \sin R = \dfrac{PQ \sin 120°}{PR}$

$$= \frac{(4.0) \sin 120°}{10.0} = 0.3464$$

Hence $\angle R = \sin^{-1} 0.3464 = 20°16'$ (or $159°44'$, which is impossible in this case).

$\angle P = 180° - 120° - 20°16' = \mathbf{39°44'}$, **which is the inclination of the jib to the vertical.**

Applying the sine rule:

$$\frac{10.0}{\sin 120°} = \frac{QR}{\sin 39°44'}$$

from which, **length of tie, QR** $= \dfrac{10.0 \sin 39°44'}{\sin 120°}$

$$= \mathbf{7.38\,m}$$

Now try the following exercise

Exercise 87 **Further problems on practical situations involving trigonometry (Answers on page 262)**

1. A ship P sails at a steady speed of 45 km/h in a direction of W 32° N (i.e. a bearing of 302°) from a port. At the same time another ship Q leaves the port at a steady speed of 35 km/h in a direction N 15° E (i.e. a bearing of 015°). Determine their distance apart after 4 hours.

2. Two sides of a triangular plot of land are 52.0 m and 34.0 m, respectively. If the area of the plot is 620 m² find (a) the length of fencing required to enclose the plot and (b) the angles of the triangular plot.

3. A jib crane is shown in Fig. 24.13. If the tie rod PR is 8.0 long and PQ is 4.5 m long determine (a) the length of jib RQ and (b) the angle between the jib and the tie rod.

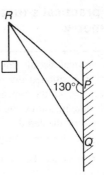

Fig. 24.13

4. A building site is in the form of a quadrilateral as shown in Fig. 24.14, and its area is 1510 m². Determine the length of the perimeter of the site.

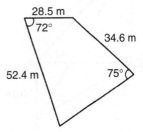

Fig. 24.14

5. Determine the length of members BF and EB in the roof truss shown in Fig. 24.15.

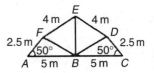

Fig. 24.15

6. A laboratory 9.0 m wide has a span roof which slopes at 36° on one side and 44° on the other. Determine the lengths of the roof slopes.

7. PQ and QR are the phasors representing the alternating currents in two branches of a circuit. Phasor PQ is 20.0 A and is horizontal. Phasor QR (which is joined to the end of PQ to form triangle PQR) is 14.0 A and is at an angle of 35° to the horizontal. Determine the resultant phasor PR and the angle it makes with phasor PQ.

24.6   Further practical situations involving trigonometry

> *Problem 11.* A vertical aerial stands on horizontal ground. A surveyor positioned due east of the aerial measures the elevation of the top as 48°. He moves due south 30.0 m and measures the elevation as 44°. Determine the height of the aerial.

In Fig. 24.16, DC represents the aerial, A is the initial position of the surveyor and B his final position. From triangle ACD, $\tan 48° = \dfrac{DC}{AC}$, from which $AC = \dfrac{DC}{\tan 48°}$

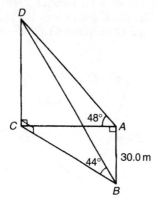

Fig. 24.16

Similarly, from triangle BCD, $BC = \dfrac{DC}{\tan 44°}$

For triangle ABC, using Pythagoras' theorem:

$$BC^2 = AB^2 + AC^2$$

$$\left(\frac{DC}{\tan 44°}\right)^2 = (30.0)^2 + \left(\frac{DC}{\tan 48°}\right)^2$$

$$DC^2\left(\frac{1}{\tan^2 44°} - \frac{1}{\tan^2 48°}\right) = 30.0^2$$

$$DC^2(1.072323 - 0.810727) = 30.0^2$$

$$DC^2 = \frac{30.0^2}{0.261596} = 3440.4$$

Hence, height of aerial, $DC = \sqrt{3340.4} = 58.65$ m.

> *Problem 12.* A crank mechanism of a petrol engine is shown in Fig. 24.17. Arm OA is 10.0 cm long and rotates clockwise about 0. The connecting rod AB is 30.0 cm long and end B is constrained to move horizontally.

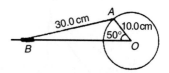

Fig. 24.17

(a) For the position shown in Fig. 24.17 determine the angle between the connecting rod AB and the horizontal and the length of OB.

(b) How far does B move when angle AOB changes from 50° to 120°?

(a) Applying the sine rule:

$$\frac{AB}{\sin 50°} = \frac{AO}{\sin B}$$

from which, $\sin B = \dfrac{AO \sin 50°}{AB} = \dfrac{10.0 \sin 50°}{30.0}$

$$= 0.2553$$

Hence $B = \sin^{-1} 0.2553 = 14°47'$ (or 165°13′, which is impossible in this case).

Hence the connecting rod AB makes an angle of 14°47′ with the horizontal.

Angle $OAB = 180° - 50° - 14°47' = 115°13'$

Applying the sine rule:

$$\frac{30.0}{\sin 50°} = \frac{OB}{\sin 115°13'}$$

from which, $\quad OB = \dfrac{30.0 \sin 115°13'}{\sin 50°} = \mathbf{35.43\ cm}$

(b) Figure 24.18 shows the initial and final positions of the crank mechanism. In triangle $OA'B'$, applying the sine rule:

$$\frac{30.0}{\sin 120°} = \frac{10.0}{\sin A'B'O}$$

from which, $\sin A'B'O = \dfrac{10.0 \sin 120°}{30.0} = 0.2887$

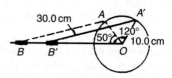

Fig. 24.18

Hence $A'B'O = \sin^{-1} 0.2887 = 16°47'$ (or 163°13′ which is impossible in this case).

Angle $OA'B' = 180° - 120° - 16°47' = 43°13'$

Applying the sine rule:

$$\frac{30.0}{\sin 120°} = \frac{OB'}{\sin 43°13'}$$

from which, $\quad OB' = \dfrac{30.0 \sin 43°13'}{\sin 120°} = 23.72\,\text{cm}$

Since $OB = 35.43\,\text{cm}$ and $OB' = 23.72\,\text{cm}$ then

$$BB' = 35.43 - 23.72 = 11.71\,\text{cm}$$

Hence B moves 11.71 cm when angle AOB changes from $50°$ to $120°$

Problem 13. The area of a field is in the form of a quadrilateral $ABCD$ as shown in Fig. 24.19. Determine its area.

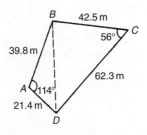

Fig. 24.19

A diagonal drawn from B to D divides the quadrilateral into two triangles.

Area of quadrilateral $ABCD$

$= $ area of triangle $ABD + $ area of triangle BCD

$= \frac{1}{2}(39.8)(21.4)\sin 114° + \frac{1}{2}(42.5)(62.3)\sin 56°$

$= 389.04 + 1097.5 = \mathbf{1487\,m^2}$

Now try the following exercise

Exercise 88   Further problems on practical situations involving trigonometry (Answers on page 262)

1. Three forces acting on a fixed point are represented by the sides of a triangle of dimensions 7.2 cm, 9.6 cm and 11.0 cm. Determine the angles between the lines of action and the three forces.
2. A vertical aerial AB, 9.60 m high, stands on ground which is inclined $12°$ to the horizontal. A stay connects the top of the aerial A to a point C on the ground 10.0 m downhill from B, the foot of the aerial. Determine (a) the length of the stay, and (b) the angle the stay makes with the ground.
3. A reciprocating engine mechanism is shown in Fig. 24.20. The crank AB is 12.0 cm long and the

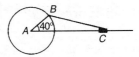

Fig. 24.20

connecting rod BC is 32.0 cm long. For the position shown determine the length of AC and the angle between the crank and the connecting rod.
4. From Fig. 22.20, determine how far C moves, correct to the nearest millimetre when angle CAB changes from $40°$ to $160°$, B moving in an anticlockwise direction.
5. A surveyor, standing W $25°$ S of a tower measures the angle of elevation of the top of the tower as $46°30'$. From a position E $23°$ S from the tower the elevation of the top is $37°15'$. Determine the height of the tower if the distance between the two observations is 75 m.
6. Calculate, correct to 3 significant figures, the co-ordinates x and y to locate the hole centre at P shown in Fig. 24.21.

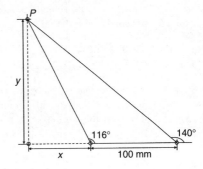

Fig. 24.21

7. An idler gear, 30 mm in diameter, has to be fitted between a 70 mm diameter driving gear and a 90 mm diameter driven gear as shown in Fig. 24.22. Determine the value of angle θ between the centre lines.

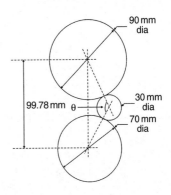

Fig. 24.22

Assignment 12

This assignment covers the material in Chapters 23 and 24. The marks for each question are shown in brackets at the end of each question.

1. Plot a graph of $y = 3x^2 + 5$ from $x = 1$ to $x = 4$. Estimate, correct to 2 decimal places, using 6 intervals, the area enclosed by the curve, the ordinates $x = 1$ and $x = 4$, and the x-axis by (a) the trapezoidal rule, (b) the mid-ordinate rule, and (c) Simpson's rule. (12)

2. A circular cooling tower is 20 m high. The inside diameter of the tower at different heights is given in the following table:

Height (m)	0	5.0	10.0	15.0	20.0
Diameter (m)	16.0	13.3	10.7	8.6	8.0

Determine the area corresponding to each diameter and hence estimate the capacity of the tower in cubic metres. (7)

3. A vehicle starts from rest and its velocity is measured every second for 6 seconds, with the following results:

Time t (s)	0	1	2	3	4	5	6
Velocity v (m/s)	0	1.2	2.4	3.7	5.2	6.0	9.2

Using Simpson's rule, calculate (a) the distance travelled in 6 s (i.e. the area under the v/t graph) and (b) the average speed over this period. (6)

4. A triangular plot of land ABC is shown in Fig. A12.1. Solve the triangle and determine its area. (10)

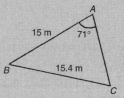

Fig. A12.1

5. A car is travelling 20 m above sea level. It then travels 500 m up a steady slope of 17°. Determine, correct to the nearest metre, how high the car is now above see level. (3)

6. Figure A12.2 shows a roof truss PQR with rafter $PQ = 3$ m. Calculate the length of (a) the roof rise PP', (b) rafter PR, and (c) the roof span QR. Find also (d) the cross-sectional area of the roof truss. (12)

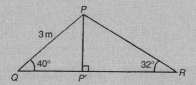

Fig. A12.2

25

Vectors

25.1 Introduction

Some physical quantities are entirely defined by a numerical value and are called **scalar quantities** or **scalars**. Examples of scalars include time, mass, temperature, energy and volume. Other physical quantities are defined by both a numerical value and a direction in space and these are called **vector quantities** or **vectors**. Examples of vectors include force, velocity, moment and displacement.

25.2 Vector addition

A vector may be represented by a straight line, the length of line being directly proportional to the magnitude of the quantity and the direction of the line being in the same direction as the line of action of the quantity. An arrow is used to denote the sense of the vector, that is, for a horizontal vector, say, whether it acts from left to right or vice-versa. The arrow is positioned at the end of the vector and this position is called the 'nose' of the vector. Figure 25.1 shows a velocity of 20 m/s at an angle of 45° to the horizontal and may be depicted by $oa = 20$ m/s at 45° to the horizontal.

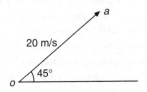

Fig. 25.1

To distinguish between vector and scalar quantities, various ways are used. These include:

(i) **bold print**,

(ii) two capital letters with an arrow above them to denote the sense of direction, e.g. $\overrightarrow{AB}$, where A is the starting point and B the end point of the vector,

(iii) a line over the top of letters, e.g. $\overline{AB}$ or $\bar{a}$,

(iv) letters with an arrow above, e.g. $\vec{a}$, $\vec{A}$,

(v) underlined letters, e.g. $\underline{a}$,

(vi) $xi + jy$, where i and j are axes at right-angles to each other; for example, $3i + 4j$ means 3 units in the i direction and 4 units in the j direction, as shown in Fig. 25.2.

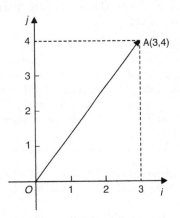

Fig. 25.2

(vii) a column matrix $\begin{pmatrix} a \\ b \end{pmatrix}$; for example, the vector OA shown in Fig. 25.2 could be represented by $\begin{pmatrix} 3 \\ 4 \end{pmatrix}$

Thus, in Fig. 25.2, $OA \equiv \overrightarrow{OA} \equiv \overline{OA} \equiv 3i + 4j \equiv \begin{pmatrix} 3 \\ 4 \end{pmatrix}$

The one adopted in this text is to denote vector quantities in **bold print**. Thus, *oa* represents a vector quantity, but *oa* is the magnitude of the vector *oa*. Also, positive angles are measured in an anticlockwise direction from a horizontal, right facing line and negative angles in a clockwise direction from this line – as with graphical work. Thus 90° is a line vertically upwards and −90° is a line vertically downwards.

The resultant of adding two vectors together, say V_1 at an angle θ_1 and V_2 at angle $(-\theta_2)$, as shown in Fig. 25.3(a), can be obtained by drawing *oa* to represent V_1 and then drawing *ar* to represent V_2. The resultant of $V_1 + V_2$ is given by *or*. This is shown in Fig. 25.3(b), the vector equation being *oa + ar = or*. This is called the **'nose-to-tail' method** of vector addition.

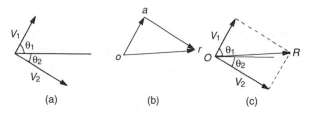

(a)　　　　(b)　　　　(c)

Fig. 25.3

Alternatively, by drawing lines parallel to V_1 and V_2 from the noses of V_2 and V_1, respectively, and letting the point of intersection of these parallel lines be *R*, gives *OR* as the magnitude and direction of the resultant of adding V_1 and V_2, as shown in Fig. 25.3(c). This is called the **'parallelogram' method** of vector addition.

> *Problem 1.* A force of 4 N is inclined at an angle of 45° to a second force of 7 N, both forces acting at a point. Find the magnitude of the resultant of these two forces and the direction of the resultant with respect to the 7 N force by both the 'triangle' and the 'parallelogram' methods.

The forces are shown in Fig. 25.4(a). Although the 7 N force is shown as a horizontal line, it could have been drawn in any direction.

Using the **'nose-to-tail' method**, a line 7 units long is drawn horizontally to give vector *oa* in Fig. 25.4(b). To the nose of this vector *ar* is drawn 4 units long at an angle of 45° to *oa*. The resultant of vector addition is *or* and by measurement is **10.2 units long and at an angle of 16° to the 7 N force**.

Figure 25.4(c) uses the **'parallelogram' method** in which lines are drawn parallel to the 7 N and 4 N forces from the noses of the 4 N and 7 N forces, respectively. These intersect

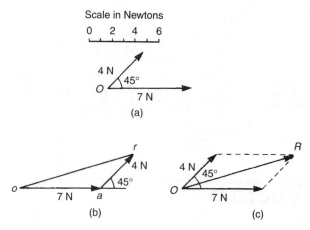

(a)

(b)　　　　　　(c)

Fig. 25.4

at *R*. Vector *OR* gives the magnitude and direction of the resultant of vector addition and as obtained by the 'nose-to-tail' method is **10.2 units long at an angle of 16° to the 7 N force**.

> *Problem 2.* Use a graphical method to determine the magnitude and direction of the resultant of the three velocities shown in Fig. 25.5.

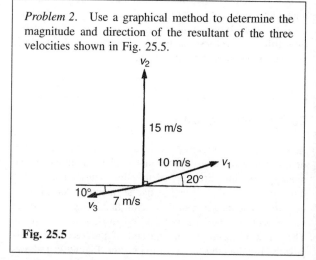

Fig. 25.5

Often it is easier to use the 'nose-to-tail' method when more than two vectors are being added. The order in which the vectors are added is immaterial. In this case the order taken is v_1, then v_2, then v_3 but just the same result would have been obtained if the order had been, say, v_1, v_3 and finally v_2. v_1 is drawn 10 units long at an angle of 20° to the horizontal, shown by *oa* in Fig. 25.6. v_2 is added to v_1 by drawing a line 15 units long vertically upwards from *a*, shown as *ab*. Finally, v_3 is added to $v_1 + v_2$ by drawing a line 7 units long at an angle at 190° from *b*, shown as *br*. The resultant of vector addition is *or* and by measurement is 17.5 units long at an angle of 82° to the horizontal.

Thus $v_1 + v_2 + v_3 = $ **17.5 m/s at 82° to the horizontal**.

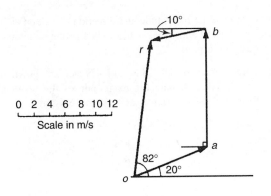

Fig. 25.6

25.3 Resolution of vectors

A vector can be resolved into two component parts such that the vector addition of the component parts is equal to the original vector. The two components usually taken are a horizontal component and a vertical component. For the vector shown as F in Fig. 25.7, the horizontal component is $F\cos\theta$ and the vertical component is $F\sin\theta$.

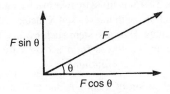

Fig. 25.7

For the vectors F_1 and F_2 shown in Fig. 25.8, the horizontal component of vector addition is:

$$H = F_1\cos\theta_1 + F_2\cos\theta_2$$

and the vertical component of vector addition is:

$$V = F_1\sin\theta_1 + F_2\sin\theta_2$$

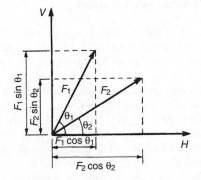

Fig. 25.8

Having obtained H and V, the magnitude of the resultant vector R is given by $\sqrt{H^2 + V^2}$ and its angle to the horizontal is given by $\tan^{-1}(V/H)$

Problem 3. Resolve the acceleration vector of $17\,\text{m/s}^2$ at an angle of $120°$ to the horizontal into a horizontal and a vertical component.

For a vector A at angle θ to the horizontal, the horizontal component is given by $A\cos\theta$ and the vertical component by $A\sin\theta$. Any convention of signs may be adopted, in this case horizontally from left to right is taken as positive and vertically upwards is taken as positive.

Horizontal component $H = 17\cos120° = -8.50\,\text{m/s}^2$, acting from left to right.

Vertical component $V = 17\sin120° = 14.72\,\text{m/s}^2$, acting vertically upwards.

These component vectors are shown in Fig. 25.9.

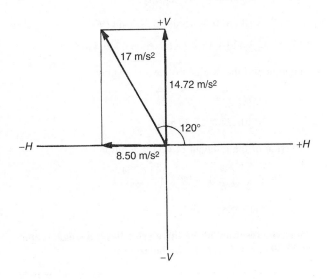

Fig. 25.9

Problem 4. Calculate the resultant force of the two forces given in Problem 1.

With reference to Fig. 25.4(a):

Horizontal component of force, $H = 7\cos0° + 4\cos45°$

$$= 7 + 2.828 = 9.828\,\text{N}$$

Vertical component of force, $V = 7\sin0° + 4\sin45°$

$$= 0 + 2.828 = 2.828\,\text{N}$$

The magnitude of the resultant of vector addition

$$= \sqrt{H^2 + V^2} = \sqrt{9.828^2 + 2.828^2}$$

$$= \sqrt{104.59} = \mathbf{10.23\,N}$$

The direction of the resultant of vector addition

$$= \tan^{-1}\left(\frac{V}{H}\right) = \tan^{-1}\left(\frac{2.828}{9.828}\right) = \mathbf{16.05°}$$

Thus, the resultant of the two forces is a single vector of 10.23 N at 16.05° to the 7 N vector.

Problem 5. Calculate the resultant velocity of the three velocities given in Problem 2.

With reference to Fig. 25.5:

Horizontal component of the velocity,

$$H = 10\cos 20° + 15\cos 90° + 7\cos 190°$$

$$= 9.397 + 0 + (-6.894) = \mathbf{2.503\,m/s}$$

Vertical component of the velocity,

$$V = 10\sin 20° + 15\sin 90° + 7\sin 190°$$

$$= 3.420 + 15 + (-1.216) = \mathbf{17.204\,m/s}$$

Magnitude of the resultant of vector addition

$$= \sqrt{H^2 + V^2} = \sqrt{2.503^2 + 17.204^2}$$

$$= \sqrt{302.24} = \mathbf{17.39\,m/s}$$

Direction of the resultant of vector addition

$$= \tan^{-1}\left(\frac{V}{H}\right) = \tan^{-1}\left(\frac{17.204}{2.503}\right)$$

$$= \tan^{-1} 6.8734 = \mathbf{81.72°}$$

Thus, the resultant of the three velocities is a single vector of 17.39 m/s at 81.72° to the horizontal

---

Now try the following exercise

Exercise 89   Further problems on vectors (Answers on page 262)

1. Forces of 23 N and 41 N act at a point and are inclined at 90° to each other. Find, by drawing, the resultant force and its direction relative to the 41 N force.

2. Forces A, B and C are coplanar and act at a point. Force A is 12 kN at 90°, B is 5 kN at 180° and C is 13 kN at 293°. Determine graphically the resultant force.

3. Calculate the magnitude and direction of velocities of 3 m/s at 18° and 7 m/s at 115° when acting simultaneously on a point.

4. Three forces of 2 N, 3 N and 4 N act as shown in Fig. 25.10. Calculate the magnitude of the resultant force and its direction relative to the 2 N force.

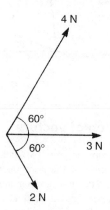

Fig. 25.10

5. A load of 5.89 N is lifted by two strings, making angles of 20° and 35° with the vertical. Calculate the tensions in the strings. [For a system such as this, the vectors representing the forces form a closed triangle when the system is in equilibrium].

6. The acceleration of a body is due to four component, coplanar accelerations. These are 2 m/s² due north, 3 m/s² due east, 4 m/s² to the south-west and 5 m/s² to the south-east. Calculate the resultant acceleration and its direction.

7. A current phasor i_1 is 5 A and horizontal. A second phasor i_2 is 8 A and is at 50° to the horizontal. Determine the resultant of the two phasors, $i_1 + i_2$, and the angle the resultant makes with current i_1.

25.4   Vector subtraction

In Fig. 25.11, a force vector F is represented by oa. The vector $(-oa)$ can be obtained by drawing a vector from o in the opposite sense to oa but having the same magnitude, shown as ob in Fig. 25.11, i.e. $ob = (-oa)$.

For two vectors acting at a point, as shown in Fig. 25.12(a), the resultant of vector addition is $os = oa + ob$. Fig. 25.12(b) shows vectors $ob + (-oa)$, that is, $ob - oa$ and the vector equation is $ob - oa = od$. Comparing od in Fig. 25.12(b) with the broken line ab in Fig. 25.12(a) shows that the second diagonal of the 'parallelogram' method of vector addition gives the magnitude and direction of vector subtraction of oa from ob.

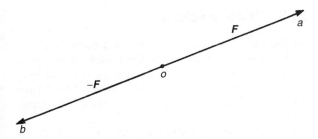

Fig. 25.11

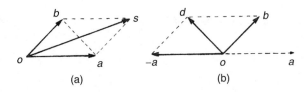

(a)   (b)

Fig. 25.12

Problem 6. Accelerations of $a_1 = 1.5\,\text{m/s}^2$ at $90°$ and $a_2 = 2.6\,\text{m/s}^2$ at $145°$ act at a point. Find $a_1 + a_2$ and $a_1 - a_2$ by (i) drawing a scale vector diagram and (ii) by calculation.

(i) The scale vector diagram is shown in Fig. 25.13. By measurement,

$$a_1 + a_2 = 3.7\,\text{m/s}^2 \text{ at } 126°$$

$$a_1 - a_2 = 2.1\,\text{m/s}^2 \text{ at } 0°$$

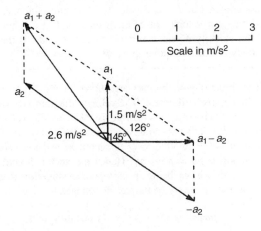

Fig. 25.13

(ii) Resolving horizontally and vertically gives:

Horizontal component of $a_1 + a_2$,

$$H = 1.5\cos 90° + 2.6\cos 145° = -2.13$$

Vertical component of $a_1 + a_2$,

$$V = 1.5\sin 90° + 2.6\sin 145° = 2.99$$

Magnitude of

$$a_1 + a_2 = \sqrt{(-2.13)^2 + 2.99^2} = \textbf{3.67 m/s}^2$$

Direction of $a_1 + a_2 = \tan^{-1}\left(\dfrac{2.99}{-2.13}\right)$ and must lie in the second quadrant since H is negative and V is positive.

$\tan^{-1}\left(\dfrac{2.99}{-2.13}\right) = -54.53°$, and for this to be in the second quadrant, the true angle is $180°$ displaced, i.e. $180° - 54.53°$ or $125.47°$

Thus $a_1 + a_2 = \textbf{3.67 m/s}^2 \text{ at } \textbf{125.47}°$.

Horizontal component of $a_1 - a_2$, that is, $a_1 + (-a_2)$

$$= 1.5\cos 90° + 2.6\cos(145° - 180°)$$

$$= 2.6\cos(-35°) = 2.13$$

Vertical component of $a_1 - a_2$, that is, $a_1 + (-a_2)$

$$= 1.5\sin 90° + 2.6\sin(-35°) = 0$$

Magnitude of $a_1 - a_2 = \sqrt{2.13^2 + 0^2} = 2.13$ m/s^2

Direction of $a_1 - a_2 = \tan^{-1}\left(\dfrac{0}{2.13}\right) = 0°$

Thus $a_1 - a_2 = \textbf{2.13 m/s}^2 \text{ at } \textbf{0}°$

Problem 7. Calculate the resultant of (i) $v_1 - v_2 + v_3$ and (ii) $v_2 - v_1 - v_3$ when $v_1 = 22$ units at $140°$, $v_2 = 40$ units at $190°$ and $v_3 = 15$ units at $290°$

(i) The vectors are shown in Fig. 25.14.

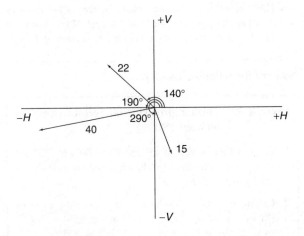

Fig. 25.14

The horizontal component of $v_1 - v_2 + v_3$

$$= (22 \cos 140°) - (40 \cos 190°) + (15 \cos 290°)$$

$$= (-16.85) - (-39.39) + (5.13) = \textbf{27.67 units}$$

The vertical component of $v_1 - v_2 + v_3$

$$= (22 \sin 140°) - (40 \sin 190°) + (15 \sin 290°)$$

$$= (14.14) - (-6.95) + (-14.10) = \textbf{6.99 units}$$

The magnitude of the resultant, R, which can be represented by the mathematical symbol for 'the **modulus** of' as $|v_1 - v_2 + v_3|$ is given by:

$$|R| = \sqrt{27.67^2 + 6.99^2} = \textbf{28.54 units}$$

The direction of the resultant, R, which can be represented by the mathematical symbol for 'the **argument** of' as arg $(v_1 - v_2 + v_3)$ is given by:

$$\textbf{arg } R = \tan^{-1}\left(\frac{6.99}{27.67}\right) = 14.18°$$

Thus $v_1 - v_2 + v_3 = \textbf{28.54 units at 14.18°}$

(ii) The horizontal component of $v_2 - v_1 - v_3$

$$= (40 \cos 190°) - (22 \cos 140°) - (15 \cos 290°)$$

$$= (-39.39) - (-16.85) - (5.13) = \textbf{-27.67 units}$$

The vertical component of $v_2 - v_1 - v_3$

$$= (40 \sin 190°) - (22 \sin 140°) - (15 \sin 290°)$$

$$= (-6.95) - (14.14) - (-14.10) = \textbf{-6.99 units}$$

Let $R = v_2 - v_1 - v_3$ then $|R| = \sqrt{(-27.67)^2 + (-6.99)^2}$ $= 28.54$ units and $\textbf{arg } R = \tan^{-1}\left(\dfrac{-6.99}{-27.67}\right)$ and must lie in the third quadrant since both H and V are negative quantities.

$\text{Tan}^{-1}\left(\dfrac{-6.99}{-27.67}\right) = 14.18°$, hence the required angle is $180° + 14.18° = 194.18°$

Thus $v_2 - v_1 - v_3 = \textbf{28.54 units at 194.18°}$

This result is as expected, since $v_2 - v_1 - v_3 = -(v_1 - v_2 + v_3)$ and the vector 28.54 units at 194.18° is minus times the vector 28.54 units at 14.18°

Now try the following exercise

Exercise 90   Further problems on vectors (Answers on page 262)

1. Forces of $F_1 = 40$ N at 45° and $F_2 = 30$ N at 125° act at a point. Determine by drawing and by calculation (a) $F_1 + F_2$ (b) $F_1 - F_2$

2. Calculate the resultant of (a) $v_1 + v_2 - v_3$ (b) $v_3 - v_2 + v_1$ when $v_1 = 15$ m/s at 85°, $v_2 = 25$ m/s at 175° and $v_3 = 12$ m/s at 235°

25.5   Relative velocity

For relative velocity problems, some fixed datum point needs to be selected. This is often a fixed point on the earth's surface. In any vector equation, only the start and finish points affect the resultant vector of a system. Two different systems are shown in Fig. 25.15, but in each of the systems, the resultant vector is ad.

The vector equation of the system shown in Fig. 25.15(a) is:

$$ad = ab + bd$$

and that for the system shown in Fig. 25.15(b) is:

$$ad = ab + bc + cd$$

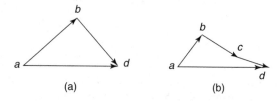

(a)   (b)

Fig. 25.15

Thus in vector equations of this form, only the first and last letters, a and d, respectively, fix the magnitude and direction of the resultant vector. This principle is used in relative velocity problems.

Problem 8.   Two cars, P and Q, are travelling towards the junction of two roads which are at right angles to one another. Car P has a velocity of 45 km/h due east and car Q a velocity of 55 km/h due south. Calculate (i) the velocity of car P relative to car Q, and (ii) the velocity of car Q relative to car P.

(i) The directions of the cars are shown in Fig. 25.16 (a), called a **space diagram**. The velocity diagram is shown in Fig. 25.16 (b), in which pe is taken as the velocity of car P relative to point e on the earth's surface. The velocity of P relative to Q is vector pq and the vector equation is $pq = pe + eq$. Hence the vector directions are as shown, eq being in the opposite direction to qe. From the geometry of the vector triangle,

$$|pq| = \sqrt{45^2 + 55^2} = 71.06 \text{ km/h and}$$

$$\text{arg } pq = \tan^{-1}\left(\frac{55}{45}\right) = 50.71°$$

i.e., the velocity of car P relative to car Q is 71.06 km/h at 50.71°

(ii) The velocity of car Q relative to car P is given by the vector equation $qp = qe + ep$ and the vector diagram

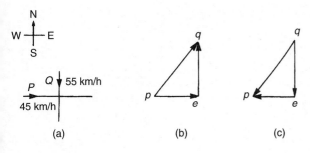

Fig. 25.16

is as shown in Fig. 25.16(c), having **ep** opposite in direction to **pe**. From the geometry of this vector triangle:

$$|qp| = \sqrt{45^2 + 55^2} = 71.06 \, \text{km/h and}$$

$$\arg qp = \tan^{-1}\left(\frac{55}{45}\right) = 50.71°$$

but must lie in the third quadrant, i.e., the required angle is $180° + 50.71° = 230.71°$

Thus the velocity of car Q relative to car P is 71.06 km/h at 230.71°

Now try the following exercise

Exercise 91   Further problems on relative velocity (Answers on page 263)

1. A car is moving along a straight horizontal road at 79.2 km/h and rain is falling vertically downwards at 26.4 km/h. Find the velocity of the rain relative to the driver of the car.

2. Calculate the time needed to swim across a river 142 m wide when the swimmer can swim at 2 km/h in still water and the river is flowing at 1 km/h.

26

Number sequences

26.1 Simple sequences

A set of numbers which are connected by a definite law is called a **series** or a **sequence of numbers**. Each of the numbers in the series is called a **term** of the series.

For example, 1, 3, 5, 7, ... is a series obtained by adding 2 to the previous term, and 2, 8, 32, 128, ... is a sequence obtained by multiplying the previous term by 4.

Problem 1. Determine the next two terms in the series: 3, 6, 9, 12, ...

We notice that the sequence 3, 6, 9, 12, ... progressively increases by 3, thus the next two terms will be **15 and 18**

Problem 2. Find the next three terms in the series: 9, 5, 1, ...

We notice that each term in the series 9, 5, 1, ... progressively decreases by 4, thus the next two terms will be $1 - 4$, i.e. **−3** and $-3 - 4$, i.e. **−7**

Problem 3. Determine the next two terms in the series: 2, 6, 18, 54, ...

We notice that the second term, 6, is three times the first term, the third term, 18, is three times the second term, and that the fourth term, 54, is three times the third term. Hence the fifth term will be $3 \times 54 = $ **162** and the sixth term will be $3 \times 162 = $ **486**

Now try the following exercise

Exercise 92 Further problems on simple sequences (Answers on page 263)

Determine the next two terms in each of the following series:

1. 5, 9, 13, 17, ...   2. 3, 6, 12, 24, ...

3. 112, 56, 28, ...   4. 12, 7, 2, ...

5. 2, 5, 10, 17, 26, 37, ... 6. 1, 0.1, 0.01, ...

7. 4, 9, 19, 34, ...

26.2 The n'th term of a series

If a series is represented by a general expression, say, $2n + 1$, where n is an integer (i.e. a whole number), then by substituting $n = 1, 2, 3, \ldots$ the terms of the series can be determined; in this example, the first three terms will be:

$$2(1) + 1, 2(2) + 1, 2(3) + 1, \ldots, \quad \text{i.e.} \quad 3, 5, 7, \ldots$$

What is the n'th term of the sequence 1, 3, 5, 7, ... ? Firstly, we notice that the gap between each term is 2, hence the law relating the numbers is:

$$\text{`}2n + \text{something'}$$

The second term, $3 = 2n + \text{something}$,

hence when $n = 2$ (i.e. the second term of the series), then $3 = 4 + \text{something}$ and the 'something' must be -1. Thus **the n'th term of 1, 3, 5, 7, ... is $2n - 1$**. Hence the fifth term is given by $2(5) - 1 = 9$, and the twentieth term is $2(20) - 1 = 39$, and so on.

> *Problem 4.* The n'th term of a sequence is given by $3n + 1$. Write down the first four terms.

The first four terms of the series $3n + 1$ will be:

$$3(1) + 1, 3(2) + 1, 3(3) + 1 \quad \text{and} \quad 3(4) + 1$$

i.e. **4, 7, 10 and 13**

> *Problem 5.* The n'th term of a series is given by $4n - 1$. Write down the first four terms.

The first four terms of the series $4n - 1$ will be:

$$4(1) - 1, 4(2) - 1, 4(3) - 1 \quad \text{and} \quad 4(4) - 1$$

i.e. **3, 7, 11 and 15**

> *Problem 6.* Find the n'th term of the series: $1, 4, 7, \ldots$.

We notice that the gap between each of the given three terms is 3, hence the law relating the numbers is:

$$\text{'}3n + \text{something'}$$

The second term, $\quad 4 = 3n + \text{something}$,

so when $n = 2$, then $4 = 6 + \text{something}$,

so the 'something' must be -2 (from simple equations).

Thus **the n'th term of the series 1, 4, 7, $\ldots$ is: $3n - 2$**

> *Problem 7.* Find the n'th term of the sequence: 3, 9, 15, 21, $\ldots$. Hence determine the 15th term of the series.

We notice that the gap between each of the given four terms is 6, hence the law relating the numbers is:

$$\text{'}6n + \text{something'}$$

The second term, $\quad 9 = 6n + \text{something}$,

so when $n = 2$, then $9 = 12 + \text{something}$,

so the 'something' must be -3

Thus **the n'th term of the series 3, 9, 15, 21, $\ldots$ is: $6n - 3$**

The 15th term of the series is given by $6n - 3$ when $n = 15$.

Hence the 15th term of the series 3, 9, 15, 21, $\ldots$ is: $6(15) - 3 = 87$

> *Problem 8.* Find the n'th term of the series: 1, 4, 9, 16, 25, $\ldots$.

This is a special series and does not follow the pattern of the previous examples. Each of the terms in the given series are square numbers,

i.e. $\quad 1, 4, 9, 16, 25, \ldots \equiv 1^2, 2^2, 3^2, 4^2, 5^2, \ldots$

Hence the n'th term is: n^2

Now try the following exercise

> **Exercise 93** **Further problems on the n'th term of a series (Answers on page 263)**
>
> 1. The n'th term of a sequence is given by $2n - 1$. Write down the first four terms.
> 2. The n'th term of a sequence is given by $3n + 4$. Write down the first five terms.
> 3. Write down the first four terms of the sequence given by $5n + 1$
>
> Find the n'th term in the following series:
>
> 4. $5, 10, 15, 20, \ldots$ 5. $4, 10, 16, 22, \ldots$
> 6. $3, 5, 7, 9, \ldots$ 7. $2, 6, 10, 14, \ldots$
> 8. $9, 12, 15, 18, \ldots$ 9. $1, 8, 27, 64, 125, \ldots$

26.3 Arithmetic progressions

When a sequence has a constant difference between successive terms it is called an **arithmetic progression** (often abbreviated to *AP*).

Examples include:

(i) $1, 4, 7, 10, 13, \ldots$ where the **common difference** is 3,

and (ii) $a, a + d, a + 2d, a + 3d, \ldots$ where the common difference is d.

If the first term of an *AP* is 'a' and the common difference is 'd' then

> the n'th term is : $a + (n - 1)d$

In example (i) above, the 7th term is given by $1 + (7 - 1)3 = $ **19**, which may be readily checked.

The sum S of an *AP* can be obtained by multiplying the average of all the terms by the number of terms.

The average of all the terms $= \dfrac{a + 1}{2}$, where 'a' is the first term and l is the last term, i.e. $l = a + (n - 1)d$, for n terms.

Hence the sum of n terms,

$$S_n = n\left(\frac{a+1}{2}\right) = \frac{n}{2}\{a + [a + (n-1)d]\}$$

i.e.

$$\boxed{S_n = \frac{n}{2}[2a + (n-1)d]}$$

For example, the sum of the first 7 terms of the series 1, 4, 7, 10, 13, ... is given by

$$S_7 = \frac{7}{2}[2(1) + (7-1)3], \text{ since } a = 1 \text{ and } d = 3$$
$$= \frac{7}{2}[2 + 18] = \frac{7}{2}[20] = \textbf{70}$$

26.4 Worked problems on arithmetic progression

Problem 9. Determine (a) the ninth, and (b) the sixteenth term of the series 2, 7, 12, 17, ...

2, 7, 12, 17, ... is an arithmetic progression with a common difference, d, of 5

(a) The n'th term of an AP is given by $a + (n-1)d$

Since the first term $a = 2, d = 5$ and $n = 9$

then the 9th term is:

$$2 + (9-1)5 = 2 + (8)(5) = 2 + 40 = \textbf{42}$$

(b) The 16th term is:

$$2 + (16-1)5 = 2 + (15)(5) = 2 + 75 = \textbf{77}$$

Problem 10. The 6th term of an AP is 17 and the 13th term is 38. Determine the 19th term.

The n'th term of an AP is $a + (n-1)d$

The 6th term is: $\quad a + 5d = 17 \qquad\qquad (1)$

The 13th term is: $\quad a + 12d = 38 \qquad\qquad (2)$

Equation (2) − equation (1) gives: $7d = 21$, from which, $d = \dfrac{21}{7} = 3$

Substituting in equation (1) gives: $a + 15 = 17$, from which, $a = 2$

Hence the 19th term is:

$$a + (n-1)d = 2 + (19-1)3 = 2 + (18)(3)$$
$$= 2 + 54 = \textbf{56}$$

Problem 11. Determine the number of the term whose value is 22 in the series $2\frac{1}{2}, 4, 5\frac{1}{2}, 7, \ldots$

$2\frac{1}{2}, 4, 5\frac{1}{2}, 7, \ldots$ is an AP where $a = 2\frac{1}{2}$ and $d = 1\frac{1}{2}$

Hence if the n'th term is 22 then: $a + (n-1)d = 22$

i.e. $\quad 2\frac{1}{2} + (n-1)\left(1\frac{1}{2}\right) = 22$

$$(n-1)\left(1\frac{1}{2}\right) = 22 - 2\frac{1}{2} = 19\frac{1}{2}$$

$$n - 1 = \frac{19\frac{1}{2}}{1\frac{1}{2}} = 13 \text{ and } n = 13 + 1 = 14$$

i.e. **the 14th term of the AP is 22**

Problem 12. Find the sum of the first 12 terms of the series 5, 9, 13, 17, ...

5, 9, 13, 17, ... is an AP where $a = 5$ and $d = 4$

The sum of n terms of an AP,

$$S_n = \frac{n}{2}[2a + (n-1)d]$$

Hence the sum of the first 12 terms,

$$S_{12} = \frac{12}{2}[2(5) + (12-1)4]$$
$$= 6[10 + 44] = 6(54) = \textbf{324}$$

Problem 13. Find the sum of the first 21 terms of the series 3.5, 4.1, 4.7, 5.3, ...

3.5, 4.1, 4.7, 5.3, ... is an AP where $a = 3.5$ and $d = 0.6$

The sum of the first 21 terms,

$$S_{21} = \frac{21}{2}[2a + (n-1)d]$$
$$= \frac{21}{2}[2(3.5) + (21-1)0.6] = \frac{21}{2}[7 + 12]$$
$$= \frac{21}{2}(19) = \frac{399}{2} = \textbf{199.5}$$

Now try the following exercise

Exercise 94 Further problems on arithmetic progressions (Answers on page 263)

1. Find the 11th term of the series 8, 14, 20, 26, ...

2. Find the 17th term of the series 11, 10.7, 10.4, 10.1, ...

3. The seventh term of a series is 29 and the eleventh term is 54. Determine the sixteenth term.

4. Find the 15th term of an arithmetic progression of which the first term is $2\frac{1}{2}$ and the tenth term is 16.

5. Determine the number of the term which is 29 in the series 7, 9.2, 11.4, 13.6, ...

6. Find the sum of the first 11 terms of the series 4, 7, 10, 13, ...

7. Determine the sum of the series 6.5, 8.0, 9.5, 11.0, ..., 32

26.5 Further worked problems on arithmetic progressions

Problem 14. The sum of 7 terms of an AP is 35 and the common difference is 1.2. Determine the first term of the series.

$n = 7, d = 1.2$ and $S_7 = 35$

Since the sum of n terms of an AP is given by

$$S_n = \frac{n}{2}[2a + (n-1)d], \text{ then}$$

$$35 = \frac{7}{2}[2a + (7-1)1.2] = \frac{7}{2}[2a + 7.2]$$

Hence $\dfrac{35 \times 2}{7} = 2a + 7.2$

$$10 = 2a + 7.2$$

Thus $2a = 10 - 7.2 = 2.8$, from which $a = \dfrac{2.8}{2} = 1.4$

i.e. **the first term, $a = 1.4$**

Problem 15. Three numbers are in arithmetic progression. Their sum is 15 and their product is 80. Determine the three numbers.

Let the three numbers be $(a-d)$, a and $(a+d)$

Then $(a-d) + a + (a+d) = 15$, i.e. $3a = 15$, from which, $a = 5$

Also, $a(a-d)(a+d) = 80$, i.e. $a(a^2 - d^2) = 80$

Since $a = 5, 5(5^2 - d^2) = 80$

$$125 - 5d^2 = 80$$

$$125 - 80 = 5d^2$$

$$45 = 5d^2$$

from which, $d^2 = \dfrac{45}{5} = 9$. Hence $d = \sqrt{9} = \pm 3$

The three numbers are thus $(5-3)$, 5 and $(5+3)$, i.e. **2, 5 and 8**

Problem 16. Find the sum of all the numbers between 0 and 207 which are exactly divisible by 3.

The series 3, 6, 9, 12, ... 207 is an AP whose first term $a = 3$ and common difference $d = 3$

The last term is $a + (n-1)d = 207$

i.e. $\quad 3 + (n-1)3 = 207$, from which

$$(n-1) = \frac{207 - 3}{3} = 68$$

Hence $n = 68 + 1 = 69$

The sum of all 69 terms is given by

$$S_{69} = \frac{n}{2}[2a + (n-1)d]$$

$$= \frac{69}{2}[2(3) + (69-1)3]$$

$$= \frac{69}{2}[6 + 204] = \frac{69}{2}(210) = \mathbf{7245}$$

Problem 17. The first, twelfth and last term of an arithmetic progression are 4, $31\frac{1}{2}$, and $376\frac{1}{2}$ respectively. Determine (a) the number of terms in the series, (b) the sum of all the terms and (c) the 80'th term.

(a) Let the AP be $a, a+d, a+2d, \ldots, a+(n-1)d$, where $a = 4$

The 12th term is: $a + (12-1)d = 31\frac{1}{2}$

i.e. $\quad 4 + 11d = 31\frac{1}{2}$, from which,

$$11d = 31\frac{1}{2} - 4 = 27\frac{1}{2}$$

Hence $d = \dfrac{27\frac{1}{2}}{11} = 2\dfrac{1}{2}$

The last term is $a + (n-1)d$

i.e. $\quad 4 + (n-1)\left(2\frac{1}{2}\right) = 376\frac{1}{2}$

$$(n-1) = \frac{376\frac{1}{2} - 4}{2\frac{1}{2}} = \frac{372\frac{1}{2}}{2\frac{1}{2}} = 149$$

Hence the number of terms in the series,

$$n = 149 + 1 = 150$$

(b) Sum of all the terms,

$$S_{150} = \frac{n}{2}[2a + (n-1)d]$$

$$= \frac{150}{2}\left[2(4) + (150-1)\left(2\frac{1}{2}\right)\right]$$

$$= 75\left[8 + (149)\left(2\frac{1}{2}\right)\right] = 85[8 + 372.5)$$

$$= 75(380.5) = \mathbf{28537.5}$$

(c) The 80th term is:

$$a + (n - 1)d = 4 + (80 - 1)\left(2\tfrac{1}{2}\right)$$

$$= 4 + (79)\left(2\tfrac{1}{2}\right)$$

$$= 4 + 197.5 = \mathbf{201\tfrac{1}{2}}$$

Now try the following exercise

Exercise 95 Further problems on arithmetic progressions (Answers on page 263)

1. The sum of 15 terms of an arithmetic progression is 202.5 and the common difference is 2. Find the first term of the series.
2. Three numbers are in arithmetic progression. Their sum is 9 and their product is $20\tfrac{1}{4}$. Determine the three numbers.
3. Find the sum of all the numbers between 5 and 250 which are exactly divisible by 4
4. Find the number of terms of the series 5, 8, 11, ... of which the sum is 1025
5. Insert four terms between 5 and $22\tfrac{1}{2}$ to form an arithmetic progression.
6. The first, tenth and last terms of an arithmetic progression are 9, 40.5, and 425.5 respectively. Find (a) the number of terms, (b) the sum of all the terms and (c) the 70th term.
7. On commencing employment a man is paid a salary of £7200 per annum and receives annual increments of £350. Determine his salary in the 9th year and calculate the total he will have received in the first 12 years.
8. An oil company bores a hole 80 m deep. Estimate the cost of boring if the cost is £30 for drilling the first metre with an increase in cost of £2 per metre for each succeeding metre.

26.6 Geometric progressions

When a sequence has a constant ratio between successive terms it is called a **geometric progression** (often abbreviated to *GP*). The constant is called the. **common ratio, r**

Examples include

(i) 1, 2, 4, 8, ... where the common ratio is 2,

and (ii) $a, ar, ar^2, ar^3, \ldots$ where the common ratio is r

If the first term of a *GP* is '*a*' and the common ratio is r, then

the n'th term is : $\quad ar^{n-1}$

which can be readily checked from the above examples.

For example, the 8th term of the *GP* 1, 2, 4, 8, ... is $(1)(2)^7 = \mathbf{128}$, since $a = 1$ and $r = 2$

Let a *GP* be $a, ar, ar^2, ar^3, \ldots ar^{n-1}$

then the sum of n terms,

$$S_n = a + ar + ar^2 + ar^3 + \ldots + ar^{n-1} \ldots \quad (1)$$

Multiplying throughout by r gives:

$$rS_n = ar + ar^2 + ar^3 + ar^4 + \ldots ar^{n-1} + ar^n \ldots \quad (2)$$

Subtracting equation (2) from equation (1) gives:

$$S_n - rS_n = a - ar^n$$

i.e. $\quad S_n(1 - r) = a(1 - r^n)$

Thus the sum of n terms, $\quad \boxed{S_n = \dfrac{a(1 - r^n)}{(1 - r)}} \quad$ which is valid

when $r < 1$

Subtracting equation (1) from equation (2) gives

$$\boxed{S_n = \frac{a(r^n - 1)}{(r - 1)}} \quad \text{which is valid when } r > 1$$

For example, the sum of the first 8 terms of the *GP* 1, 2, 4, 8, 16, ... is given by

$$S_8 = \frac{1(2^8 - 1)}{(2 - 1)}, \quad \text{since } a = 1 \text{ and } r = 2$$

i.e. $\quad S_8 = \dfrac{1(256 - 1)}{1} = \mathbf{255}$

When the common ratio r of a *GP* is less than unity, the sum of n terms,

$$S_n = \frac{a(1 - r^n)}{(1 - r)}, \quad \text{which may be written as}$$

$$S_n = \frac{a}{(1 - r)} - \frac{ar^n}{(1 - r)}$$

Since $r < 1$, r^n becomes less as n increases,

i.e. $\quad r^n \to 0 \quad$ as $\quad n \to \infty$

Hence $\quad \dfrac{ar^n}{(1 - r)} \to 0 \quad$ as $\quad n \to \infty$

Thus $\quad S_n \to \dfrac{a}{(1 - r)} \quad$ as $\quad n \to \infty$

The quantity $\dfrac{a}{(1 - r)}$ is called the **sum to infinity**, S_∞, and is the limiting value of the sum of an infinite number of terms,

i.e. $\quad \boxed{S_\infty = \dfrac{a}{(1 - r)}} \quad$ which is valid when $-1 < r < 1$

For example, the sum to infinity of the *GP* $1, \frac{1}{2}, \frac{1}{4}, \ldots$ is

$$S_\infty = \frac{1}{1 - \frac{1}{2}}, \text{ since } a = 1 \text{ and } r = \frac{1}{2}, \text{ i.e. } S_\infty = 2$$

26.7 Worked problems on geometric progressions

> **Problem 18.** Determine the tenth term of the series 3, 6, 12, 24, ...

3, 6, 12, 24, ... is a geometric progression with a common ratio *r* of 2

The *n*'th term of a *GP* is ar^{n-1}, where *a* is the first term. Hence the 10th term is:

$$(3)(2)^{10-1} = (3)(2)^9 = 3(512) = \mathbf{1536}$$

> **Problem 19.** Find the sum of the first 7 terms of the series, $\frac{1}{2}, 1\frac{1}{2}, 4\frac{1}{2}, 13\frac{1}{2}, \ldots$.

$\frac{1}{2}, 1\frac{1}{2}, 4\frac{1}{2}, 13\frac{1}{2}, \ldots$ is a *GP* with a common ratio $r = 3$

The sum of *n* terms, $S_n = \dfrac{a(r^n - 1)}{(r - 1)}$

Hence $S_7 = \dfrac{\frac{1}{2}(3^7 - 1)}{(3 - 1)} = \dfrac{\frac{1}{2}(2187 - 1)}{2} = \mathbf{546\frac{1}{2}}$

> **Problem 20.** The first term of a geometric progression is 12 and the fifth term is 55. Determine the 8'th term and the 11'th term.

The 5th term is given by $ar^4 = 55$, where the first term $a = 12$

Hence $r^4 = \dfrac{55}{a} = \dfrac{55}{12}$ and $r = \sqrt[4]{\left(\dfrac{55}{12}\right)} = 1.4631719\ldots$

The 8th term is

$$ar^7 = (12)(1.4631719 \ldots)^7 = \mathbf{172.3}$$

The 11th term is

$$ar^{10} = (12)(1.4631719 \ldots)^{10} = \mathbf{539.7}$$

> **Problem 21.** Which term of the series 2187, 729, 243, ... is $\frac{1}{9}$?

2187, 729, 243, ... is a *GP* with a common ratio $r = \frac{1}{3}$ and first term $a = 2187$

The *n*'th term of a *GP* is given by: ar^{n-1}

Hence $\dfrac{1}{9} = (2187)\left(\dfrac{1}{3}\right)^{n-1}$ from which

$$\left(\frac{1}{3}\right)^{n-1} = \frac{1}{(9)(2187)} = \frac{1}{3^2 3^7} = \frac{1}{3^9} = \left(\frac{1}{3}\right)^9$$

Thus $(n - 1) = 9$, from which, $n = 9 + 1 = 10$

i.e. $\dfrac{1}{9}$ **is the 10th term of the *GP***

> **Problem 22.** Find the sum of the first 9 terms of the series 72.0, 57.6, 46.08, ...

The common ratio,

$$r = \frac{ar}{a} = \frac{57.6}{72.0} = 0.8 \left(\text{also } \frac{ar^2}{ar} = \frac{46.08}{57.6} = 0.8\right)$$

The sum of 9 terms,

$$S_9 = \frac{a(1 - r^n)}{(1 - r)} = \frac{72.0(1 - 0.8^9)}{(1 - 0.8)}$$

$$= \frac{72.0(1 - 0.1342)}{0.2} = \mathbf{311.7}$$

> **Problem 23.** Find the sum to infinity of the series 3, 1, $\frac{1}{3}$, ...

3, 1, $\frac{1}{3}$, ... is a *GP* of common ratio, $r = \frac{1}{3}$

The sum to infinity,

$$S_\infty = \frac{a}{1 - r} = \frac{3}{1 - \frac{1}{3}} = \frac{3}{\frac{2}{3}} = \frac{9}{2} = \mathbf{4\frac{1}{2}}$$

Now try the following exercise

> **Exercise 96 Further problems on geometric progressions (Answers on page 263)**
>
> 1. Find the 10th term of the series 5, 10, 20, 40, ...
>
> 2. Determine the sum of the first 7 terms of the series $\frac{1}{4}, \frac{3}{4}, 2\frac{1}{4}, 6\frac{3}{4}, \ldots$.
>
> 3. The first term of a geometric progression is 4 and the 6th term is 128. Determine the 8th and 11th terms.
>
> 4. Which term of the series 3, 9, 27, ... is 59049?
>
> 5. Find the sum of the first 7 terms of the series 2, 5, $12\frac{1}{2}$, ... (correct to 4 significant figures).
>
> 6. Determine the sum to infinity of the series 4, 2, 1, ...
>
> 7. Find the sum to infinity of the series $2\frac{1}{2}, -1\frac{1}{4}, \frac{5}{8}, \ldots$.

26.8 Further worked problems on geometric progressions

Problem 24. In a geometric progression the sixth term is 8 times the third term and the sum of the seventh and eighth terms is 192. Determine (a) the common ratio, (b) the first term, and (c) the sum of the fifth to eleventh terms, inclusive.

(a) Let the GP be $a, ar, ar^2, ar^3, \ldots, ar^{n-1}$

The 3rd term $= ar^2$ and the sixth term $= ar^5$

The 6th term is 8 times the 3rd

Hence $ar^5 = 8ar^2$ from which, $r^3 = 8, r = \sqrt[3]{8}$

i.e. **the common ratio $r = 2$**

(b) The sum of the 7th and 8th terms is 192. Hence $ar^6 + ar^7 = 192$

Since $r = 2$, then

$$64a + 128a = 192$$
$$192a = 192$$

from which, a, **the first term $= 1$**

(c) The sum of the 5th to 11th terms (inclusive) is given by:

$$S_{11} - S_4 = \frac{a(r^{11} - 1)}{(r - 1)} - \frac{a(r^4 - 1)}{(r - 1)}$$

$$= \frac{1(2^{11} - 1)}{(2 - 1)} - \frac{1(2^4 - 1)}{(2 - 1)}$$

$$= (2^{11} - 1) - (2^4 - 1)$$

$$= 2^{11} - 2^4 = 2408 - 16 = \mathbf{2032}$$

Problem 25. A hire tool firm finds that their net return from hiring tools is decreasing by 10% per annum. If their net gain on a certain tool this year is £400, find the possible total of all future profits from this tool (assuming the tool lasts for ever).

The net gain forms a series:

$$£400 + £400 \times 0.9 + £400 \times 0.9^2 + \ldots,$$

which is a GP with $a = 400$ and $r = 0.9$

The sum to infinity,

$$S_\infty = \frac{a}{(1 - r)} = \frac{400}{(1 - 0.9)}$$

$$= \mathbf{£4000} = \textbf{total future profits}$$

Problem 26. If £100 is invested at compound interest of 8% per annum, determine (a) the value after

10 years, (b) the time, correct to the nearest year, it takes to reach more than £300

(a) Let the GP be $a, ar, ar^2, \ldots ar^n$

The first term $a = £100$

The common ratio $r = 1.08$

Hence the second term is $ar = (100)(1.08) = £108$, which is the value after 1 year, the third term is $ar^2 = (100)(1.08)^2 = £116.64$, which is the value after 2 years, and so on.

Thus the value after 10 years $= ar^{10} = (100)(1.08)^{10}$

$$= \mathbf{£215.89}$$

(b) When £300 has been reached, $300 = ar^n$

i.e. $300 = 100(1.08)^n$

and $3 = (1.08)^n$

Taking logarithms to base 10 of both sides gives:

$$\lg 3 = \lg(1.08)^n = n \lg(1.08),$$

by the laws of logarithms from which,

$$n = \frac{\lg 3}{\lg 1.08} = 14.3$$

Hence it will take 15 years to reach more than £300

Problem 27. A drilling machine is to have 6 speeds ranging from 50 rev/min to 750 rev/min. If the speeds form a geometric progression determine their values, each correct to the nearest whole number.

Let the GP of n terms be given by $a, ar, ar^2, \ldots ar^{n-1}$

The first term $a = 50$ rev/min.

The 6th term is given by ar^{6-1}, which is 750 rev/min, i.e., $ar^5 = 750$

from which $r^5 = \frac{750}{a} = \frac{750}{50} = 15$

Thus the common ratio, $r = \sqrt[5]{15} = 1.7188$

The first term is $a = 50$ rev/min

the second term is $ar = (50)(1.7188) = 85.94$,

the third term is $ar^2 = (50)(1.7188)^2 = 147.71$,

the fourth term is $ar^3 = (50)(1.7188)^3 = 253.89$,

the fifth term is $ar^4 = (50)(1.7188)^4 = 436.39$,

the sixth term is $ar^5 = (50)(1.7188)^5 = 750.06$

Hence, correct to the nearest whole number, the 6 speeds of the drilling machine are **50, 86, 148, 254, 436 and 750 rev/min.**

Now try the following exercise

Exercise 97 Further problems on geometric progressions (Answers on page 263)

1. In a geometric progression the 5th term is 9 times the 3rd term and the sum of the 6th and 7th terms is 1944. Determine (a) the common ratio, (b) the first term and (c) the sum of the 4th to 10th terms inclusive.

2. The value of a lathe originally valued at £3000 depreciates 15% per annum. Calculate its value after 4 years. The machine is sold when its value is less than £550. After how many years is the lathe sold?

3. If the population of Great Britain is 55 million and is decreasing at 2.4% per annum, what will be the population in 5 years time?

4. 100 g of a radioactive substance disintegrates at a rate of 3% per annum. How much of the substance is left after 11 years?

5. If £250 is invested at compound interest of 6% per annum determine (a) the value after 15 years, (b) the time, correct to the nearest year, it takes to reach £750

6. A drilling machine is to have 8 speeds ranging from 100 rev/min to 1000 rev/min. If the speeds form a geometric progression determine their values, each correct to the nearest whole number.

Assignment 13

This assignment covers the material in chapters 25 and 26.

The marks for each question are shown in brackets at the end of each question.

1. Forces of 10 N, 16 N and 20 N act as shown in Figure A13.1. Determine the magnitude of the resultant force and its direction relative to the 16 N force (a) by scaled drawing, and (b) by calculation. (9)

2. For the three forces shown in Fig. A13.1, calculate the resultant of $F_1 - F_2 - F_3$ and its direction relative to force F_2 (6)

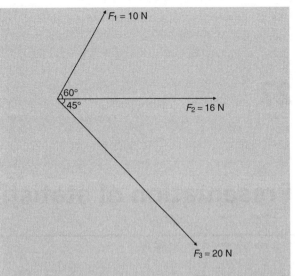

Fig. A13.1

3. Two cars, A and B, are travelling towards crossroads. A has a velocity of 60 km/h due south and B a velocity of 75 km/h due west. Calculate the velocity of A relative to B. (6)

4. Determine the 20th term of the series 15.6, 15, 14.4, 13.8, ... (3)

5. The sum of 13 terms of an arithmetic progression is 286 and the common difference is 3. Determine the first term of the series. (4)

6. An engineer earns £21000 per annum and receives annual increments of £600. Determine the salary in the 9th year and calculate the total earnings in the first 11 years. (5)

7. Determine the 11th term of the series 1.5, 3, 6, 12, ... (2)

8. Find the sum of the first eight terms of the series 1, $2\frac{1}{2}$, $6\frac{1}{4}$, ..., correct to 1 decimal place. (4)

9. Determine the sum to infinity of the series 5, 1, $\frac{1}{5}$, ... (3)

10. A machine is to have seven speeds ranging from 25 rev/min to 500 rev/min. If the speeds form a geometric progression, determine their value, each correct to the nearest whole number. (8)

27

Presentation of statistical data

27.1 Some statistical terminology

Data are obtained largely by two methods:

(a) by counting – for example, the number of stamps sold by a post office in equal periods of time, and
(b) by measurement – for example, the heights of a group of people.

When data are obtained by counting and only whole numbers are possible, the data are called **discrete**. Measured data can have any value within certain limits and are called **continuous** (see Problem 1).

A **set** is a group of data and an individual value within the set is called a **member** of the set. Thus, if the masses of five people are measured correct to the nearest 0.1 kilogram and are found to be 53.1 kg, 59.4 kg, 62.1 kg, 77.8 kg and 64.4 kg, then the set of masses in kilograms for these five people is: {53.1, 59.4, 62.1, 77.8, 64.4} and one of the members of the set is 59.4

A set containing all the members is called a **population**. Some members selected at random from a population are called a **sample**. Thus all car registration numbers form a population, but the registration numbers of, say, 20 cars taken at random throughout the country are a sample drawn from that population.

The number of times that the value of a member occurs in a set is called the **frequency** of that member. Thus in the set: {2, 3, 4, 5, 4, 2, 4, 7, 9}, member 4 has a frequency of three, member 2 has a frequency of 2 and the other members have a frequency of one.

The **relative frequency** with which any member of a set occurs is given by the ratio:

$$\frac{\text{frequency of member}}{\text{total frequency of all members}}$$

For the set: {2, 3, 5, 4, 7, 5, 6, 2, 8}, the relative frequency of member 5 is $\frac{2}{9}$.

Often, relative frequency is expressed as a percentage and the **percentage relative frequency** is:

(relative frequency × 100)%

Problem 1. Data are obtained on the topics given below. State whether they are discrete or continuous data.

(a) The number of days on which rain falls in a month for each month of the year.
(b) The mileage travelled by each of a number of salesmen.
(c) The time that each of a batch of similar batteries lasts.
(d) The amount of money spent by each of several families on food.

(a) The number of days on which rain falls in a given month must be an integer value and is obtained by **counting** the number of days. Hence, these data are **discrete**.

(b) A salesman can travel any number of miles (and parts of a mile) between certain limits and these data are **measured**. Hence the data are **continuous**.

(c) The time that a battery lasts is **measured** and can have any value between certain limits. Hence these data are **continuous**.

(d) The amount of money spent on food can only be expressed correct to the nearest pence, the amount being **counted**. Hence, these data are **discrete**.

Now try the following exercise

Exercise 98 Further problems on discrete and continuous data (Answers on page 263)

In Problems 1 and 2, state whether data relating to the topics given are discrete or continuous.

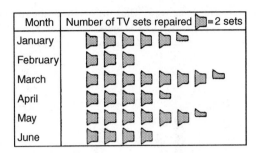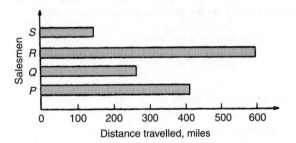

1. (a) The amount of petrol produced daily, for each of 31 days, by a refinery.
   (b) The amount of coal produced daily by each of 15 miners.
   (c) The number of bottles of milk delivered daily by each of 20 milkmen.
   (d) The size of 10 samples of rivets produced by a machine.

2. (a) The number of people visiting an exhibition on each of 5 days.
   (b) The time taken by each of 12 athletes to run 100 metres.
   (c) The value of stamps sold in a day by each of 20 post offices.
   (d) The number of defective items produced in each of 10 one-hour periods by a machine.

27.2 Presentation of ungrouped data

Ungrouped data can be presented diagrammatically in several ways and these include:

(a) **pictograms**, in which pictorial symbols are used to represent quantities (see Problem 2),

(b) **horizontal bar charts**, having data represented by equally spaced horizontal rectangles (see Problem 3), and

(c) **vertical bar charts**, in which data are represented by equally spaced vertical rectangles (see Problem 4).

Trends in ungrouped data over equal periods of time can be presented diagrammatically by a **percentage component bar chart**. In such a chart, equally spaced rectangles of any width, but whose height corresponds to 100%, are constructed. The rectangles are then subdivided into values corresponding to the percentage relative frequencies of the members (see Problem 5).

A **pie diagram** is used to show diagrammatically the parts making up the whole. In a pie diagram, the area of a circle represents the whole, and the areas of the sectors of the circle are made proportional to the parts which make up the whole (see Problem 6).

Problem 2. The number of television sets repaired in a workshop by a technician in six, one-month periods is as shown below. Present these data as a pictogram.

Month	January	February	March	April	May	June
Number repaired	11	6	15	9	13	8

Each symbol shown in Fig. 27.1 represents two television sets repaired. Thus, in January, $5\frac{1}{2}$ symbols are used to

Fig. 27.1

represent the 11 sets repaired, in February, 3 symbols are used to represent the 6 sets repaired, and so on.

Problem 3. The distance in miles travelled by four salesmen in a week are as shown below.

Salesmen	P	Q	R	S
Distance traveled (miles)	413	264	597	143

Use a horizontal bar chart to represent these data diagrammatically.

Equally spaced horizontal rectangles of any width, but whose length is proportional to the distance travelled, are used. Thus, the length of the rectangle for salesman *P* is proportional to 413 miles, and so on. The horizontal bar chart depicting these data is shown in Fig. 27.2.

Fig. 27.2

Problem 4. The number of issues of tools or materials from a store in a factory is observed for seven, one-hour periods in a day, and the results of the survey are as follows:

Period	1	2	3	4	5	6	7
Number of issues	34	17	9	5	27	13	6

Present these data on a vertical bar chart.

In a vertical bar chart, equally spaced vertical rectangles of any width, but whose height is proportional to the quantity being represented, are used. Thus the height of the rectangle for period 1 is proportional to 34 units, and so on. The vertical bar chart depicting these data is shown in Fig. 27.3.

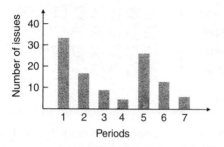

Fig. 27.3

Problem 5.   The numbers of various types of dwellings sold by a company annually over a three-year period are as shown below. Draw percentage component bar charts to present these data.

	Year 1	Year 2	Year 3
4-roomed bungalows	24	17	7
5-roomed bungalows	38	71	118
4-roomed houses	44	50	53
5-roomed houses	64	82	147
6-roomed houses	30	30	25

A table of percentage relative frequency values, correct to the nearest 1%, is the first requirement. Since,

percentage relative frequency

$$= \frac{\text{frequency of member} \times 100}{\text{total frequency}}$$

then for 4-roomed bungalows in year 1:

$$\text{percentage relative frequency} = \frac{24 \times 100}{24 + 38 + 44 + 64 + 30}$$

$$= 12\%$$

The percentage relative frequencies of the other types of dwellings for each of the three years are similarly calculated and the results are as shown in the table below.

	Year 1	Year 2	Year 3
4-roomed bungalows	12%	7%	2%
5-roomed bungalows	19%	28%	34%
4-roomed houses	22%	20%	15%
5-roomed houses	32%	33%	42%
6-roomed houses	15%	12%	7%

The percentage component bar chart is produced by constructing three equally spaced rectangles of any width,

corresponding to the three years. The heights of the rectangles correspond to 100% relative frequency, and are subdivided into the values in the table of percentages shown above. A key is used (different types of shading or different colour schemes) to indicate corresponding percentage values in the rows of the table of percentages. The percentage component bar chart is shown in Fig. 27.4.

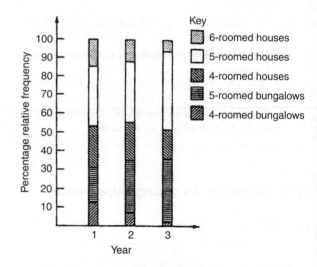

Fig. 27.4

Problem 6.   The retail price of a product costing £2 is made up as follows: materials 10p, labour 20p, research and development 40p, overheads 70p, profit 60p. Present these data on a pie diagram.

A circle of any radius is drawn, and the area of the circle represents the whole, which in this case is £2. The circle is subdivided into sectors so that the areas of the sectors are proportional to the parts, i.e. the parts which make up the total retail price. For the area of a sector to be proportional to a part, the angle at the centre of the circle must be proportional to that part. The whole, £2 or 200p, corresponds to 360°. Therefore,

$$10\text{p corresponds to } 360 \times \frac{10}{200} \text{ degrees, i.e. } 18°$$

$$20\text{p corresponds to } 360 \times \frac{20}{200} \text{ degrees, i.e. } 36°$$

and so on, giving the angles at the centre of the circle for the parts of the retail price as: 18°, 36°, 72°, 126° and 108°, respectively.

The pie diagram is shown in Fig. 27.5

Problem 7.   (a) Using the data given in Fig. 27.2 only, calculate the amount of money paid to each salesman for travelling expenses, if they are paid an allowance of 37p per mile.

(b) Using the data presented in Fig. 27.4, comment on the housing trends over the three-year period.

(c) Determine the profit made by selling 700 units of the product shown in Fig. 27.5

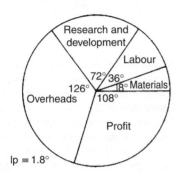

Fig. 27.5

(a) By measuring the length of rectangle P the mileage covered by salesman P is equivalent to 413 miles. Hence salesman P receives a travelling allowance of

$$\frac{£413 \times 37}{100} \quad \text{i.e.} \quad £152.81$$

Similarly, for salesman Q, the miles travelled are 264 and his allowance is

$$\frac{£264 \times 37}{100} \quad \text{i.e.} \quad £97.68$$

Salesman R travels 597 miles and he receives

$$\frac{£597 \times 37}{100} \quad \text{i.e.} \quad £220.89$$

Finally, salesman S receives

$$\frac{£143 \times 37}{100} \quad \text{i.e.} \quad £52.91$$

(b) An analysis of Fig. 27.4 shows that 5-roomed bungalows and 5-roomed houses are becoming more popular, the greatest change in the three years being a 15% increase in the sales of 5-roomed bungalows.

(c) Since 1.8° corresponds to 1p and the profit occupies 108° of the pie diagram, then the profit per unit is

$$\frac{108 \times 1}{1.8} \quad \text{i.e.} \quad 60\text{p}$$

The profit when selling 700 units of the product is

$$£\frac{700 \times 60}{100} \quad \text{i.e.} \quad £420$$

Now try the following exercise

Exercise 99   Further problems on presentation of ungrouped data (Answers on page 263)

1. The number of vehicles passing a stationary observer on a road in six ten-minute intervals is as shown. Draw a pictogram to represent these data.

Period of Time	1	2	3	4	5	6
Number of Vehicles	35	44	62	68	49	41

2. The number of components produced by a factory in a week is as shown below:

Day	Mon	Tues	Wed	Thur	Fri
Number of Components	1580	2190	1840	2385	1280

Show these data on a pictogram.

3. For the data given in Problem 1 above, draw a horizontal bar chart.

4. Present the data given in Problem 2 above on a horizontal bar chart.

5. For the data given in Problem 1 above, construct a vertical bar chart.

6. Depict the data given in Problem 2 above on a vertical bar chart.

7. A factory produces three different types of components. The percentages of each of these components produced for three, one-month periods are as shown below. Show this information on percentage component bar charts and comment on the changing trend in the percentages of the types of component produced.

Month	1	2	3
Component P	20	35	40
Component Q	45	40	35
Component R	35	25	25

8. A company has five distribution centres and the mass of goods in tonnes sent to each centre during four, one-week periods, is as shown.

Week	1	2	3	4
Centre A	147	160	174	158
Centre B	54	63	77	69
Centre C	283	251	237	211
Centre D	97	104	117	144
Centre E	224	218	203	194

Use a percentage component bar chart to present these data and comment on any trends.

9. The employees in a company can be split into the following categories: managerial 3, supervisory 9, craftsmen 21, semi-skilled 67, others 44. Show these data on a pie diagram.

10. The way in which an apprentice spent his time over a one-month period is as follows:

> drawing office 44 hours, production 64 hours, training 12 hours, at college 28 hours.

Use a pie diagram to depict this information.

11. (a) With reference to Fig. 27.5, determine the amount spent on labour and materials to produce 1650 units of the product.

(b) If in year 2 of Fig. 27.4, 1% corresponds to 2.5 dwellings, how many bungalows are sold in that year.

12. (a) If the company sell 23 500 units per annum of the product depicted in Fig. 27.5, determine the cost of their overheads per annum.

(b) If 1% of the dwellings represented in year 1 of Fig. 27.4 corresponds to 2 dwellings, find the total number of houses sold in that year.

27.3   Presentation of grouped data

When the number of members in a set is small, say ten or less, the data can be represented diagrammatically without further analysis, by means of pictograms, bar charts, percentage components bar charts or pie diagrams (as shown in Section 27.2).

For sets having more than ten members, those members having similar values are grouped together in **classes** to form a **frequency distribution**. To assist in accurately counting members in the various classes, a **tally diagram** is used (see Problems 8 and 12).

A frequency distribution is merely a table showing classes and their corresponding frequencies (see Problems 8 and 12).

The new set of values obtained by forming a frequency distribution is called **grouped data**.

The terms used in connection with grouped data are shown in Fig. 27.6(a). The size or range of a class is given by the **upper class boundary value** minus the **lower class boundary value**, and in Fig. 27.6 is $7.65 - 7.35$, i.e. 0.30. The **class interval** for the class shown in Fig. 27.6(b) is 7.4 to 7.6 and the class mid-point value is given by

$$\frac{\text{(upper class boundary value)} + \text{(lower class boundary value)}}{2}$$

and in Fig. 27.6 is $\dfrac{7.65 + 7.35}{2}$, i.e. 7.5

One of the principal ways of presenting grouped data diagrammatically is by using a **histogram**, in which the **areas**

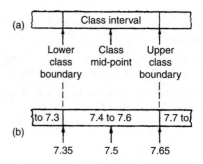

Fig. 27.6

of vertical, adjacent rectangles are made proportional to frequencies of the classes (see Problem 9). When class intervals are equal, the heights of the rectangles of a histogram are equal to the frequencies of the classes. For histograms having unequal class intervals, the area must be proportional to the frequency. Hence, if the class interval of class A is twice the class interval of class B, then for equal frequencies, the height of the rectangle representing A is half that of B (see Problem 11).

Another method of presenting grouped data diagrammatically is by using a **frequency polygon**, which is the graph produced by plotting frequency against class mid-point values and joining the co-ordinates with straight lines (see Problem 12).

A **cumulative frequency distribution** is a table showing the cumulative frequency for each value of upper class boundary. The cumulative frequency for a particular value of upper class boundary is obtained by adding the frequency of the class to the sum of the previous frequencies. A cumulative frequency distribution is formed in Problem 13.

The curve obtained by joining the co-ordinates of cumulative frequency (vertically) against upper class boundary (horizontally) is called an **ogive** or a **cumulative frequency distribution curve** (see Problem 13).

Problem 8.   The data given below refer to the gain of each of a batch of 40 transistors, expressed correct to the nearest whole number. Form a frequency distribution for these data having seven classes.

81	83	87	74	76	89	82	84
86	76	77	71	86	85	87	88
84	81	80	81	73	89	82	79
81	79	78	80	85	77	84	78
83	79	80	83	82	79	80	77

The **range** of the data is the value obtained by taking the value of the smallest member from that of the largest member. Inspection of the set of data shows that, range $= 89 - 71 = 18$. The size of each class is given approximately by range divided by the number of classes. Since 7 classes are required, the size of each class is 18/7, that is, approximately 3. To achieve seven equal classes spanning a range of values from 71 to 89, the class intervals are selected as: 70–72, 73–75, and so on.

To assist with accurately determining the number in each class, a **tally diagram** is produced, as shown in Table 27.1(a). This is obtained by listing the classes in the left-hand column, and then inspecting each of the 40 members of the set in turn and allocating them to the appropriate classes by putting '1s' in the appropriate rows. Every fifth '1' allocated to a particular row is shown as an oblique line crossing the four previous '1s', to help with final counting.

A **frequency distribution** for the data is shown in Table 27.1(b) and lists classes and their corresponding frequencies, obtained from the tally diagram. (Class mid-point values are also shown in the table, since they are used for constructing the histogram for these data (see Problem 9)).

Table 27.1(a)

Class	Tally
70–72	1
73–75	11
76–78	̶L̶H̶T̶ 11
79–81	̶L̶H̶T̶ ̶L̶H̶T̶ 11
82–84	̶L̶H̶T̶ 1111
85–87	̶L̶H̶T̶ 1
88–90	111

Table 27.1(b)

Class	Class mid-point	Frequency
70–72	71	1
73–75	74	2
76–78	77	7
79–81	80	12
82–84	83	9
85–87	86	6
88–90	89	3

Problem 9. Construct a histogram for the data given in Table 27.1(b).

The histogram is shown in Fig. 27.7. The width of the rectangles correspond to the upper class boundary values minus the lower class boundary values and the heights of the rectangles correspond to the class frequencies. The easiest way to draw a histogram is to mark the class mid-point values on the horizontal scale and draw the rectangles symmetrically about the appropriate class mid-point values and touching one another.

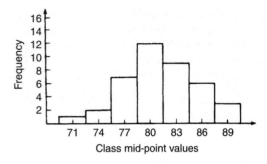

Fig. 27.7

Problem 10. The amount of money earned weekly by 40 people working part-time in a factory, correct to the nearest £10, is shown below. Form a frequency distribution having 6 classes for these data.

80	90	70	110	90	160	110	80
140	30	90	50	100	110	60	100
80	90	110	80	100	90	120	70
130	170	80	120	100	110	40	110
50	100	110	90	100	70	110	80

Inspection of the set given shows that the majority of the members of the set lie between £80 and £110 and that there are a much smaller number of extreme values ranging from £30 to £170. If equal class intervals are selected, the frequency distribution obtained does not give as much information as one with unequal class intervals. Since the majority of members are between £80 and £100, the class intervals in this range are selected to be smaller than those outside of this range. There is no unique solution and one possible solution is shown in Table 27.2.

Table 27.2

Class	Frequency
20–40	2
50–70	6
80–90	12
100–110	14
120–140	4
150–170	2

Problem 11. Draw a histogram for the data given in Table 27.2.

When dealing with unequal class intervals, the histogram must be drawn so that the areas, (and not the heights), of the rectangles are proportional to the frequencies of the classes. The data given are shown in columns 1 and 2 of Table 27.3. Columns 3 and 4 give the upper and lower class boundaries, respectively. In column 5, the class ranges (i.e. upper class boundary minus lower class boundary values) are listed. The heights of the rectangles are proportional to the ratio $\dfrac{\text{frequency}}{\text{class range}}$, as shown in column 6. The histogram is shown in Fig. 27.8

Table 27.3

1 Class	2 Frequency	3 Upper class boundary	4 Lower class boundary	5 Class range	6 Height of rectangle
20–40	2	45	15	30	$\dfrac{2}{30} = \dfrac{1}{15}$
50–70	6	75	45	30	$\dfrac{6}{30} = \dfrac{3}{15}$
80–90	12	95	75	20	$\dfrac{12}{20} = \dfrac{9}{15}$
100–110	14	115	95	20	$\dfrac{14}{20} = \dfrac{10\frac{1}{2}}{15}$
120–140	4	145	115	30	$\dfrac{4}{30} = \dfrac{2}{15}$
150–170	2	175	145	30	$\dfrac{2}{30} = \dfrac{1}{15}$

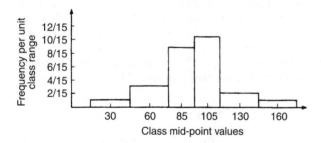

Fig. 27.8

Problem 12. The masses of 50 ingots in kilograms are measured correct to the nearest 0.1 kg and the results are as shown below. Produce a frequency distribution having about 7 classes for these data and then present the grouped data as (a) a frequency polygon and (b) a histogram.

8.0	8.6	8.2	7.5	8.0	9.1	8.5	7.6	8.2	7.8
8.3	7.1	8.1	8.3	8.7	7.8	8.7	8.5	8.4	8.5
7.7	8.4	7.9	8.8	7.2	8.1	7.8	8.2	7.7	7.5
8.1	7.4	8.8	8.0	8.4	8.5	8.1	7.3	9.0	8.6
7.4	8.2	8.4	7.7	8.3	8.2	7.9	8.5	7.9	8.0

The **range** of the data is the member having the largest value minus the member having the smallest value. Inspection of the set of data shows that:

$$\text{range} = 9.1 - 7.1 = 2.0$$

The size of each class is given approximately by

$$\frac{\text{range}}{\text{number of classes}}$$

Since about seven classes are required, the size of each class is 2.0/7, that is approximately 0.3, and thus the **class limits** are selected as 7.1 to 7.3, 7.4 to 7.6, 7.7 to 7.9, and so on.

The **class mid-point** for the 7.1 to 7.3 class is $\dfrac{7.35 + 7.05}{2}$, i.e. 7.2, for the 7.4 to 7.6 class is $\dfrac{7.65 + 7.35}{2}$, i.e. 7.5, and so on.

To assist with accurately determining the number in each class, a **tally diagram** is produced as shown in Table 27.4. This is obtained by listing the classes in the left-hand column and then inspecting each of the 50 members of the set of data in turn and allocating it to the appropriate class by putting a '1' in the appropriate row. Each fifth '1' allocated to a particular row is marked as an oblique line to help with final counting.

Table 27.4

Class	Tally
7.1 to 7.3	111
7.4 to 7.6	ℍℍ
7.7 to 7.9	ℍℍ 1111
8.0 to 8.2	ℍℍ ℍℍ 1111
8.3 to 8.5	ℍℍ ℍℍ 1
8.6 to 8.8	ℍℍ 1
8.9 to 9.1	11

A **frequency distribution** for the data is shown in Table 27.5 and lists classes and their corresponding frequencies. Class mid-points are also shown in this table, since they are used when constructing the frequency polygon and histogram.

Table 27.5

Class	Class mid-point	Frequency
7.1 to 7.3	7.2	3
7.4 to 7.6	7.5	5
7.7 to 7.9	7.8	9
8.0 to 8.2	8.1	14
8.3 to 8.5	8.4	11
8.6 to 8.8	8.7	6
8.9 to 9.1	9.0	2

A **frequency polygon** is shown in Fig. 27.9, the co-ordinates corresponding to the class mid-point/frequency values, given in Table 27.5. The co-ordinates are joined by straight lines and the polygon is 'anchored-down' at each end by joining to the next class mid-point value and zero frequency.

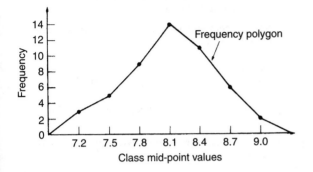

Fig. 27.9

A **histogram** is shown in Fig. 27.10, the width of a rectangle corresponding to (upper class boundary value – lower class boundary value) and height corresponding to the class frequency. The easiest way to draw a histogram is to mark class mid-point values on the horizontal scale and to draw the rectangles symmetrically about the appropriate class mid-point values and touching one another. A histogram for the data given in Table 27.5 is shown in Fig. 27.10.

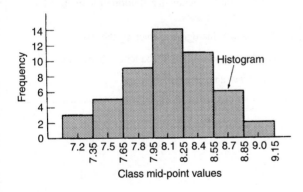

Fig. 27.10

Problem 13. The frequency distribution for the masses in kilograms of 50 ingots is:

7.1 to 7.3	3	7.4 to 7.6	5	7.7 to 7.9	9
8.0 to 8.2	14	8.3 to 8.5	11	8.6 to 8.8	6
8.9 to 9.1	2				

Form a cumulative frequency distribution for these data and draw the corresponding ogive.

A **cumulative frequency distribution** is a table giving values of cumulative frequency for the values of upper class boundaries, and is shown in Table 27.6. Columns 1 and 2 show the classes and their frequencies. Column 3 lists the upper class boundary values for the classes given in column 1. Column 4 gives the cumulative frequency values for all frequencies less than the upper class boundary values given in column 3. Thus, for example, for the 7.7 to 7.9 class shown in row 3, the cumulative frequency value is the sum of all frequencies having values of less than 7.95, i.e. $3 + 5 + 9 = 17$, and so on. The **ogive** for the cumulative frequency distribution given in Table 27.6 is shown in Fig. 27.11. The co-ordinates corresponding to each upper class boundary/cumulative frequency value are plotted and the co-ordinates are joined by straight lines (– not the best curve drawn through the co-ordinates as in experimental work.) The ogive is 'anchored' at its start by adding the co-ordinate (7.05, 0).

Table 27.6

1 Class	2 Frequency	3 Upper class boundary	4 Cumulative frequency
		Less than	
7.1–7.3	3	7.35	3
7.4–7.6	5	7.65	8
7.7–7.9	9	7.95	17
8.0–8.2	14	8.25	31
8.3–8.5	11	8.55	42
8.6–8.8	6	8.85	48
8.9–9.1	2	9.15	50

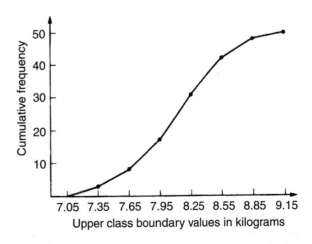

Fig. 27.11

Now try the following exercise

Exercise 100   Further problems on presentation of grouped data (Answers on page 264)

1. The mass in kilograms, correct to the nearest one-tenth of a kilogram, of 60 bars of metal are as shown. Form a frequency distribution of about 8 classes for these data.

```
39.8 40.1 40.3 40.0 40.6 39.7 40.0 40.4 39.6 39.3
39.6 40.7 40.2 39.9 40.3 40.2 40.4 39.9 39.8 40.0
40.2 40.1 40.3 39.7 39.9 40.5 39.9 40.5 40.0 39.9
40.1 40.8 40.0 40.0 40.1 40.2 40.1 40.0 40.2 39.9
39.7 39.8 40.4 39.7 39.9 39.5 40.1 40.1 39.9 40.2
39.5 40.6 40.0 40.1 39.8 39.7 39.5 40.2 39.9 40.3
```

2. Draw a histogram for the frequency distribution given in the solution of Problem 1.

3. The information given below refers to the value of resistance in ohms of a batch of 48 resistors of similar value. Form a frequency distribution for the data, having about 6 classes and draw a frequency polygon and histogram to represent these data diagramatically.

```
21.0 22.4 22.8 21.5 22.6 21.1 21.6 22.3
22.9 20.5 21.8 22.2 21.0 21.7 22.5 20.7
23.2 22.9 21.7 21.4 22.1 22.2 22.3 21.3
22.1 21.8 22.0 22.7 21.7 21.9 21.1 22.6
21.4 22.4 22.3 20.9 22.8 21.2 22.7 21.6
22.2 21.6 21.3 22.1 21.5 22.0 23.4 21.2
```

4. The time taken in hours to the failure of 50 specimens of a metal subjected to fatigue failure tests are as shown. Form a frequency distribution, having about 8 classes and unequal class intervals, for these data.

```
28 22 23 20 12 24 37 28 21 25
21 14 30 23 27 13 23 7  26 19
24 22 26 3  21 24 28 40 27 24
```
```
20 25 23 26 47 21 29 26 22 33
27 9  13 35 20 16 20 25 18 22
```

5. Form a cumulative frequency distribution and hence draw the ogive for the frequency distribution given in the solution to Problem 3.

6. Draw a histogram for the frequency distribution given in the solution to Problem 4.

7. The frequency distribution for a batch of 48 resistors of similar value, measured in ohms, is:

```
20.5–20.9  3   21.0–21.4  10   21.5–21.9  11
22.0–22.4  13  22.5–22.9  9    23.0–23.4  2
```

Form a cumulative frequency distribution for these data.

8. Draw an ogive for the data given in the solution of Problem 7.

9. The diameter in millimetres of a reel of wire is measured in 48 places and the results are as shown.

```
2.10 2.29 2.32 2.21 2.14 2.22
2.28 2.18 2.17 2.20 2.23 2.13
2.26 2.10 2.21 2.17 2.28 2.15
2.16 2.25 2.23 2.11 2.27 2.34
2.24 2.05 2.29 2.18 2.24 2.16
2.15 2.22 2.14 2.27 2.09 2.21
2.11 2.17 2.22 2.19 2.12 2.20
2.23 2.07 2.13 2.26 2.16 2.12
```

(a) Form a frequency distribution of diameters having about 6 classes.

(b) Draw a histogram depicting the data.

(c) Form a cumulative frequency distribution.

(d) Draw an ogive for the data.

28

Measures of central tendency and dispersion

28.1 Measures of central tendency

A single value, which is representative of a set of values, may be used to give an indication of the general size of the members in a set, the word **'average'** often being used to indicate the single value.

The statistical term used for 'average' is the arithmetic mean or just the **mean**.

Other measures of central tendency may be used and these include the **median** and the **modal** values.

28.2 Mean, median and mode for discrete data

Mean

The **arithmetic mean value** is found by adding together the values of the members of a set and dividing by the number of members in the set. Thus, the mean of the set of numbers: $\{4, 5, 6, 9\}$ is:

$$\frac{4+5+6+9}{4}, \quad \text{i.e.} \quad 6$$

In general, the mean of the set: $\{x_1, x_2, x_3, \ldots x_n\}$ is

$$\bar{x} = \frac{x_1 + x_2 + x_3 + \cdots + x_n}{n}, \text{ written as } \frac{\sum x}{n}$$

where $\sum$ is the Greek letter 'sigma' and means 'the sum of', and $\bar{x}$ (called x-bar) is used to signify a mean value.

Median

The **median value** often gives a better indication of the general size of a set containing extreme values. The set: $\{7, 5, 74, 10\}$ has a mean value of 24, which is not really representative of any of the values of the members of the set. The median value is obtained by:

(a) **ranking** the set in ascending order of magnitude, and

(b) selecting the value of the **middle member** for sets containing an odd number of members, or finding the value of the mean of the two middle members for sets containing an even number of members.

For example, the set: $\{7, 5, 74, 10\}$ is ranked as $\{5, 7, 10, 74\}$, and since it contains an even number of members (four in this case), the mean of 7 and 10 is taken, giving a median value of 8.5. Similarly, the set: $\{3, 81, 15, 7, 14\}$ is ranked as $\{3, 7, 14, 15, 81\}$ and the median value is the value of the middle member, i.e. 14

Mode

The **modal value**, or **mode**, is the most commonly occurring value in a set. If two values occur with the same frequency, the set is 'bi-modal'. The set: $\{5, 6, 8, 2, 5, 4, 6, 5, 3\}$ has a modal value of 5, since the member having a value of 5 occurs three times.

Problem 1. Determine the mean, median and mode for the set:

$$\{2, 3, 7, 5, 5, 13, 1, 7, 4, 8, 3, 4, 3\}$$

The mean value is obtained by adding together the values of the members of the set and dividing by the number of members in the set.

Thus, **mean value**,

$$\bar{x} = \frac{\begin{array}{c}2+3+7+5+5+13+1\\+7+4+8+3+4+3\end{array}}{13} = \frac{65}{13} = 5$$

To obtain the median value the set is ranked, that is, placed in ascending order of magnitude, and since the set contains an odd number of members the value of the middle member is the median value. Ranking the set gives:

$$\{1, 2, 3, 3, 3, 4, 4, 5, 5, 7, 7, 8, 13\}$$

The middle term is the seventh member, i.e. 4, thus the **median value is 4**. The **modal value** is the value of the most commonly occurring member and is **3**, which occurs three times, all other members only occurring once or twice.

Problem 2. The following set of data refers to the amount of money in £s taken by a news vendor for 6 days. Determine the mean, median and modal values of the set:

$$\{27.90, 34.70, 54.40, 18.92, 47.60, 39.68\}$$

Mean value

$$= \frac{27.90 + 34.70 + 54.40 + 18.92 + 47.60 + 39.68}{6}$$

$$= \textbf{£37.20}$$

The ranked set is:

$$\{18.92, 27.90, 34.70, 39.68, 47.60, 54.40\}$$

Since the set has an even number of members, the mean of the middle two members is taken to give the median value, i.e.

$$\text{median value} = \frac{34.70 + 39.68}{2} = \textbf{£37.19}$$

Since no two members have the same value, this set has **no mode**.

Now try the following exercise

Exercise 101   Further problems on mean, median and mode for discrete data (Answers on page 264)

In Problems 1 to 4, determine the mean, median and modal values for the sets given.

1. $\{3, 8, 10, 7, 5, 14, 2, 9, 8\}$
2. $\{26, 31, 21, 29, 32, 26, 25, 28\}$
3. $\{4.72, 4.71, 4.74, 4.73, 4.72, 4.71, 4.73, 4.72\}$
4. $\{73.8, 126.4, 40.7, 141.7, 28.5, 237.4, 157.9\}$

28.3   Mean, median and mode for grouped data

The mean value for a set of grouped data is found by determining the sum of the (frequency × class mid-point

values) and dividing by the sum of the frequencies,

i.e. mean value $\bar{x} = \dfrac{f_1 x_1 + f_2 x_2 + \cdots f_n x_n}{f_1 + f_2 + \cdots + f_n} = \dfrac{\sum (f x)}{\sum f}$

where f is the frequency of the class having a mid-point value of x, and so on.

Problem 3. The frequency distribution for the value of resistance in ohms of 48 resistors is as shown. Determine the mean value of resistance.

20.5–20.9   3,   21.0–21.4   10,   21.5–21.9   11,

22.0–22.4   13,   22.5–22.9   9,   23.0–23.4   2

The class mid-point/frequency values are:

20.7   3, 21.2   10, 21.7   11, 22.2   13, 22.7   9   and   23.2   2

For grouped data, the mean value is given by:

$$\bar{x} = \frac{\sum (f x)}{\sum f}$$

where f is the class frequency and x is the class mid-point value. Hence mean value,

$$\bar{x} = \frac{\begin{array}{l}(3 \times 20.7) + (10 \times 21.2) + (11 \times 21.7) \\ + (13 \times 22.2) + (9 \times 22.7) + (2 \times 23.2)\end{array}}{48}$$

$$= \frac{1052.1}{48} = 21.919\ldots$$

i.e. **the mean value is 21.9 ohms**, correct to 3 significant figures.

Histogram

The mean, median and modal values for grouped data may be determined from a **histogram**. In a histogram, frequency values are represented vertically and variable values horizontally. The mean value is given by the value of the variable corresponding to a vertical line drawn through the centroid of the histogram. The median value is obtained by selecting a variable value such that the area of the histogram to the left of a vertical line drawn through the selected variable value is equal to the area of the histogram on the right of the line. The modal value is the variable value obtained by dividing the width of the highest rectangle in the histogram in proportion to the heights of the adjacent rectangles. The method of determining the mean, median and modal values from a histogram is shown in Problem 4.

Problem 4. The time taken in minutes to assemble a device is measured 50 times and the results are as shown. Draw a histogram depicting this data and hence determine the mean, median and modal values of the distribution.

14.5–15.5 5,   16.5–17.5 8,   18.5–19.5 16,

20.5–21.5 12,   22.5–23.5 6,   24.5–25.5 3

The histogram is shown in Fig. 28.1. The mean value lies at the centroid of the histogram. With reference to any arbitrary axis, say YY shown at a time of 14 minutes, the position of the horizontal value of the centroid can be obtained from the relationship $AM = \sum(am)$, where A is the area of the histogram, M is the horizontal distance of the centroid from the axis YY, a is the area of a rectangle of the histogram and m is the distance of the centroid of the rectangle from YY. The areas of the individual rectangles are shown circled on the histogram giving a total area of 100 square units. The positions, m, of the centroids of the individual rectangles are $1, 3, 5, \ldots$ units from YY. Thus

$$100M = (10 \times 1) + (16 \times 3) + (32 \times 5) + (24 \times 7)$$
$$+ (12 \times 9) + (6 \times 11)$$

i.e. $M = \dfrac{560}{100} = 5.6$ units from YY

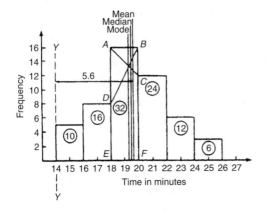

Fig. 28.1

Thus the position of the **mean** with reference to the time scale is $14 + 5.6$, i.e. **19.6 minutes.**

The median is the value of time corresponding to a vertical line dividing the total area of the histogram into two equal parts. The total area is 100 square units, hence the vertical line must be drawn to give 50 units of area on each side. To achieve this with reference to Fig. 28.1, rectangle $ABFE$ must be split so that $50 - (10 + 16)$ units of area lie on one side and $50 - (24 + 12 + 6)$ units of area lie on the other. This shows that the area of $ABFE$ is split so that 24 units of area lie to the left of the line and 8 units of area lie to the

right, i.e. the vertical line must pass through 19.5 minutes. Thus the **median value** of the distribution is **19.5 minutes.**

The mode is obtained by dividing the line AB, which is the height of the highest rectangle, proportionally to the heights of the adjacent rectangles. With reference to Fig. 28.1, this is done by joining AC and BD and drawing a vertical line through the point of intersection of these two lines. This gives the **mode** of the distribution and is **19.3 minutes**.

Now try the following exercise

Exercise 102 Further problems on mean, median and mode for grouped data (Answers on page 264)

1. The frequency distribution given below refers to the heights in centimetres of 100 people. Determine the mean value of the distribution, correct to the nearest millimetre.

   150–156 5,   157–163 18,   164–170 20

   171–177 27,   178–184 22,   185–191   8

2. The gain of 90 similar transistors is measured and the results are as shown.

   83.5–85.5   6,   86.5–88.5 39,   89.5–91.5 27,

   92.5–94.5 15,   95.5–97.5   3

   By drawing a histogram of this frequency distribution, determine the mean, median and modal values of the distribution.

3. The diameters, in centimetres, of 60 holes bored in engine castings are measured and the results are as shown. Draw a histogram depicting these results and hence determine the mean, median and modal values of the distribution.

   2.011–2.014   7,   2.016–2.019 16,

   2.021–2.024 23,   2.026–2.029   9,

   2.031–2.034   5

28.4 Standard deviation

(a) Discrete data

The standard deviation of a set of data gives an indication of the amount of dispersion, or the scatter, of members of the set from the measure of central tendency. Its value is the root-mean-square value of the members of the set and for discrete data is obtained as follows:

(a) determine the measure of central tendency, usually the mean value, (occasionally the median or modal values are specified),

(b) calculate the deviation of each member of the set from the mean, giving

$$(x_1 - \bar{x}), (x_2 - \bar{x}), (x_3 - \bar{x}), \ldots,$$

(c) determine the squares of these deviations, i.e.

$$(x_1 - \bar{x})^2, (x_2 - \bar{x})^2, (x_3 - \bar{x})^2, \ldots,$$

(d) find the sum of the squares of the deviations, that is

$$(x_1 - \bar{x})^2 + (x_2 - \bar{x})^2 + (x_3 - \bar{x})^2, \ldots,$$

(e) divide by the number of members in the set, n, giving

$$\frac{(x_1 - \bar{x})^2 + (x_2 - \bar{x})^2 + (x^3 - \bar{x})^2 + \cdots}{n}$$

(f) determine the square root of (e).

The standard deviation is indicated by σ (the Greek letter small 'sigma') and is written mathematically as:

$$\textbf{standard deviation, } \sigma = \sqrt{\left\{ \frac{\sum(x - \bar{x})^2}{n} \right\}}$$

where x is a member of the set, $\bar{x}$ is the mean value of the set and n is the number of members in the set. The value of standard deviation gives an indication of the distance of the members of a set from the mean value. The set: $\{1, 4, 7, 10, 13\}$ has a mean value of 7 and a standard deviation of about 4.2. The set $\{5, 6, 7, 8, 9\}$ also has a mean value of 7, but the standard deviation is about 1.4. This shows that the members of the second set are mainly much closer to the mean value than the members of the first set. The method of determining the standard deviation for a set of discrete data is shown in Problem 5.

Problem 5. Determine the standard deviation from the mean of the set of numbers: $\{5, 6, 8, 4, 10, 3\}$, correct to 4 significant figures.

The arithmetic mean, $\bar{x} = \dfrac{\sum x}{n}$

$$= \frac{5 + 6 + 8 + 4 + 10 + 3}{6} = 6$$

Standard deviation, $\sigma = \sqrt{\left\{ \dfrac{\sum(x - \bar{x})^2}{n} \right\}}$

The $(x - \bar{x})^2$ values are: $(5 - 6)^2, (6 - 6)^2, (8 - 6)^2, (4 - 6)^2,$ $(10 - 6)^2$ and $(3 - 6)^2$.

The sum of the $(x - \bar{x})^2$ values,

i.e. $\displaystyle\sum(x - \bar{x})^2 = 1 + 0 + 4 + 4 + 16 + 9 = 34$

and $\dfrac{\sum(x - \bar{x})^2}{n} = \dfrac{34}{6} = 5.\dot{6}$

since there are 6 members in the set.

Hence, **standard deviation,**

$$\sigma = \sqrt{\left\{ \frac{\sum(x - \bar{x}^2}{n} \right\}} = \sqrt{5.\dot{6}} = \textbf{2.380}$$

correct to 4 significant figures.

(b) Grouped data

For **grouped data, standard deviation**

$$\sigma = \sqrt{\left\{ \frac{\sum\{f(x - \bar{x})^2\}}{\sum f} \right\}}$$

where f is the class frequency value, x is the class mid-point value and $\bar{x}$ is the mean value of the grouped data. The method of determining the standard deviation for a set of grouped data is shown in Problem 6.

Problem 6. The frequency distribution for the values of resistance in ohms of 48 resistors is as shown. Calculate the standard deviation from the mean of the resistors, correct to 3 significant figures.

20.5–20.9 3, 21.0–21.4 10, 21.5–21.9 11,

22.0–22.4 13, 22.5–22.9 9, 23.0–23.4 2

The standard deviation for grouped data is given by:

$$\sigma = \sqrt{\left\{ \frac{\sum\{f(x - \bar{x})^2\}}{\sum f} \right\}}$$

From Problem 3, the distribution mean value, $\bar{x} = 21.92$, correct to 4 significant figures.

The 'x-values' are the class mid-point values, i.e. 20.7, 21.2, 21.7, ...

Thus the $(x - \bar{x})^2$ values are $(20.7 - 21.92)^2$, $(21.2 - 21.92)^2$, $(21.7 - 21.92)^2, \ldots,$

and the $f(x - \bar{x})^2$ values are $3(20.7 - 21.92)^2$, $10(21.2 - 21.92)^2$, $11(21.7 - 21.92)^2, \ldots.$

The $\sum f(x - \bar{x})^2$ values are

$$4.4652 + 5.1840 + 0.5324 + 1.0192$$

$$+ 5.4756 + 3.2768 = 19.9532$$

$$\frac{\sum\{f(x - \bar{x})^2\}}{\sum f} = \frac{19.9532}{48} = 0.41569$$

and **standard deviation,**

$$\sigma = \sqrt{\left\{ \frac{\sum\{f(x-\bar{x})^2\}}{\sum f} \right\}} = \sqrt{0.41569}$$

$$= \mathbf{0.645}, \text{ correct to 3 significant figures}$$

Now try the following exercise

Exercise 103   Further problems on standard devia-
              tion (Answers on page 264)

1. Determine the standard deviation from the mean of the set of numbers:

   {35, 22, 25, 23, 28, 33, 30}

   correct to 3 significant figures.

2. The values of capacitances, in microfarads, of ten capacitors selected at random from a large batch of similar capacitors are:

   34.3, 25.0, 30.4, 34.6, 29.6, 28.7,

   33.4, 32.7, 29.0 and 31.3

   Determine the standard deviation from the mean for these capacitors, correct to 3 significant figures.

3. The tensile strength in megapascals for 15 samples of tin were determined and found to be:

   34.61, 34.57, 34.40, 34.63, 34.63, 34.51, 34.49, 34.61, 34.52, 34.55, 34.58, 34.53, 34.44, 34.48 and 34.40

   Calculate the mean and standard deviation from the mean for these 15 values, correct to 4 significant figures.

4. Determine the standard deviation from the mean, correct to 4 significant figures, for the heights of the 100 people given in Problem 1 of Exercise 102, page 219.

5. Calculate the standard deviation from the mean for the data given in Problem 2 of Exercise 102, page 219, correct to 3 decimal places.

28.5   Quartiles, deciles and percentiles

Other measures of dispersion which are sometimes used are the quartile, decile and percentile values. The **quartile values** of a set of discrete data are obtained by selecting the values of members which divide the set into four equal parts. Thus for the set: {2, 3, 4, 5, 5, 7, 9, 11, 13, 14, 17} there are 11 members and the values of the members dividing the set into four equal parts are 4, 7, and 13. These values are signified by Q_1, Q_2 and Q_3 and called the first, second and third quartile values, respectively. It can be seen that the second quartile value, Q_2, is the value of the middle member and hence is the median value of the set.

For grouped data the ogive may be used to determine the quartile values. In this case, points are selected on the vertical cumulative frequency values of the ogive, such that they divide the total value of cumulative frequency into four equal parts. Horizontal lines are drawn from these values to cut the ogive. The values of the variable corresponding to these cutting points on the ogive give the quartile values (see Problem 7).

When a set contains a large number of members, the set can be split into ten parts, each containing an equal number of members. These ten parts are then called **deciles**. For sets containing a very large number of members, the set may be split into one hundred parts, each containing an equal number of members. One of these parts is called a **percentile**.

Problem 7. The frequency distribution given below refers to the overtime worked by a group of craftsmen during each of 48 working weeks in a year.

25–29   5,   30–34 4,   35–39 7,   40–44 11,

45–49 12,   50–54 8,   55–59 1

Draw an ogive for this data and hence determine the quartile values.

The cumulative frequency distribution (i.e. upper class boundary/cumulative frequency values) is:

29.5   5,   34.5   9,   39.5 16,   44.5 27,

49.5 39,   54.5 47,   59.5 48

The ogive is formed by plotting these values on a graph, as shown in Fig. 26.2. The total frequency is divided into four equal parts, each having a range of 48/4, i.e. 12. This gives cumulative frequency values of 0 to 12 corresponding to the first quartile, 12 to 24 corresponding to the second quartile, 24 to 36 corresponding to the third quartile and 36 to 48 corresponding to the fourth quartile of the distribution, i.e. the distribution is divided into four equal parts. The quartile values are those of the variable corresponding to cumulative frequency values of 12, 24 and 36, marked Q_1, Q_2 and Q_3 in Fig. 28.2. These values, correct to the nearest hour, are **37 hours, 43 hours and 48 hours**, respectively. The Q_2 value is also equal to the median value of the distribution. One measure of the dispersion of a distribution is called the **semi-interquartile range** and is given by $(Q_3 - Q_1)/2$, and is $(48-37)/2$ in this case, i.e. $\mathbf{5\frac{1}{2}}$ **hours.**

Problem 8. Determine the numbers contained in the (a) 41st to 50th percentile group, and (b) 8th decile group of the set of numbers shown below:

14 22   17 21   30 28   37   7   23 32

24 17   20 22   27 19   26 21   15 29

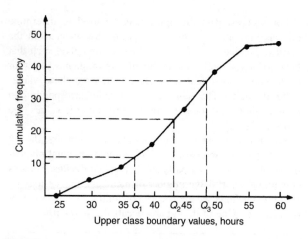

Fig. 28.2

The set is ranked, giving:

7 14 15 17 17 19 20 21 21 22

22 23 24 26 27 28 29 30 32 37

(a) There are 20 numbers in the set, hence the first 10% will be the two numbers 7 and 14, the second 10% will be 15 and 17, and so on. Thus the 41st to 50th percentile group will be the numbers **21 and 22**.

(b) The first decile group is obtained by splitting the ranked set into 10 equal groups and selecting the first group, i.e. the numbers 7 and 14. The second decile group are the numbers 15 and 17, and so on. Thus the 8th decile group contains the numbers **27 and 28**.

Now try the following exercise

Exercise 104   Further problems on quartiles, deciles and percentiles (Answers on page 264)

1. The number of working days lost due to accidents for each of 12 one-monthly periods are as shown. Determine the median and first and third quartile values for this data.

   27 37   40 28   23 30   35 24   30 32   31 28

2. The number of faults occurring on a production line in a nine-week period are as shown below. Determine the median and quartile values for the data.

   30 27   25 24   27 37   31 27   35

3. Determine the quartile values and semi-interquartile range for the frequency distribution given in Problem 1 of Exercise 102, page 219.

4. Determine the numbers contained in the 5th decile group and in the 61st to 70th percentile groups for the set of numbers:

   40 46   28 32   37 42   50 31   48 45
   32 38   27 33   40 35   25 42   38 41

5. Determine the numbers in the 6th decile group and in the 81st to 90th percentile group for the set of numbers:

   43 47 30 25 15 51 17 21 37 33 44 56 40 49 22
   36 44 33 17 35 58 51 35 44 40 31 41 55 50 16

29

Probability

29.1 Introduction to probability

Probability

The **probability** of something happening is the likelihood or chance of it happening. Values of probability lie between 0 and 1, where 0 represents an absolute impossibility and 1 represents an absolute certainty. The probability of an event happening usually lies somewhere between these two extreme values and is expressed either as a proper or decimal fraction. Examples of probability are:

that a length of copper wire has zero resistance at 100°C	0
that a fair, six-sided dice will stop with a 3 upwards	$\frac{1}{6}$ or 0.1667
that a fair coin will land with a head upwards	$\frac{1}{2}$ or 0.5
that a length of copper wire has some resistance at 100°C	1

If p is the probability of an event happening and q is the probability of the same event not happening, then the total probability is $p+q$ and is equal to unity, since it is an absolute certainty that the event either does or does not occur, i.e. $p + q = 1$

Expectation

The **expectation**, E, of an event happening is defined in general terms as the product of the probability p of an event happening and the number of attempts made, n, i.e. $E = pn$.

Thus, since the probability of obtaining a 3 upwards when rolling a fair dice is $\frac{1}{6}$, the expectation of getting a 3 upwards on four throws of the dice is $\frac{1}{6} \times 4$, i.e. $\frac{2}{3}$

Thus expectation is the average occurrence of an event.

Dependent event

A **dependent event** is one in which the probability of an event happening affects the probability of another ever happening. Let 5 transistors be taken at random from a batch of 100 transistors for test purposes, and the probability of there being a defective transistor, p_1, be determined. At some later time, let another 5 transistors be taken at random from the 95 remaining transistors in the batch and the probability of there being a defective transistor, p_2, be determined. The value of p_2 is different from p_1 since batch size has effectively altered from 100 to 95, i.e. probability p_2 is dependent on probability p_1. Since transistors are drawn, and then another 5 transistors drawn without replacing the first 5, the second random selection is said to be **without replacement**.

Independent event

An independent event is one in which the probability of an event happening does not affect the probability of another event happening. If 5 transistors are taken at random from a batch of transistors and the probability of a defective transistor p_1 is determined and the process is repeated after the original 5 have been replaced in the batch to give p_2, then p_1 is equal to p_2. Since the 5 transistors are replaced between draws, the second selection is said to be **with replacement**.

29.2 Laws of probability

The addition law of probability

The addition law of probability is recognised by the word **'or'** joining the probabilities. If p_A is the probability of event A happening and p_B is the probability of event B happening, the probability of **event A or event B** happening is given by

$p_A + p_B$. Similarly, the probability of events **A or B or C or ... N** happening is given by

$$p_A + p_B + p_C + \cdots + p_N$$

The multiplication law of probability

The multiplication law of probability is recognised by the word **'and'** joining the probabilities. If p_A is the probability of event A happening and p_B is the probability of event B happening, the probability of **event A and event B** happening is given by $p_A \times p_B$. Similarly, the probability of events **A and B and C and ... N** happening is given by

$$p_A \times p_B \times p_C \times \cdots \times p_N$$

29.3   Worked problems on probability

Problem 1. Determine the probabilities of selecting at random (a) a man, and (b) a woman from a crowd containing 20 men and 33 women.

(a) The probability of selecting at random a man, p, is given by the ratio

$$\frac{\text{number of men}}{\text{number in crowd}} \quad \text{i.e.} \quad p = \frac{20}{20 + 33}$$

$$= \frac{20}{53} \quad \text{or} \quad \textbf{0.3774}$$

(b) The probability of selecting at random a women, q, is given by the ratio

$$\frac{\text{number of women}}{\text{number in crowd}} \quad \text{i.e.} \quad q = \frac{33}{20 + 33}$$

$$= \frac{33}{53} \quad \text{or} \quad \textbf{0.6226}$$

(Check: the total probability should be equal to 1;

$$p = \frac{20}{53} \quad \text{and} \quad q = \frac{33}{53}$$

thus the total probability,

$$p + q = \frac{20}{53} + \frac{33}{53} = 1$$

hence no obvious error has been made).

Problem 2. Find the expectation of obtaining a 4 upwards with 3 throws of a fair dice.

Expectation is the average occurrence of an event and is defined as the probability times the number of attempts. The probability, p, of obtaining a 4 upwards for one throw of the dice is $\frac{1}{6}$.

Also, 3 attempts are made, hence $n = 3$ and the expectation, E, is pn, i.e. $E = \frac{1}{6} \times 3 = \frac{1}{2}$ or **0.50**

Problem 3. Calculate the probabilities of selecting at random:

(a) the winning horse in a race in which 10 horses are running,

(b) the winning horses in both the first and second races if there are 10 horses in each race.

(a) Since only one of the ten horses can win, the probability of selecting at random the winning horse is $\frac{\text{number of winners}}{\text{number of horses}}$, i.e. $\frac{1}{10}$ or **0.10**

(b) The probability of selecting the winning horse in the first race is $\frac{1}{10}$. The probability of selecting the winning horse in the second race is $\frac{1}{10}$. The probability of selecting the winning horses in the first **and** second race is given by the multiplication law of probability,

i.e.   **probability** $= \dfrac{1}{10} \times \dfrac{1}{10} = \dfrac{1}{100}$ or   **0.01**

Problem 4. The probability of a component failing in one year due to excessive temperature is $\frac{1}{20}$, due to excessive vibration is $\frac{1}{25}$ and due to excessive humidity is $\frac{1}{50}$. Determine the probabilities that during a one-year period a component: (a) fails due to excessive temperature and excessive vibration, (b) fails due to excessive vibration or excessive humidity, and (c) will not fail because of both excessive temperature and excessive humidity.

Let p_A be the probability of failure due to excessive temperature, then

$$p_A = \frac{1}{20} \quad \text{and} \quad \overline{p_A} = \frac{19}{20}$$

(where $\overline{p_A}$ is the probability of not failing.)

Let p_B be the probability of failure due to excessive vibration, then

$$p_B = \frac{1}{25} \quad \text{and} \quad \overline{p_B} = \frac{24}{25}$$

Let p_C be the probability of failure due to excessive humidity, then

$$p_C = \frac{1}{50} \quad \text{and} \quad \overline{p_C} = \frac{49}{50}$$

(a) The probability of a component failing due to excessive temperature **and** excessive vibration is given by:

$$p_A \times p_B = \frac{1}{20} \times \frac{1}{25} = \frac{1}{500} \quad \text{or} \quad \mathbf{0.002}$$

(b) The probability of a component failing due to excessive vibration **or** excessive humidity is:

$$p_B + p_C = \frac{1}{25} + \frac{1}{50} = \frac{3}{50} \quad \text{or} \quad \mathbf{0.06}$$

(c) The probability that a component will not fail due excessive temperature **and** will not fail due to excess humidity is:

$$\overline{p_A} \times \overline{p_C} = \frac{19}{20} \times \frac{49}{50} = \frac{931}{1000} \quad \text{or} \quad \mathbf{0.931}$$

Problem 5. A batch of 100 capacitors contains 73 which are within the required tolerance values, 17 which are below the required tolerance values, and the remainder are above the required tolerance values. Determine the probabilities that when randomly selecting a capacitor and then a second capacitor: (a) both are within the required tolerance values when selecting with replacement, and (b) the first one drawn is below and the second one drawn is above the required tolerance value, when selection is without replacement.

(a) The probability of selecting a capacitor within the required tolerance values is $\frac{73}{100}$. The first capacitor drawn is now replaced and a second one is drawn from the batch of 100. The probability of this capacitor being within the required tolerance values is also $\frac{73}{100}$.

Thus, the probability of selecting a capacitor within the required tolerance values for both the first **and** the second draw is

$$\frac{73}{100} \times \frac{73}{100} = \frac{5329}{10\,000} \quad \text{or} \quad \mathbf{0.5329}$$

(b) The probability of obtaining a capacitor below the required tolerance values on the first draw is $\frac{17}{100}$. There are now only 99 capacitors left in the batch, since the first capacitor is not replaced. The probability of drawing a capacitor above the required tolerance values on the second draw is $\frac{10}{99}$, since there are $(100 - 73 - 17)$, i.e. 10 capacitors above the required tolerance value. Thus, the probability of randomly selecting a capacitor below the required tolerance values and followed by randomly selecting a capacitor above the tolerance values is

$$\frac{17}{100} \times \frac{10}{99} = \frac{170}{9900} = \frac{17}{990} \quad \text{or} \quad \mathbf{0.0172}$$

Now try the following exercise

Exercise 105   Further problems on probability (Answers on page 264)

1. In a batch of 45 lamps there are 10 faulty lamps. If one lamp is drawn at random, find the probability of it being (a) faulty and (b) satisfactory.

2. A box of fuses are all of the same shape and size and comprises 23 2 A fuses, 47 5 A fuses and 69 13 A fuses. Determine the probability of selecting at random (a) a 2 A fuse, (b) a 5 A fuse and (c) a 13 A fuse.

3. (a) Find the probability of having a 2 upwards when throwing a fair 6-sided dice. (b) Find the probability of having a 5 upwards when throwing a fair 6-sided dice. (c) Determine the probability of having a 2 and then a 5 on two successive throws of a fair 6-sided dice.

4. The probability of event A happening is $\frac{3}{5}$ and the probability of event B happening is $\frac{2}{3}$. Calculate the probabilities of (a) both A and B happening, (b) only event A happening, i.e. event A happening and event B not happening, (c) only event B happening, and (d) either A, or B, or A and B happening.

5. When testing 1000 soldered joints, 4 failed during a vibration test and 5 failed due to having a high resistance. Determine the probability of a joint failing due to (a) vibration, (b) high resistance, (c) vibration or high resistance and (d) vibration and high resistance.

29.4 Further worked problems on probability

Problem 6. A batch of 40 components contains 5 which are defective. A component is drawn at random from the batch and tested and then a second component is drawn. Determine the probability that neither of the components is defective when drawn (a) with replacement, and (b) without replacement.

(a) With replacement

The probability that the component selected on the first draw is satisfactory is $\frac{35}{40}$, i.e. $\frac{7}{8}$. The component is now replaced and a second draw is made. The probability that this component is also satisfactory is $\frac{7}{8}$. Hence, the probability that both the first component drawn **and** the second component drawn are satisfactory is:

$$\frac{7}{8} \times \frac{7}{8} = \frac{49}{64} \quad \text{or} \quad \mathbf{0.7656}$$

(b) Without replacement

The probability that the first component drawn is satisfactory is $\frac{7}{8}$. There are now only 34 satisfactory components left in the batch and the batch number is 39. Hence, the probability of drawing a satisfactory component on the second draw is $\frac{34}{39}$. Thus the probability that the first component drawn **and** the second component drawn are satisfactory, i.e. neither is defective, is:

$$\frac{7}{8} \times \frac{34}{39} = \frac{238}{312} \quad \text{or} \quad \mathbf{0.7628}$$

Problem 7. A batch of 40 components contains 5 which are defective. If a component is drawn at random from the batch and tested and then a second component is drawn at random, calculate the probability of having one defective component, both with and without replacement.

The probability of having one defective component can be achieved in two ways. If p is the probability of drawing a defective component and q is the probability of drawing a satisfactory component, then the probability of having one defective component is given by drawing a satisfactory component and then a defective component **or** by drawing a defective component and then a satisfactory one, i.e. by $q \times p + p \times q$

With replacement:

$$p = \frac{5}{40} = \frac{1}{8} \quad \text{and} \quad q = \frac{35}{40} = \frac{7}{8}$$

Hence, probability of having one defective component is:

$$\frac{1}{8} \times \frac{7}{8} + \frac{7}{8} \times \frac{1}{8}$$

i.e.
$$\frac{7}{64} + \frac{7}{64} = \frac{7}{32} \quad \text{or} \quad \mathbf{0.2188}$$

Without replacement:

$p_1 = \frac{1}{8}$ and $q_1 = \frac{7}{8}$ on the first of the two draws. The batch number is now 39 for the second draw, thus,

$$p_2 = \frac{5}{39} \quad \text{and} \quad q_2 = \frac{35}{39}$$

$$p_1 q_2 + q_1 p_2 = \frac{1}{8} \times \frac{35}{39} + \frac{7}{8} \times \frac{5}{39} = \frac{35 + 35}{312}$$

$$= \frac{70}{312} \quad \text{or} \quad \mathbf{0.2244}$$

Problem 8. A box contains 74 brass washers, 86 steel washers and 40 aluminium washers. Three washers

---

are drawn at random from the box without replacement. Determine the probability that all three are steel washers.

Assume, for clarity of explanation, that a washer is drawn at random, then a second, then a third (although this assumption does not affect the results obtained). The total number of washers is $74 + 86 + 40$, i.e. 200.

The probability of randomly selecting a steel washer on the first draw is $\frac{86}{200}$. There are now 85 steel washers in a batch of 199. The probability of randomly selecting a steel washer on the second draw is $\frac{85}{199}$. There are now 84 steel washers in a batch of 198. The probability of randomly selecting a steel washer on the third draw is $\frac{84}{198}$. Hence the probability of selecting a steel washer on the first draw **and** the second draw **and** the third draw is:

$$\frac{86}{200} \times \frac{85}{199} \times \frac{84}{198} = \frac{614\,040}{7\,880\,400} = 0.0779$$

Problem 9. For the box of washers given in Problem 8 above, determine the probability that there are no aluminium washers drawn, when three washers are drawn at random from the box without replacement.

The probability of not drawing an aluminium washer on the first draw is $1 - \left(\frac{40}{200}\right)$, i.e. $\frac{160}{200}$. There are now 199 washers in the batch of which 159 are not aluminium washers. Hence, the probability of not drawing an aluminium washer on the second draw is $\frac{159}{199}$. Similarly, the probability of not drawing an aluminium washer on the third draw is $\frac{158}{198}$. Hence the probability of not drawing an aluminium washer on the first **and** second **and** third draws is

$$\frac{160}{200} \times \frac{159}{199} \times \frac{158}{198} = \frac{4\,019\,520}{7\,880\,400} = 0.5101$$

Problem 10. For the box of washers in Problem 8 above, find the probability that there are two brass washers and either a steel or an aluminium washer when three are drawn at random, without replacement.

Two brass washers (A) and one steel washer (B) can be obtained in any of the following ways:

1st draw	2nd draw	3rd draw
A	A	B
A	B	A
B	A	A

Two brass washers and one aluminium washer (*C*) can also be obtained in any of the following ways:

1st draw	2nd draw	3rd draw
A	*A*	*C*
A	*C*	*A*
C	*A*	*A*

Thus there are six possible ways of achieving the combinations specified. If *A* represents a brass washer, *B* a steel washer and *C* an aluminium washer, then the combinations and their probabilities are as shown:

First	Second	Third	PROBABILITY
A	*A*	*B*	$\frac{74}{200} \times \frac{73}{199} \times \frac{86}{198} = 0.0590$
A	*B*	*A*	$\frac{74}{200} \times \frac{86}{199} \times \frac{73}{198} = 0.0590$
B	*A*	*A*	$\frac{86}{200} \times \frac{74}{199} \times \frac{73}{198} = 0.0590$
A	*A*	*C*	$\frac{74}{200} \times \frac{73}{199} \times \frac{40}{198} = 0.0274$
A	*C*	*A*	$\frac{74}{200} \times \frac{40}{199} \times \frac{73}{198} = 0.0274$
C	*A*	*A*	$\frac{40}{200} \times \frac{74}{199} \times \frac{73}{198} = 0.0274$

The probability of having the first combination **or** the second, **or** the third, and so on, is given by the sum of the probabilities,

i.e. by $3 \times 0.0590 + 3 \times 0.0274$, that is, **0.2592**

Now try the following exercise

Exercise 106   Further problems on probability
(Answers on page 264)

1. The probability that component *A* will operate satisfactorily for 5 years is 0.8 and that *B* will operate satisfactorily over that same period of time is 0.75. Find the probabilities that in a 5 year period: (a) both components operate satisfactorily, (b) only component *A* will operate satisfactorily, and (c) only component *B* will operate satisfactorily.

2. In a particular street, 80% of the houses have telephones. If two houses selected at random are visited, calculate the probabilities that (a) they both have a telephone and (b) one has a telephone but the other does not have telephone.

3. Veroboard pins are packed in packets of 20 by a machine. In a thousand packets, 40 have less than 20 pins. Find the probability that if 2 packets are chosen at random, one will contain less than 20 pins and the other will contain 20 pins or more.

4. A batch of 1 kW fire elements contains 16 which are within a power tolerance and 4 which are not. If 3 elements are selected at random from the batch, calculate the probabilities that (a) all three are within the power tolerance and (b) two are within but one is not within the power tolerance.

5. An amplifier is made up of three transistors, *A*, *B* and *C*. The probabilities of *A*, *B* or *C* being defective are $\frac{1}{20}$, $\frac{1}{25}$ and $\frac{1}{50}$, respectively. Calculate the percentage of amplifiers produced (a) which work satisfactorily and (b) which have just one defective transistor.

6. A box contains 14 40 W lamps, 28 60 W lamps and 58 25 W lamps, all the lamps being of the same shape and size. Three lamps are drawn at random from the box, first one, then a second, then a third. Determine the probabilities of: (a) getting one 25 W, one 40 W and one 60 W lamp, with replacement, (b) getting one 25 W, one 40 W and one 60 W lamp without replacement, and (c) getting either one 25 W and two 40 W or one 60 W and two 40 W lamps with replacement.

Assignment 14

This assignment covers the material in Chapters 27 to 29. The marks for each question are shown in brackets at the end of each question.

1. A company produces five products in the following proportions:

Product *A* 24   Product *B*   6   Product *C* 15

Product *D*   9   Product *E* 18

Draw (a) a horizontal bar chart, and (b) a pie diagram to represent these data visually. (7)

2. State whether the data obtained on the following topics are likely to be discrete or continuous:

(a) the number of books in a library

(b) the speed of a car

(c) the time to failure of a light bulb (3)

3. Draw a histogram, frequency polygon and ogive for the data given below which refers to the diameter of 50 components produced by a machine.

Class intervals	Frequency
1.30–1.32 mm	4
1.33–1.35 mm	7
1.36–1.38 mm	10
1.39–1.41 mm	12
1.42–1.44 mm	8
1.45–1.47 mm	5
1.48–1.50 mm	4

(10)

4. Determine the mean, median and modal values for the lengths given in metres:

28, 20, 44, 30, 32, 30, 28, 34, 26, 28 (3)

5. The length in millimetres of 100 bolts is as shown below:

50–56 6 57–63 16 64–70 22

71–77 30 78–84 19 85–91 7

Determine for the sample (a) the mean value, and (b) the standard deviation, correct to 4 significant figures. (9)

6. The number of faulty components in a factory in a 12 week period is:

14 12 16 15 10 13 15 11 16 19 17 19

Determine the median and the first and third quartile values. (3)

7. Determine the probability of winning a prize in a lottery by buying 10 tickets when there are 10 prizes and a total of 5000 tickets sold. (2)

8. A sample of 50 resistors contains 44 which are within the required tolerance value, 4 which are below and the remainder which are above. Determine the probability of selecting from the sample a resistor which is (a) below the required tolerance, and (b) above the required tolerance. Now two resistors are selected at random from the sample. Determine the probability, correct to 3 decimal places, that neither resistor is defective when drawn (c) with replacement, and (d) without replacement. (e) If a resistor is drawn at random from the batch and tested, and then a second resistor is drawn from those left, calculate the probability of having one defective component when selection is without replacement. (13)

30

Introduction to differentiation

30.1 Introduction to calculus

Calculus is a branch of mathematics involving or leading to calculations dealing with continuously varying functions. Calculus is a subject that falls into two parts:

 (i) **differential calculus** (or **differentiation**) and

(ii) **integral calculus** (or **integration**).

Differentiation is used in calculations involving rates of change (see section 30.10), velocity and acceleration, and maximum and minimum values of curves (see 'Engineering Mathematics').

30.2 Functional notation

In an equation such as $y = 3x^2 + 2x - 5$, y is said to be a function of x and may be written as $y = f(x)$.

An equation written in the form $f(x) = 3x^2 + 2x - 5$ is termed **functional notation**.

The value of $f(x)$ when $x = 0$ is denoted by $f(0)$, and the value of $f(x)$ when $x = 2$ is denoted by $f(2)$ and so on. Thus when $f(x) = 3x^2 + 2x - 5$, then

$$f(0) = 3(0)^2 + 2(0) - 5 = -5$$

and $\qquad f(2) = 3(2)^2 + 2(2) - 5 = 11$ and so on.

Problem 1. If $f(x) = 4x^2 - 3x + 2$ find: $f(0)$, $f(3)$, $f(-1)$ and $f(3) - f(-1)$

$$f(x) = 4x^2 - 3x + 2$$

$$f(0) = 4(0)^2 - 3(0) + 2 = \mathbf{2}$$

$$f(3) = 4(3)^2 - 3(3) + 2 = 36 - 9 + 2 = \mathbf{29}$$

$$f(-1) = 4(-1)^2 - 3(-1) + 2 = 4 + 3 + 2 = \mathbf{9}$$

$$f(3) - f(-1) = 29 - 9 = \mathbf{20}$$

Problem 2. Given that $f(x) = 5x^2 + x - 7$ determine:

(i) $f(2) \div f(1)$ $\qquad$ (ii) $f(3 + a)$

(iii) $f(3 + a) - f(3)$ $\qquad$ (iv) $\dfrac{f(3 + a) - f(3)}{a}$

$$f(x) = 5x^2 + x - 7$$

(i) $f(2) = 5(2)^2 + 2 - 7 = 15$

$\qquad f(1) = 5(1)^2 + 1 - 7 = -1$

$\qquad f(2) \div f(1) = \dfrac{15}{-1} = \mathbf{-15}$

(ii) $f(3 + a) = 5(3 + a)^2 + (3 + a) - 7$

$\qquad\qquad = 5(9 + 6a + a^2) + (3 + a) - 7$

$\qquad\qquad = 45 + 30a + 5a^2 + 3 + a - 7 = \mathbf{41 + 31a + 5a^2}$

(iii) $f(3) = 5(3)^2 + 3 - 7 = 41$

$\qquad f(3 + a) - f(3) = (41 + 31a + 5a^2) - (41) = \mathbf{31a + 5a^2}$

(iv) $\dfrac{f(3 + a) - f(3)}{a} = \dfrac{31a + 5a^2}{a} = \mathbf{31 + 5a}$

Now try the following exercise

Exercise 107 Further problems on functional notation (Answers on page 265)

1. If $f(x) = 6x^2 - 2x + 1$ find $f(0)$, $f(1)$, $f(2)$, $f(-1)$ and $f(-3)$

2. If $f(x) = 2x^2 + 5x - 7$ find $f(1)$, $f(2)$, $f(-1)$, $f(2) - f(-1)$

3. Given $f(x) = 3x^3 + 2x^2 - 3x + 2$ prove that
$$f(1) = \frac{1}{7}f(2)$$

4. If $f(x) = -x^2 + 3x + 6$ find $f(2)$, $f(2+a)$,
$f(2+a) - f(2)$ and $\dfrac{f(2+a) - f(2)}{a}$

30.3 The gradient of a curve

(a) If a tangent is drawn at a point P on a curve, then the
gradient of this tangent is said to be the **gradient of the
curve** at P. In Figure 30.1, the gradient of the curve at
P is equal to the gradient of the tangent PQ.

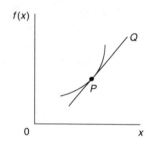

Fig. 30.1

(b) For the curve shown in Figure 30.2, let the points A and
B have co-ordinates (x_1, y_1) and (x_2, y_2), respectively.
In functional notation, $y_1 = f(x_1)$ and $y_2 = f(x_2)$ as
shown.

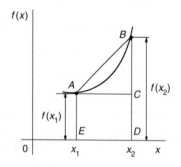

Fig. 30.2

The gradient of the chord AB
$$= \frac{BC}{AC} = \frac{BD - CD}{ED}$$
$$= \frac{f(x_2) - f(x_1)}{(x_2 - x_1)}$$

(c) For the curve $f(x) = x^2$ shown in Figure 30.3:

(i) the gradient of chord AB
$$= \frac{f(3) - f(1)}{3 - 1} = \frac{9 - 1}{2} = 4$$

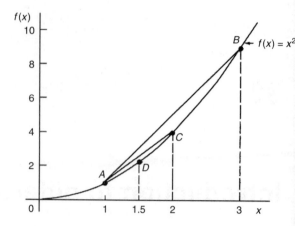

Fig. 30.3

(ii) the gradient of chord AC
$$= \frac{f(2) - f(1)}{2 - 1} = \frac{4 - 1}{1} = 3$$

(iii) the gradient of chord AD
$$= \frac{f(1.5) - f(1)}{1.5 - 1} = \frac{2.25 - 1}{0.5} = 2.5$$

(iv) if E is the point on the curve
$(1.1, f(1.1))$ then the gradient of chord AE
$$= \frac{f(1.1) - f(1)}{1.1 - 1}$$
$$= \frac{1.21 - 1}{0.1} = 2.1$$

(v) if F is the point on the curve $(1.01, f(1.01))$ then
the gradient of chord AF
$$= \frac{f(1.01) - f(1)}{1.01 - 1}$$
$$= \frac{1.0201 - 1}{0.01} = 2.01$$

Thus as point B moves closer and closer to point A the
gradient of the chord approaches nearer and nearer to the
value 2. This is called the **limiting value** of the gradient of
the chord AB and when B coincides with A the chord becomes
the tangent to the curve.

Now try the following exercise

**Exercise 108   A further problem on the gradient of a
curve (Answers on page 265)**

1. Plot the curve $f(x) = 4x^2 - 1$ for values of x from
$x = -1$ to $x = +4$. Label the co-ordinates $(3, f(3))$ and
$(1, f(1))$ as J and K, respectively. Join points J and K

to form the chord JK. Determine the gradient of chord JK. By moving J nearer and nearer to K determine the gradient of the tangent of the curve at K.

30.4 Differentiation from first principles

(i) In Figure 30.4, A and B are two points very close together on a curve, δx (delta x) and δy (delta y) representing small increments in the x and y directions, respectively.

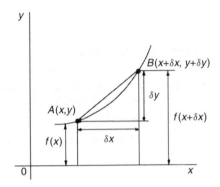

Fig. 30.4

Gradient of chord $AB = \dfrac{\delta y}{\delta x}$

however, $\delta y = f(x + \delta x) - f(x)$

Hence $\dfrac{\delta y}{\delta x} = \dfrac{f(x + \delta x) - f(x)}{\delta x}$

As δx approaches zero, $\dfrac{\delta y}{\delta x}$ approaches a limiting value and the gradient of the chord approaches the gradient of the tangent at A.

(ii) When determining the gradient of a tangent to a curve there are two notations used. The gradient of the curve at A in Figure 30.4 can either be written as:

$$\lim_{\delta x \to 0} \frac{\delta y}{\delta x} \ \text{or} \ \lim_{\delta x \to 0} \left\{ \frac{f(x + \delta x) - f(x)}{\delta x} \right\}$$

In **Leibniz notation**, $\dfrac{dy}{dx} = \lim\limits_{\delta x \to 0} \dfrac{\delta y}{\delta x}$

In **functional notation**,

$$f'(x) = \lim_{\delta x \to 0} \left\{ \frac{f(x + \delta x) - f(x)}{\delta x} \right\}$$

(iii) $\dfrac{dy}{dx}$ is the same as $f'(x)$ and is called the **differential coefficient** or the **derivative**. The process of finding the differential coefficient is called **differentiation**.

Summarising, the differential coefficient,

$$\frac{dy}{dx} = f'(x) = \lim_{\delta x \to 0} \frac{\delta y}{\delta x} = \lim_{\delta x \to 0} \left\{ \frac{f(x + \delta x) - f(x)}{\delta x} \right\}$$

Problem 3. Differentiate from first principles $f(x) = x^2$ and determine the value of the gradient of the curve at $x = 2$

To 'differentiate from first principles' means 'to find $f'(x)$' by using the expression

$$f'(x) = \lim_{\delta x \to 0} \left\{ \frac{f(x + \delta x) - f(x)}{\delta x} \right\}$$

$$f(x) = x^2$$

Substituting $(x + \delta x)$ for x gives
$f(x + \delta x) = (x + \delta x)^2 = x^2 + 2x\delta x + \delta x^2$, hence

$$f'(x) = \lim_{\delta x \to 0} \left\{ \frac{(x^2 + 2x\delta x + \delta x^2) - (x^2)}{\delta x} \right\}$$

$$= \lim_{\delta x \to 0} \left\{ \frac{2x\delta x + \delta x^2}{\delta x} \right\} = \lim_{\delta x \to 0} \{2x + \delta x\}$$

As $\delta x \to 0$, $\{2x + \delta x\} \to \{2x + 0\}$. Thus $f'(x) = 2x$, i.e. **the differential coefficient of x^2 is $2x$.**

At $x = 2$, the gradient of the curve, $f'(x) = 2(2) = 4$

Problem 4. Find the differential coefficient of $y = 5x$

By definition, $\dfrac{dy}{dx} = f'(x) = \lim\limits_{\delta x \to 0} \left\{ \dfrac{f(x + \delta x) - f(x)}{\delta x} \right\}$

The function being differentiated is $y = f(x) = 5x$. Substituting $(x + \delta x)$ for x gives: $f(x + \delta x) = 5(x + \delta x) = 5x + 5\delta x$. Hence

$$\frac{dy}{dx} = f'(x) = \lim_{\delta x \to 0} \left\{ \frac{(5x + 5\delta x) - (5x)}{\delta x} \right\}$$

$$= \lim_{\delta x \to 0} \left\{ \frac{5\delta x}{\delta x} \right\} = \lim_{\delta x \to 0} \{5\}$$

Since the term δx does not appear in $\{5\}$ the limiting value as $\delta x \to 0$ of $\{5\}$ is 5.

Thus $\dfrac{dy}{dx} = 5$, i.e. **the differential coefficient of $5x$ is 5.**

The equation $y = 5x$ represents a straight line of gradient 5 (see Chapter 11).

The 'differential coefficient' (i.e. $\dfrac{dy}{dx}$ or $f'(x)$) means 'the gradient of the curve', and since the gradient of the line $y = 5x$ is 5 this result can be obtained by inspection. Hence, in general, **if $y = kx$** (where k is a constant), then the gradient of the line is k and $\dfrac{dy}{dx}$ or $f'(x) = k$.

Problem 5. Find the derivative of $y = 8$

$y = f(x) = 8$. Since there are no x-values in the original equation, substituting $(x + \delta x)$ for x still gives $f(x + \delta x) = 8$. Hence

$$\frac{dy}{dx} = f'(x) = \lim_{\delta x \to 0} \left\{ \frac{f(x + \delta x) - f(x)}{\delta x} \right\}$$

$$= \lim_{\delta x \to 0} \left\{ \frac{8 - 8}{\delta x} \right\} = 0$$

Thus, when $y = 8$, $\dfrac{dy}{dx} = 0$

The equation $y = 8$ represents a straight horizontal line and the gradient of a horizontal line is zero, hence the result could have been determined by inspection. 'Finding the derivative' means 'finding the gradient', hence, in general, for any horizontal line **if $y = k$** (where k is a constant) then $\dfrac{dy}{dx} = 0$.

Problem 6. Differentiate from first principles $f(x) = 2x^3$

Substituting $(x + \delta x)$ for x gives

$$f(x + \delta x) = 2(x + \delta x)^3 = 2(x + \delta x)(x^2 + 2x\delta x + \delta x^2)$$

$$= 2(x^3 + 3x^2\delta x + 3x\delta x^2 + \delta x^3)$$

$$= 2x^3 + 6x^2\delta x + 6x\delta x^2 + 2\delta x^3$$

$$\frac{dy}{dx} = f'(x) = \lim_{\delta x \to 0} \left\{ \frac{f(x + \delta x) - f(x)}{\delta x} \right\}$$

$$= \lim_{\delta x \to 0} \left\{ \frac{(2x^3 + 6x^2\delta x + 6x\delta x^2 + 2\delta x^3) - (2x^3)}{\delta x} \right\}$$

$$= \lim_{\delta x \to 0} \left\{ \frac{6x^2\delta x + 6x\delta x^2 + 2\delta x^3}{\delta x} \right\}$$

$$= \lim_{\delta x \to 0} \left\{ 6x^2 + 6x\delta x + 2\delta x^2 \right\}$$

Hence $f'(x) = 6x^2$, i.e. **the differential coefficient of $2x^3$ is $6x^2$**.

Problem 7. Find the differential coefficient of $y = 4x^2 + 5x - 3$ and determine the gradient of the curve at $x = -3$

$$y = f(x) = 4x^2 + 5x - 3$$

$$f(x + \delta x) = 4(x + \delta x)^2 + 5(x + \delta x) - 3$$

$$= 4(x^2 + 2x\delta x + \delta x^2) + 5x + 5\delta x - 3$$

$$= 4x^2 + 8x\delta x + 4\delta x^2 + 5x + 5\delta x - 3$$

$$\frac{dy}{dx} = f'(x) = \lim_{\delta x \to 0} \left\{ \frac{f(x + \delta x) - f(x)}{\delta x} \right\}$$

$$= \lim_{\delta x \to 0} \left\{ \frac{\begin{array}{c}(4x^2 + 8x\delta x + 4\delta x^2 + 5x + 5\delta x - 3) \\ -(4x^2 + 5x - 3)\end{array}}{\delta x} \right\}$$

$$= \lim_{\delta x \to 0} \left\{ \frac{8x\delta x + 4\delta x^2 + 5\delta x}{\delta x} \right\} = \lim_{\delta x \to 0} \left\{ 8x + 4\delta x + 5 \right\}$$

i.e. $\dfrac{dy}{dx} = f'(x) = \mathbf{8x + 5}$

At $x = -3$, the gradient of the curve

$$= \frac{dy}{dx} = f'(x) = 8(-3) + 5 = \mathbf{-19}$$

Now try the following exercise

Exercise 109 **Further problems on differentiation from first principles (Answers on page 265)**

In Problems 1 to 12, differentiate from first principles.

1. $y = x$
2. $y = 7x$
3. $y = 4x^2$
4. $y = 5x^3$
5. $y = -2x^2 + 3x - 12$
6. $y = 23$
7. $f(x) = 9x$
8. $f(x) = \dfrac{2x}{3}$
9. $f(x) = 9x^2$
10. $f(x) = -7x^3$
11. $f(x) = x^2 + 15x - 4$
12. $f(x) = 4$
13. Determine $\dfrac{d}{dx}(4x^3 - 1)$ from first principles
14. Find $\dfrac{d}{dx}(3x^2 + 5)$ from first principles

30.5 Differentiation of $y = ax^n$ by the general rule

From differentiation by first principles, a general rule for differentiating ax^n emerges where a and n are any constants. This rule is:

if	$y = ax^n$	then	$\dfrac{dy}{dx} = anx^{n-1}$
or, if	$f(x) = ax^n$	then	$f'(x) = anx^{n-1}$

(Each of the results obtained in worked problems 3 and 7 may be deduced by using this general rule).

When differentiating, results can be expressed in a number of ways.

For example: (i) if $y = 3x^2$ then $\dfrac{dy}{dx} = 6x$,

              (ii) if $f(x) = 3x^2$ then $f'(x) = 6x$,

              (iii) the differential coefficient of $3x^2$ is $6x$,

              (iv) the derivative of $3x^2$ is $6x$, and

              (v) $\dfrac{d}{dx}(3x^2) = 6x$

Problem 8. Using the general rule, differentiate the following with respect to x:

(a) $y = 5x^7$ (b) $y = 3\sqrt{x}$ (c) $y = \dfrac{4}{x^2}$

(a) Comparing $y = 5x^7$ with $y = ax^n$ shows that $a = 5$ and $n = 7$. Using the general rule,

$$\frac{dy}{dx} = anx^{n-1} = (5)(7)x^{7-1} = 35x^6$$

(b) $y = 3\sqrt{x} = 3x^{\frac{1}{2}}$. Hence $a = 3$ and $n = \dfrac{1}{2}$

$$\frac{dy}{dx} = anx^{n-1} = (3)\frac{1}{2}x^{\frac{1}{2}-1}$$

$$= \frac{3}{2}x^{-\frac{1}{2}} = \frac{3}{2x^{\frac{1}{2}}} = \frac{3}{2\sqrt{x}}$$

(c) $y = \dfrac{4}{x^2} = 4x^{-2}$. Hence $a = 4$ and $n = -2$

$$\frac{dy}{dx} = anx^{n-1} = (4)(-2)x^{-2-1}$$

$$= -8x^{-3} = -\frac{8}{x^3}$$

Problem 9. Find the differential coefficient of
$y = \dfrac{2}{5}x^3 - \dfrac{4}{x^3} + 4\sqrt{x^5} + 7$

$$y = \frac{2}{5}x^3 - \frac{4}{x^3} + 4\sqrt{x^5} + 7$$

i.e. $y = \dfrac{2}{5}x^3 - 4x^{-3} + 4x^{5/2} + 7$ from the laws of

indices (see Chapter 3)

$$\frac{dy}{dx} = \left(\frac{2}{5}\right)(3)x^{3-1} - (4)(-3)x^{-3-1}$$

$$+ (4)\left(\frac{5}{2}\right)x^{(5/2)-1} + 0$$

$$= \frac{6}{5}x^2 + 12x^{-4} + 10x^{3/2}$$

i.e. $\dfrac{dy}{dx} = \dfrac{6}{5}x^2 + \dfrac{12}{x^4} + 10\sqrt{x^3}$

Problem 10. If $f(t) = 5t + \dfrac{1}{\sqrt{t^3}}$ find $f'(t)$

$$f(t) = 5t + \frac{1}{\sqrt{t^3}} = 5t + \frac{1}{t^{\frac{3}{2}}} = 5t^1 + t^{-\frac{3}{2}}$$

Hence $f'(t) = (5)(1)t^{1-1} + \left(-\dfrac{3}{2}\right)t^{-\frac{3}{2}-1} = 5t^0 - \dfrac{3}{2}t^{-\frac{5}{2}}$

i.e. $f'(t) = 5 - \dfrac{3}{2t^{\frac{5}{2}}} = 5 - \dfrac{3}{2\sqrt{t^5}}$ (since $t^0 = 1$)

Problem 11. Differentiate $y = \dfrac{(x+2)^2}{x}$ with respect to x

---

$$y = \frac{(x+2)^2}{x} = \frac{x^2 + 4x + 4}{x} = \frac{x^2}{x} + \frac{4x}{x} + \frac{4}{x}$$

i.e. $y = x^1 + 4 + 4x^{-1}$

Hence $\dfrac{dy}{dx} = 1x^{1-1} + 0 + (4)(-1)x^{-1-1} = x^0 - 4x^{-2}$

$$= 1 - \frac{4}{x^2} \quad \text{(since } x^0 = 1)$$

Now try the following exercise

Exercise 110   **Further problems on differentiation of $y = ax^n$ by the general rule (Answers on page 265)**

In Problems 1 to 8, determine the differential coefficients with respect to the variable.

1. $y = 7x^4$ 2. $y = \sqrt{x}$

3. $y = \sqrt{t^3}$ 4. $y = 6 + \dfrac{1}{x^3}$

5. $y = 3x - \dfrac{1}{\sqrt{x}} + \dfrac{1}{x}$ 6. $y = \dfrac{5}{x^2} - \dfrac{1}{\sqrt{x^7}} + 2$

7. $y = 3(t-2)^2$ 8. $y = (x+1)^3$

9. Using the general rule for ax^n check the results of Problems 1 to 12 of Exercise 109, page 232.

10. Differentiate $f(x) = 6x^2 - 3x + 5$ and find the gradient of the curve at (a) $x = -1$, and (b) $x = 2$.

11. Find the differential coefficient of $y = 2x^3 + 3x^2 - 4x - 1$ and determine the gradient. of the curve at $x = 2$.

12. Determine the derivative of $y = -2x^3 + 4x + 7$ and determine the gradient of the curve at $x = -1.5$.

30.6 Differentiation of sine and cosine functions

Figure 30.5(a) shows a graph of $y = \sin\theta$. The gradient is continually changing as the curve moves from 0 to A to B to C to D. The gradient, given by $\dfrac{dy}{d\theta}$, may be plotted in a corresponding position below $y = \sin\theta$, as shown in Figure 30.5(b).

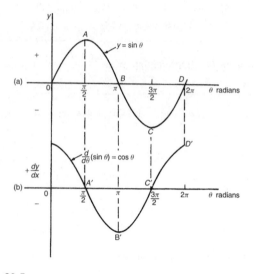

Fig. 30.5

(i) At 0, the gradient is positive and is at its steepest. Hence $0'$ is a maximum positive value.

(ii) Between 0 and A the gradient is positive but is decreasing in value until at A the gradient is zero, shown as A′.

(iii) Between A and B the gradient is negative but is increasing in value until at B the gradient is at its steepest. Hence B′ is a maximum negative value.

(iv) If the gradient of $y = \sin\theta$ is further investigated between B and C and C and D then the resulting graph of $\dfrac{dy}{d\theta}$ is seen to be a cosine wave.

Hence the rate of change of $\sin\theta$ is $\cos\theta$, i.e.

$$\text{if} \quad y = \sin\theta \quad \text{then} \quad \frac{dy}{d\theta} = \cos\theta$$

It may also be shown that:

$$\text{if} \quad y = \sin a\theta, \qquad \frac{dy}{d\theta} = a\,\cos a\theta \qquad (1)$$

(where a is a constant)

$$\text{and if} \quad y = \sin(a\theta + \alpha), \qquad \frac{dy}{d\theta} = a\,\cos(a\theta + \alpha) \qquad (2)$$

(where a and α are constants)

If a similar exercise is followed for $y = \cos\theta$ then the graphs of Figure 30.6 result, showing $\dfrac{dy}{d\theta}$ to be a graph of $\sin\theta$, but displaced by π radians.

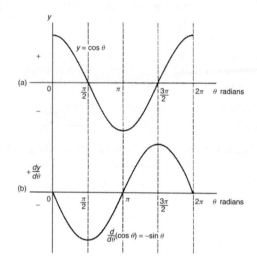

Fig. 30.6

If each point on the curve $y = \sin\theta$ (as shown in Figure 30.5(a)) were to be made negative, (i.e. $+\dfrac{\pi}{2}$ is made $-\dfrac{\pi}{2}$, $-\dfrac{3\pi}{2}$ is made $+\dfrac{3\pi}{2}$, and so on) then the graph shown in Figure 30.6(b) would result. This latter graph therefore represents the curve of $-\sin\theta$.

Thus, if $y = \cos\theta$, $\dfrac{dy}{d\theta} = -\sin\theta$

It may also be shown that:

$$\text{if} \quad y = \cos a\theta, \quad \frac{dy}{d\theta} = -a\sin a\theta \qquad (3)$$

(where a is a constant)

$$\text{and} \quad \text{if} \quad y = \cos(a\theta + \alpha), \quad \frac{dy}{d\theta} = -a\sin(a\theta + \alpha) \qquad (4)$$

(where a and α are constants)

Problem 12. Differentiate the following with respect to the variable:
(a) $y = 2\sin 5\theta$ (b) $f(t) = 3\cos 2t$

(a) $y = 2\sin 5\theta$

$$\frac{dy}{d\theta} = (2)(5\cos 5\theta) = \mathbf{10\cos 5\theta} \qquad \text{from equation (1)}$$

(b) $f(t) = 3\cos 2t$

$$f'(t) = (3)(-2\sin 2t) = \mathbf{-6\sin 2t} \quad \text{from equation (3)}$$

Problem 13. Find the differential coefficient of
$y = 7 \sin 2x - 3 \cos 4x$

$y = 7 \sin 2x - 3 \cos 4x$

$\dfrac{dy}{dx} = (7)(2 \cos 2x) - (3)(-4 \sin 4x)$ from

equations (1) and (3)

$= \mathbf{14 \cos 2x + 12 \sin 4x}$

Problem 14. Differentiate the following with respect to the variable:

(a) $f(\theta) = 5 \sin(100\pi\theta - 0.40)$

(b) $f(t) = 2 \cos(5t + 0.20)$

(a) If $f(\theta) = 5 \sin(100\pi\theta - 0.40)$

$f'(\theta) = 5[100\pi \cos(100\pi\theta - 0.40)]$ from equation (2),

where $a = 100\pi$

$= \mathbf{500\pi \cos(100\pi\theta - 0.40)}$

(b) If $f(t) = 2 \cos(5t + 0.20)$

$f'(t) = 2[-5 \sin(5t + 0.20)]$ from equation (4),

where $a = 5$

$= \mathbf{-10 \sin(5t + 0.20)}$

Problem 15. An alternating voltage is given by:
$v = 100 \sin 200t$ volts, where t is the time in seconds. Calculate the rate of change of voltage when
(a) $t = 0.005$ s and (b) $t = 0.01$ s

$v = 100 \sin 200t$ volts. The rate of change of v is given by $\dfrac{dv}{dt}$.

$\dfrac{dv}{dt} = (100)(200 \cos 200t) = 20000 \cos 200t$

(a) When $t = 0.005$ s,

$\dfrac{dv}{dt} = 20\,000 \cos(200)(0.005) = 20\,000 \cos 1$

cos 1 means 'the cosine of 1 radian' (make sure your calculator is on radians – not degrees).

Hence $\dfrac{dv}{dt} = \mathbf{10\,806\ volts\ per\ second}$

(b) When $t = 0.01$ s,

$\dfrac{dv}{dt} = 20\,000 \cos(200)(0.01) = 20000 \cos 2$

Hence $\dfrac{dv}{dt} = \mathbf{-8323\ volts\ per\ second}$

Now try the following exercise

Exercise 111 Further problems on the differentiation of sine and cosine functions (Answers on page 265)

1. Differentiate with respect to x:

    (a) $y = 4 \sin 3x$ (b) $y = 2 \cos 6x$

2. Given $f(\theta) = 2 \sin 3\theta - 5 \cos 2\theta$, find $f'(\theta)$

3. An alternating current is given by $i = 5 \sin 100t$ amperes, where t is the time in seconds. Determine the rate of change of current (i.e. $\dfrac{di}{dt}$) when $t = 0.01$ seconds.

4. $v = 50 \sin 40t$ volts represents an alternating voltage, v, where t is the time in seconds. At a time of 20×10^{-3} seconds, find the rate of change of voltage (i.e. $\dfrac{dv}{dt}$).

5. If $f(t) = 3 \sin(4t + 0.12) - 2 \cos(3t - 0.72)$ determine $f'(t)$.

30.7 Differentiation of e^{ax} and ln ax

A graph of $y = e^x$ is shown in Figure 30.7(a). The gradient of the curve at any point is given by $\dfrac{dy}{dx}$ and is continually changing. By drawing tangents to the curve at many points on the curve and measuring the gradient of the tangents, values of $\dfrac{dy}{dx}$ for corresponding values of x may be obtained. These values are shown graphically in Figure 30.7(b). The graph of $\dfrac{dy}{dx}$ against x is identical to the original graph of $y = e^x$. It follows that:

$$\text{if } \mathbf{y = e^x}, \quad \text{then} \quad \dfrac{dy}{dx} = \mathbf{e^x}$$

It may also be shown that

$$\text{if } \mathbf{y = e^{ax}}, \quad \text{then} \quad \dfrac{dy}{dx} = \mathbf{a\,e^{ax}}$$

Therefore if $y = 2e^{6x}$, then $\dfrac{dy}{dx} = (2)(6e^{6x}) = 12e^{6x}$

A graph of $y = \ln x$ is shown in Figure 30.8(a). The gradient of the curve at any point is given by $\dfrac{dy}{dx}$ and is continually changing. By drawing tangents to the curve at many points on the curve and measuring the gradient of the tangents, values of $\dfrac{dy}{dx}$ for corresponding values of x may be obtained. These values are shown graphically in Figure 30.8(b). The graph of $\dfrac{dy}{dx}$ against x is the graph of $\dfrac{dy}{dx} = \dfrac{1}{x}$. It follows that:

$$\text{if } \mathbf{y = \ln x}, \quad \text{then} \quad \dfrac{dy}{dx} = \dfrac{1}{x}$$

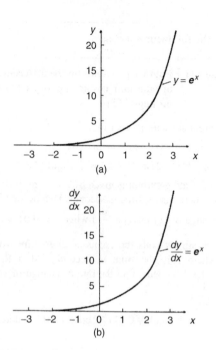

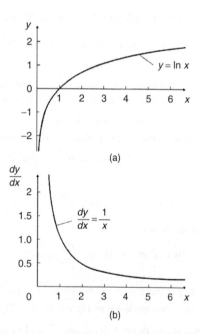

Fig. 30.7

Fig. 30.8

It may also be shown that

$$\text{if } y = \ln\ ax, \quad \text{then} \quad \frac{dy}{dx} = \frac{1}{x}$$

(Note that in the latter expression, the constant 'a' does not appear in the $\dfrac{dy}{dx}$ term).

Thus if $y = \ln\ 4x$, then $\dfrac{dy}{dx} = \dfrac{1}{x}$

Problem 16. Differentiate the following with respect to the variable:

(a) $y = 3e^{2x}$ (b) $f(t) = \dfrac{4}{3e^{5t}}$

(a) If $y = 3e^{2x}$ then $\dfrac{dy}{dx} = (3)(2e^{2x}) = 6e^{2x}$

(b) If $f(t) = \dfrac{4}{3e^{5t}} = \dfrac{4}{3}e^{-5t}$, then

$$f'(t) = \frac{4}{3}(-5e^{-5t}) = -\frac{20}{3}e^{-5t} = -\frac{20}{3e^{5t}}$$

Problem 17. Differentiate $y = 5\ln\ 3x$

If $y = 5\ln\ 3x$, then $\dfrac{dy}{dx} = (5)\left(\dfrac{1}{x}\right) = \dfrac{5}{x}$

Now try the following exercise

Exercise 112   Further problems on the differentiation of e^{ax} and ln ax (Answers on page 265)

1. Differentiate with respect to x:
   (a) $y = 5e^{3x}$ (b) $y = \dfrac{2}{7e^{2x}}$

2. Given $f(\theta) = 5\ln\ 2\theta - 4\ln\ 3\theta$, determine $f'(\theta)$

3. If $f(t) = 4\ln\ t + 2$, evaluate $f'(t)$ when $t = 0.25$

4. Evaluate $\dfrac{dy}{dx}$ when $x = 1$, given $y = 3e^{4x} - \dfrac{5}{2e^{3x}} + 8\ln 5x$. Give the answer correct to 3 significant figures.

30.8 Summary of standard derivatives

The five main differential coefficients used in this chapter are summarised in Table 30.1.

Table 30.1

y or $f(x)$	$\dfrac{dy}{dx}$ or $f'(x)$
ax^n	anx^{n-1}
$\sin\ ax$	$a\cos\ ax$
$\cos\ ax$	$-a\sin\ ax$
e^{ax}	ae^{ax}
$\ln\ ax$	$\dfrac{1}{x}$

Problem 18. Find the gradient of the curve $y = 3x^2 - 7x + 2$ at the point $(1, -2)$

If $y = 3x^2 - 7x + 2$, then gradient $= \dfrac{dy}{dx} = 6x - 7$

At the point $(1, -2)$, $x = 1$, hence **gradient** $= 6(1) - 7 = -1$

Problem 19. If $y = \dfrac{3}{x^2} - 2\sin 4x + \dfrac{2}{e^x} + \ln 5x$, determine $\dfrac{dy}{dx}$

$$y = \frac{3}{x^2} - 2\sin 4x + \frac{2}{e^x} + \ln 5x$$
$$= 3x^{-2} - 2\sin 4x + 2e^{-x} + \ln 5x$$
$$\frac{dy}{dx} = 3(-2x^{-3}) - 2(4\cos 4x) + 2\left(-e^{-x}\right) + \frac{1}{x}$$
$$= -\frac{6}{x^3} - 8\cos 4x - \frac{2}{e^x} + \frac{1}{x}$$

Now try the following exercise

Exercise 113 Further problems on standard derivatives (Answers on page 265)

1. Find the gradient of the curve $y = 2x^4 + 3x^3 - x + 4$ at the points (a) $(0, 4)$ and (b) $(1, 8)$

2. Differentiate:

$$y = \frac{2}{x^2} + 2\ln 2x - 2(\cos 5x + 3\sin 2x) - \frac{2}{e^{3x}}$$

30.9 Successive differentiation

When a function $y = f(x)$ is differentiated with respect to x the differential coefficient is written as $\dfrac{dy}{dx}$ or $f'(x)$. If the expression is differentiated again, the second differential coefficient is obtained and is written as $\dfrac{d^2y}{dx^2}$ (pronounced dee two y by dee x squared) or $f''(x)$ (pronounced f double-dash x).

By successive differentiation further higher derivatives such as $\dfrac{d^3y}{dx^3}$ and $\dfrac{d^4y}{dx^4}$ may be obtained.

Thus if $y = 5x^4$, $\dfrac{dy}{dx} = 20x^3$, $\dfrac{d^2y}{dx^2} = 60x^2$, $\dfrac{d^3y}{dx^3} = 120x$, $\dfrac{d^4y}{dx^4} = 120$ and $\dfrac{d^5y}{dx^5} = 0$

Problem 20. If $f(x) = 4x^5 - 2x^3 + x - 3$, find $f''(x)$

$$f(x) = 4x^5 - 2x^3 + x - 3$$
$$f'(x) = 20x^4 - 6x^2 + 1$$
$$f''(x) = 80x^3 - 12x \quad \text{or} \quad 4x(20x^2 - 3)$$

Problem 21. Given $y = \dfrac{2}{3}x^3 - \dfrac{4}{x^2} + \dfrac{1}{2x} - \sqrt{x}$, determine $\dfrac{d^2y}{dx^2}$

$$y = \frac{2}{3}x^3 - \frac{4}{x^2} + \frac{1}{2x} - \sqrt{x}$$
$$= \frac{2}{3}x^3 - 4x^{-2} + \frac{1}{2}x^{-1} - x^{\frac{1}{2}}$$
$$\frac{dy}{dx} = \left(\frac{2}{3}\right)(3x^2) - 4(-2x^{-3}) + \left(\frac{1}{2}\right)(-1x^{-2}) - \frac{1}{2}x^{-\frac{1}{2}}$$

i.e. $\dfrac{dy}{dx} = 2x^2 + 8x^{-3} - \dfrac{1}{2}x^{-2} - \dfrac{1}{2}x^{-\frac{1}{2}}$

$$\frac{d^2y}{dx^2} = 4x + (8)(-3x^{-4}) - \left(\frac{1}{2}\right)(-2x^{-3})$$
$$- \left(\frac{1}{2}\right)\left(-\frac{1}{2}x^{-\frac{3}{2}}\right)$$
$$= 4x - 24x^{-4} + 1x^{-3} + \frac{1}{4}x^{-\frac{3}{2}}$$

i.e. $\dfrac{d^2y}{dx^2} = 4x - \dfrac{24}{x^4} + \dfrac{1}{x^3} + \dfrac{1}{4\sqrt{x^3}}$

Now try the following exercise

Exercise 114 Further problems on successive differentiation (Answers on page 265)

1. If $y = 3x^4 + 2x^3 - 3x + 2$, find (a) $\dfrac{d^2y}{dx^2}$ (b) $\dfrac{d^3y}{dx^3}$

2. (a) Given $f(t) = \dfrac{2}{5}t^2 - \dfrac{1}{t^3} + \dfrac{3}{t} - \sqrt{t} + 1$, determine $f''(t)$

   (b) Evaluate $f''(t)$ in part (a) when $t = 1$

3. If $y = 3\sin 2t + \cos t$, find $\dfrac{d^2y}{dt^2}$

4. If $f(\theta) = 2\ln 4\theta$, show that $f''(\theta) = -\dfrac{2}{\theta^2}$

30.10 Rates of change

If a quantity y depends on and varies with a quantity x then the rate of change of y with respect to x is $\dfrac{dy}{dx}$. Thus, for example, the rate of change of pressure p with height h is $\dfrac{dp}{dh}$.

A rate of change with respect to time is usually just called 'the rate of change', the 'with respect to time' being assumed. Thus, for example, a rate of change of current, i, is $\dfrac{di}{dt}$ and a rate of change of temperature, θ, is $\dfrac{d\theta}{dt}$, and so on.

Problem 22. The length L metres of a certain metal rod at temperature $t°C$ is given by: $L = 1 + 0.00003t + 0.0000004t^2$. Determine the rate of change of length, in mm/°C, when the temperature is (a) 100°C and (b) 250°C

The rate of change of length means $\dfrac{dL}{dt}$

Since length $L = 1 + 0.00003t + 0.0000004t^2$, then

$$\frac{dl}{dt} = 0.00003 + 0.0000008t$$

(a) When $t = 100°C$, $\dfrac{dL}{dt} = 0.00003 + (0.0000008)(100)$

$$= 0.00011 \text{ m/°C} = \mathbf{0.11 \text{ mm/°C}}$$

(b) When $t = 250°C$, $\dfrac{dL}{dt} = 0.00003 + (0.0000008)(250)$

$$= 0.00023 \text{ m/°C} = \mathbf{0.23 \text{ mm/°C}}$$

Problem 23. The luminous intensity I candelas of a lamp at varying voltage V is given by: $I = 5 \times 10^{-4} V^2$. Determine the voltage at which the light is increasing at a rate of 0.4 candelas per volt.

The rate of change of light with respect to voltage is given by $\dfrac{dI}{dV}$.

Since $I = 5 \times 10^{-4} V^2$,

$$\frac{dI}{dV} = (5 \times 10^{-4})(2V) = 10 \times 10^{-4} V = 10^{-3} V$$

When the light is increasing at 0.4 candelas per volt then $+0.4 = 10^{-3} V$, from which, voltage

$$V = \frac{0.4}{10^{-3}} = 0.4 \times 10^{+3} = \mathbf{400 \text{ volts}}$$

Problem 24. Newton's law of cooling is given by: $\theta = \theta_0 e^{-kt}$, where the excess of temperature at zero time is $\theta_0°C$ and at time t seconds is $\theta°C$. Determine the rate of change of temperature after 50 s, given that $\theta_0 = 15°C$ and $k = -0.02$

The rate of change of temperature is $\dfrac{d\theta}{dt}$.

Since $\theta = \theta_0 e^{-kt}$ then $\dfrac{d\theta}{dt} = (\theta_0)(-ke^{-kt}) = -k\theta_0 e^{-kt}$

When $\theta_0 = 15$, $k = -0.02$ and $t = 50$, then

$$\frac{d\theta}{dt} = -(-0.02)(15)e^{-(-0.02)(50)} = 0.30e^1 = \mathbf{0.815°C/s}$$

Problem 25. The pressure p of the atmosphere at height h above ground level is given by $p = p_0 e^{-h/c}$, where p_0 is the pressure at ground level and c is a constant. Determine the rate of change of pressure with height when $p_0 = 10^5$ Pascals and $c = 6.2 \times 10^4$ at 1550 metres.

The rate of change of pressure with height is $\dfrac{dp}{dh}$.

Since $p = p_0 e^{-h/c}$ then

$$\frac{dp}{dh} = (p_0)\left(-\frac{1}{c}e^{-h/c}\right) = -\frac{p_0}{c}e^{-h/c}$$

When $p_0 = 10^5$, $c = 6.2 \times 10^4$ and $h = 1550$, then **rate of change of pressure,**

$$\frac{dp}{dh} = -\frac{10^5}{6.2 \times 10^4}e^{-(1550/6.2 \times 10^4)}$$

$$= -\frac{10}{6.2}e^{-0.025}$$

$$= \mathbf{-1.573 \text{ Pa/m}}$$

Now try the following exercise

Exercise 115   Further problems on rates of change (Answers on page 265)

1. An alternating current, i amperes, is given by $i = 10 \sin 2\pi ft$, where f is the frequency in hertz and t the time in seconds. Determine the rate of change of current when $t = 12$ ms, given that $f = 50$ Hz.

2. The luminous intensity, I candelas, of a lamp is given by $I = 8 \times 10^{-4} V^2$, where V is the voltage. Find (a) the rate of change of luminous intensity with voltage when $V = 100$ volts, and (b) the voltage at which the light is increasing at a rate of 0.5 candelas per volt.

3. The voltage across the plates of a capacitor at any time t seconds is given by $v = V e^{-t/CR}$, where V, C and R are constants. Given $V = 200$ volts, $C = 0.10 \times 10^{-6}$ farads and $R = 2 \times 10^6$ ohms, find (a) the initial rate of change of voltage, and (b) the rate of change of voltage after 0.2 s.

4. The pressure p of the atmosphere at height h above ground level is given by $p = p_0 e^{-h/c}$, where p_0 is the pressure at ground level and c is a constant. Determine the rate of change of pressure with height when $p_0 = 1.013 \times 10^5$ Pascals and $c = 6.05 \times 10^4$ at 1450 metres.

31

Introduction to integration

31.1 The process of integration

The process of integration reverses the process of differentiation. In differentiation, if $f(x) = 2x^2$ then $f'(x) = 4x$. Thus, the integral of $4x$ is $2x^2$, i.e. integration is the process of moving from $f'(x)$ to $f(x)$. By similar reasoning, the integral of $2t$ is t^2.

Integration is a process of summation or adding parts together and an elongated S, shown as $\int$, is used to replace the words 'the integral of'. Hence, from above, $\int 4x = 2x^2$ and $\int 2t$ is t^2.

In differentiation, the differential coefficient $\dfrac{dy}{dx}$ indicates that a function of x is being differentiated with respect to x, the dx indicating that it is 'with respect to x'. In integration the variable of integration is shown by adding d (the variable) after the function to be integrated.

Thus $\int 4x\,dx$ means 'the integral of $4x$ with respect to x', and $\int 2t\,dt$ means 'the integral of $2t$ with respect to t'

As stated above, the differential coefficient of $2x^2$ is $4x$, hence $\int 4x\,dx = 2x^2$. However, the differential coefficient of $2x^2+7$ is also $4x$. Hence $\int 4x\,dx$ could also be equal to $2x^2 + 7$. To allow for the possible presence of a constant, whenever the process of integration is performed, a constant 'c' is added to the result.

Thus $\int 4x\,dx = 2x^2 + c$ and $\int 2t\,dt = t^2 + c$

'c' is called the **arbitrary constant of integration**.

31.2 The general solution of integrals of the form ax^n

The general solution of integrals of the form $\int ax^n\,dx$, where a and n are constants and $n \neq -1$, is given by:

$$\boxed{\int ax^n\ dx = \frac{ax^{n+1}}{n+1} + c}$$

Using this rule gives:

(i) $\displaystyle\int 3x^4\ dx = \frac{3x^{4+1}}{4+1} + c = \frac{3}{5}x^5 + c$

(ii) $\displaystyle\int \frac{4}{9}t^3\ dt\ dx = \frac{4}{9}\left(\frac{t^{3+1}}{3+1}\right) + c$

$$= \frac{4}{9}\left(\frac{t^4}{4}\right) + c = \frac{1}{9}t^4 + c$$

Both of these results may be checked by differentiation.

31.3 Standard integrals

From Chapter 30, $\dfrac{d}{dx}(\sin\ ax) = a\cos\ ax$

Since integration is the reverse process of differentiation it follows that:

$$\int a\cos\ ax\ dx = \sin\ ax + c$$

or $\displaystyle\int \cos\ ax\ dx = \frac{1}{a}\sin\ ax + c$

By similar reasoning

$$\int \sin\ ax\ dx = -\frac{1}{a}\cos\ ax + c$$

$$\int e^{ax}\ dx = \frac{1}{a}e^{ax} + c$$

and $\displaystyle\int \frac{1}{x}\ dx = \ln x + c$

From above, $\int ax^n \, dx = \dfrac{ax^{n+1}}{n+1} + c$ **except when $n = -1$**

When $n = -1$, then $\int x^{-1} \, dx = \int \dfrac{1}{x} \, dx = \ln x + c$

A list of **standard integrals** is summarised in Table 31.1.

Table 31.1 Standard integrals

(i)	$\int ax^n \, dx$	$= \dfrac{ax^{n+1}}{n+1} + c$
		(except when $n = -1$)
(ii)	$\int \cos ax \, dx$	$= \dfrac{1}{a} \sin ax + c$
(iii)	$\int \sin ax \, dx$	$= -\dfrac{1}{a} \cos ax + c$
(iv)	$\int e^{ax} \, dx$	$= \dfrac{1}{a} e^{ax} + c$
(v)	$\int \dfrac{1}{x} \, dx$	$= \ln x + c$

Problem 1. Determine: (a) $\int 3x^2 \, dx$ (b) $\int 2t^3 \, dt$

The general rule is $\int ax^n \, dx = \dfrac{ax^{n+1}}{n+1} + c$

(a) When $a = 3$ and $n = 2$ then

$$\int 3x^2 \, dx = \dfrac{3x^{2+1}}{2+1} + c = x^3 + c$$

(b) When $a = 2$ and $n = 3$ then

$$\int 2t^3 \, dt = \dfrac{2t^{3+1}}{3+1} + c = \dfrac{2t^4}{4} + c = \dfrac{1}{2}t^4 + c$$

Each of these results may be checked by differentiating them.

Problem 2. Determine (a) $\int 8 \, dx$ (b) $\int 2x \, dx$

(a) $\int 8 \, dx$ is the same as $\int 8x^0 \, dx$ and, using the general rule when $a = 8$ and $n = 0$ gives:

$$\int 8x^0 \, dx = \dfrac{8x^{0+1}}{0+1} + c = 8x + c$$

In general, if k is a constant then $\int k \, dx = kx + c$

(b) When $a = 2$ and $n = 1$, then

$$\int 2x \, dx = \int 2x^1 \, dx = \dfrac{2x^{1+1}}{1+1} + c = \dfrac{2x^2}{2} + c$$
$$= x^2 + c$$

Problem 3. Determine: $\int \left(2 + \dfrac{5}{7}x - 6x^2 \right) \, dx$

$\int \left(2 + \dfrac{5}{7}x - 6x^2 \right) \, dx$ may be written as:

$$\int 2 \, dx + \int \dfrac{5}{7}x \, dx - \int 6x^2 \, dx$$

i.e. each term is integrated separately.
(This splitting up of terms only applies for addition and subtraction).

Hence $\int \left(2 + \dfrac{5}{7}x - 6x^2 \right) \, dx$

$$= 2x + \left(\dfrac{5}{7} \right) \dfrac{x^{1+1}}{1+1} - (6)\dfrac{x^{2+1}}{2+1} + c$$

$$= 2x + \left(\dfrac{5}{7} \right) \dfrac{x^2}{2} - (6)\dfrac{x^3}{3} + c$$

$$= 2x + \dfrac{5}{14}x^2 - 2x^3 + c$$

Note that when an integral contains more than one term there is no need to have an arbitrary constant for each; just a single constant at the end is sufficient.

Problem 4. Determine: $\int \dfrac{3}{x^2} \, dx$

$\int \dfrac{3}{x^2} \, dx = \int 3x^{-2} \, dx$. Using the standard integral, $\int ax^n \, dx$ when $a = 3$ and $n = -2$ gives:

$$\int 3x^{-2} \, dx = \dfrac{3x^{-2+1}}{-2+1} + c$$

$$= \dfrac{3x^{-1}}{-1} + c = -3x^{-1} + c = \dfrac{-3}{x} + c$$

Problem 5. Determine: $\int 3\sqrt{x} \, dx$

For fractional powers it is necessary to appreciate that $\sqrt[n]{a^m} = a^{\frac{m}{n}}$, from which, $\sqrt{x} = x^{\frac{1}{2}}$. Hence,

$$\int 3\sqrt{x} \, dx = \int 3x^{\frac{1}{2}} \, dx = \dfrac{3x^{\frac{1}{2}+1}}{\frac{1}{2}+1} + c = \dfrac{3x^{\frac{3}{2}}}{\frac{3}{2}} + c$$

$$= (3) \left(\dfrac{2}{3} \right) x^{\frac{3}{2}} = 2x^{\frac{3}{2}} + c = 2\sqrt{x^3} + c$$

Problem 6. Determine: $\int \dfrac{5}{\sqrt{x}} \, dx$

$$\int \frac{5}{\sqrt{x}} \, dx = \int 5x^{-\frac{1}{2}} \, dx = \frac{5x^{-\frac{1}{2}+1}}{-\frac{1}{2}+1} + c = \frac{5x^{\frac{1}{2}}}{\frac{1}{2}} + c$$

$$= (5) \left(\frac{2}{1}\right) x^{\frac{1}{2}} + c = 10\sqrt{x} + c$$

Problem 7. Determine:

(a) $\int \left(\dfrac{x^3 - 2x}{3x}\right) dx$ (b) $\int (1-x)^2 \, dx$

(a) Rearranging into standard integral form gives:

$$\int \left(\frac{x^3 - 2x}{3x}\right) dx = \int \left(\frac{x^3}{3x} - \frac{2x}{3x}\right) dx$$

$$= \int \left(\frac{x^2}{3} - \frac{2}{3}\right) dx$$

$$= \left(\frac{1}{3}\right) \frac{x^{2+1}}{2+1} - \frac{2}{3}x + c$$

$$= \left(\frac{1}{3}\right) \frac{x^3}{3} - \frac{2}{3}x + c$$

$$= \frac{1}{9}x^3 - \frac{2}{3}x + c$$

(b) Rearranging $\int (1-x)^2 \, dx$ gives:

$$\int (1 - 2x + x^2) \, dx = x - \frac{2x^{1+1}}{1+1} + \frac{x^{2+1}}{2+1} + c$$

$$= x - \frac{2x^2}{2} + \frac{x^3}{3} + c$$

$$= x - x^2 + \frac{1}{3}x^3 + c$$

This problem shows that functions often have to be rearranged into the standard form of $\int ax^n \, dx$ before it is possible to integrate them.

Problem 8. Determine:

(a) $\int 5 \cos 3x \, dx$ (b) $\int 3 \sin 2x \, dx$

(a) From Table 31.1(ii),

$$\int 5 \cos 3x \, dx = 5 \int \cos 3x \, dx$$

$$= (5) \left(\frac{1}{3} \sin 3x\right) + c$$

$$= \frac{5}{3} \sin 3x + c$$

(b) From Table 31.1(iii),

$$\int 3 \sin 2x \, dx = 3 \int \sin 2x \, dx = (3)\left(-\frac{1}{2} \cos 2x\right) + c$$

$$= -\frac{3}{2} \cos 2x + c$$

Problem 9. Determine: (a) $\int 5e^{3x} \, dx$ (b) $\int \dfrac{6}{e^{2x}} \, dx$

(a) From Table 31.1(iv),

$$\int 5e^{3x} \, dx = 5 \int e^{3x} \, dx = (5) \left(\frac{1}{3}e^{3x}\right) + c$$

$$= \frac{5}{3}e^{3x} + c$$

(b) $\int \dfrac{6}{e^{2x}} dx = \int 6e^{-2x} \, dx = 6 \int e^{-2x} \, dx$

$$= (6) \left(\frac{1}{-2}e^{-2x}\right) + c = -3e^{-2x} + c$$

$$= -\frac{3}{e^{2x}} + c$$

Problem 10. Determine:

(a) $\int \dfrac{3}{5x} \, dx$ (b) $\int \left(\dfrac{3x^2 - 1}{x}\right) dx$

(a) $\int \dfrac{3}{5x} \, dx = \int \left(\dfrac{3}{5}\right)\left(\dfrac{1}{x}\right) dx = \dfrac{3}{5} \int \left(\dfrac{1}{x}\right) dx$

$$= \frac{3}{5} \ln x + c \text{ (from Table 31.1(v))}$$

(b) $\int \left(\dfrac{3x^2 - 1}{x}\right) dx = \int \left(\dfrac{3x^2}{x} - \dfrac{1}{x}\right) dx$

$$= \int \left(3x - \frac{1}{x}\right) dx = \frac{3x^2}{2} + \ln x + c$$

$$= \frac{3}{2}x^2 + \ln x + c$$

Now try the following exercise

Exercise 116   Further problems on standard integrals (Answers on page 265)

Determine the following integrals:

1. (a) $\int 4 \, dx$ (b) $\int 7x \, dx$

2. (a) $\int \dfrac{2}{5}x^2 \, dx$ (b) $\int \dfrac{5}{6}x^3 \, dx$

3. (a) $\int (3 + 2x - 4x^2)\, dx$ (b) $3\int (x + 5x^2)\, dx$

4. (a) $\int \left(\dfrac{3x^2 - 5x}{x}\right) dx$ (b) $\int (2 + x)^2\, dx$

5. (a) $\int \dfrac{4}{3x^2}\, dx$ (b) $\int \dfrac{3}{4x^4}\, dx$

6. (a) $\int \sqrt{x}\, dx$ (b) $\int \dfrac{2}{\sqrt{x}}\, dx$

7. (a) $\int 3\cos\, 2x\, dx$ (b) $\int 7\sin\, 3x\, dx$

8. (a) $\int \dfrac{3}{4} e^{2x}\, dx$ (b) $\dfrac{2}{3}\int \dfrac{dx}{e^{5x}}$

9. (a) $\int \dfrac{2}{3x}\, dx$ (b) $\int \left(\dfrac{x^2 - 1}{x}\right) dx$

31.4 Definite integrals

Integrals containing an arbitrary constant c in their results are called **indefinite integrals** since their precise value cannot be determined without further information. **Definite integrals** are those in which limits are applied.

If an expression is written as $[x]_a^b$ 'b' is called the upper limit and 'a' the lower limit.

The operation of applying the limits is defined as: $[x]_a^b = (b) - (a)$

The increase in the value of the integral x^2 as x increases from 1 to 3 is written as $\int_1^3 x^2\, dx$

Applying the limits gives:

$$\int_1^3 x^2\, dx = \left[\dfrac{x^3}{3} + c\right]_1^3 = \left(\dfrac{3^3}{3} + c\right) - \left(\dfrac{1^3}{3} + c\right)$$

$$= (9 + c) - \left(\dfrac{1}{3} + c\right)$$

$$= 8\dfrac{2}{3}$$

Note that the 'c' term always cancels out when limits are applied and it need not be shown with definite integrals.

Problem 11.   Evaluate:

(a) $\int_1^2 3x\, dx$ (b) $\int_{-2}^3 (4 - x^2)\, dx$

(a) $\int_1^2 3x\, dx = \left[\dfrac{3x^2}{2}\right]_1^2 = \left\{\dfrac{3}{2}(2)^2\right\} - \left\{\dfrac{3}{2}(1)^2\right\}$

$$= 6 - 1\dfrac{1}{2} = 4\dfrac{1}{2} \quad \text{or} \quad \textbf{4.5}$$

(b) $\int_{-2}^3 (4 - x^2)\, dx = \left[4x - \dfrac{x^3}{3}\right]_{-2}^3$

$$= \left\{4(3) - \dfrac{(3)^3}{3}\right\} - \left\{4(-2) - \dfrac{(-2)^3}{3}\right\}$$

$$= \{12 - 9\} - \left\{-8 - \dfrac{-8}{3}\right\}$$

$$= \{3\} - \left\{-5\dfrac{1}{3}\right\} = 8\dfrac{1}{3} \quad \text{or} \quad \textbf{8.33}$$

Problem 12.   Evaluate:

(a) $\int_0^2 x(3 + 2x)\, dx$ (b) $\int_{-1}^1 \left(\dfrac{x^4 - 5x^2 + x}{x}\right) dx$

(a) $\int_0^2 x(3 + 2x)\, dx = \int_0^2 (3x + 2x^2)\, dx = \left[\dfrac{3x^2}{2} + \dfrac{2x^3}{3}\right]_0^2$

$$= \left\{\dfrac{3(2)^2}{2} + \dfrac{2(2)^3}{3}\right\} - \{0 + 0\}$$

$$= 6 + \dfrac{16}{3} = 11\dfrac{1}{3} \quad \text{or} \quad \textbf{11.33}$$

(b) $\int_{-1}^1 \left(\dfrac{x^4 - 5x^2 + x}{x}\right) dx = \int_{-1}^1 \left(\dfrac{x^4}{x} - \dfrac{5x^2}{x} + \dfrac{x}{x}\right) dx$

$$= \int_{-1}^1 (x^3 - 5x + 1)\, dx$$

$$= \left[\dfrac{x^4}{4} - \dfrac{5x^2}{2} + x\right]_{-1}^1$$

$$= \left\{\dfrac{1}{4} - \dfrac{5}{2} + 1\right\} - \left\{\dfrac{(-1)^4}{4} - \dfrac{5(-1)^2}{2} + (-1)\right\}$$

$$= \left\{\dfrac{1}{4} - \dfrac{5}{2} + 1\right\} - \left\{\dfrac{1}{4} - \dfrac{5}{2} - 1\right\} = 2$$

Problem 13.   Evaluate: $\int_0^{\pi/2} 3\sin\, 2x\, dx$

$$\int_0^{\pi/2} 3\sin 2x\, dx = \left[(3)\left(-\dfrac{1}{2}\cos\, 2x\right)\right]_0^{\pi/2}$$

$$= \left[-\dfrac{3}{2}\cos\, 2x\right]_0^{\pi/2} = -\dfrac{3}{2}[\cos\, 2x]_0^{\pi/2}$$

$$= -\dfrac{3}{2}\left[\left\{\cos\, 2\left(\dfrac{\pi}{2}\right)\right\} - \{\cos\, 2(0)\}\right]$$

$$= -\dfrac{3}{2}[(\cos\, \pi) - (\cos 0)]$$

$$= -\dfrac{3}{2}[(-1) - (1)] = -\dfrac{3}{2}(-2) = 3$$

Problem 14. Evaluate: $\int_1^2 4\cos\ 3t\ dt$

$$\int_1^2 4\cos\ 3t\ dt = \left[(4)\left(\frac{1}{3}\sin\ 3t\right)\right]_1^2 = \frac{4}{3}[\sin 3t]_1^2$$

$$= \frac{4}{3}[\sin\ 6 - \sin\ 3]$$

Note that limits of trigonometric functions are always expressed in radians – thus, for example, sin 6 means the sine of 6 radians = −0.279415..

Hence $\int_1^2 4\cos\ 3t\ dt = \frac{4}{3}[(-0.279415..) - (0.141120..)]$

$$= \frac{4}{3}[-0.420535] = -0.5607$$

Problem 15. Evaluate: (a) $\int_1^2 4e^{2x}\ dx$ (b) $\int_1^4 \frac{3}{4x}\ du$, each correct to 4 significant figures.

(a) $\int_1^2 4e^{2x}\ dx = \left[\frac{4}{2}e^{2x}\right]_1^2 = 2[e^{2x}]_1^2 = 2[e^4 - e^2]$

$$= 2[54.5982 - 7.3891] = 94.42$$

(b) $\int_1^4 \frac{3}{4x}\ du = \left[\frac{3}{4}\ln x\right]_1^4 = \frac{3}{4}[\ln 4 - \ln 1]$

$$= \frac{3}{4}[1.3863 - 0] = 1.040$$

Now try the following exercise

Exercise 117   Further problems on definite integrals (Answers on page 265)

Evaluate the following definite integrals (where necessary, correct to 4 significant figures):

1. (a) $\int_1^4 5x\ dx$ (b) $\int_{-1}^1 \frac{3}{4}t^2\ dt$

2. (a) $\int_{-1}^2 (3 - x^2)\ dx$ (b) $\int_1^3 (x^2 - 4x + 3)\ dx$

3. (a) $\int_1^2 x(1 + 4x)\ dx$ (b) $\int_{-1}^2 \left(\frac{x^3 + 2x^2 - 3x}{x}\right)\ dx$

4. (a) $\int_0^\pi \frac{3}{2}\cos\theta\ d\theta$ (b) $\int_0^{\pi/2} 4\cos\ x\ dx$

5. (a) $\int_{\pi/6}^{\pi/3} 2\sin\ 2x\ dx$ (b) $\int_0^2 3\sin\ t\ dt$

6. (a) $\int_0^1 5\cos\ 3x\ dx$ (b) $\int_{\pi/4}^{\pi/2} (3\sin\ 2x - 2\cos\ 3x)\ dx$

7. (a) $\int_0^1 3e^{3t}\ dt$ (b) $\int_{-1}^2 \frac{2}{3e^{2x}}\ dx$

8. (a) $\int_2^3 \frac{2}{3x}\ dx$ (b) $\int_1^3 \left(\frac{2x^2 + 1}{x}\right)\ dx$

31.5   Area under a curve

The area shown shaded in Figure 31.1 may be determined using approximate methods such as the trapezoidal rule, the mid-ordinate rule or Simpson's rule (see Chapter 23) or, more precisely, by using integration.

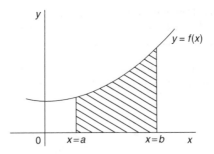

Fig. 31.1

The shaded area in Figure 31.1 is given by:

$$\text{shaded area} = \int_a^b y\ dx = \int_a^b f(x)\ dx$$

Thus, determining the area under a curve by integration merely involves evaluating a definite integral, as shown in Section 31.4.

There are several instances in engineering and science where the area beneath a curve needs to be accurately determined. For example, the areas between limits of a:

velocity/time graph gives distance travelled,
force/distance graph gives work done,
voltage/current graph gives power, and so on.

Should a curve drop below the x-axis, then $y(= f(x))$ becomes negative and $\int f(x)\ dx$ is negative. When determining such areas by integration, a negative sign is placed before the integral. For the curve shown in Figure 31.2, the total shaded area is given by (area E + area F + area G).

By integration,

$$\text{total shaded area} = \int_a^b f(x)\ dx - \int_b^c f(x)\ dx$$
$$+ \int_c^d f(x)\ dx$$

(Note that this is **not** the same as $\int_a^d f(x)\ dx$.)

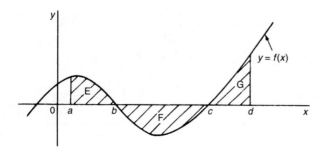

Fig. 31.2

It it usually necessary to sketch a curve in order to check whether it crosses the x-axis.

Problem 16.   Determine the area enclosed by $y = 2x + 3$, the x-axis and ordinates $x = 1$ and $x = 4$.

$y = 2x + 3$ is a straight line graph as shown in Figure 31.3, where the required area is shown shaded.

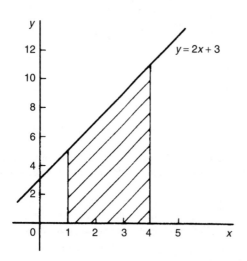

Fig. 31.3

By integration, shaded area $= \displaystyle\int_{1}^{4} y \, dx$

$$= \int_{1}^{4} (2x + 3) \, dx$$

$$= \left[\frac{2x^2}{2} + 3x \right]_{1}^{4}$$

$$= [(16 + 12) - (1 + 3)]$$

$$= \textbf{24 square units}$$

[This answer may be checked since the shaded area is a trapezium.

Area of trapezium $= \dfrac{1}{2}$ (sum of parallel sides)

(perpendicular distance between parallel sides)

$$= \frac{1}{2}(5 + 11)(3) = \textbf{24 square units}]$$

Problem 17.   The velocity v of a body t seconds after a certain instant is: $(2t^2 + 5)$m/s. Find by integration how far it moves in the interval from $t = 0$ to $t = 4$ s.

Since $2t^2 + 5$ is a quadratic expression, the curve $v = 2t^2 + 5$ is a parabola cutting the v-axis at $v = 5$, as shown in Figure 31.4.

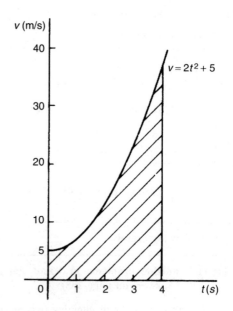

Fig. 31.4

The distance travelled is given by the area under the v/t curve (shown shaded in Figure 31.4).

By integration, shaded area $= \displaystyle\int_{0}^{4} v \, dt = \int_{0}^{4} (2t^2 + 5) \, dt$

$$= \left[\frac{2t^3}{3} + 5t \right]_{0}^{4}$$

$$= \left(\frac{2(4^3)}{3} + 5(4) \right) - (0)$$

i.e.   **distance travelled = 62.67 m**

Problem 18. Sketch the graph $y = x^3 + 2x^2 - 5x - 6$ between $x = -3$ and $x = 2$ and determine the area enclosed by the curve and the x-axis.

A table of values is produced and the graph sketched as shown in Figure 31.5 where the area enclosed by the curve and the x-axis is shown shaded.

x	-3	-2	-1	0	1	2
x^3	-27	-8	-1	0	1	8
$2x^2$	18	8	2	0	2	8
$-5x$	15	10	5	0	-5	-10
-6	-6	-6	-6	-6	-6	-6
y	0	4	0	-6	-8	0

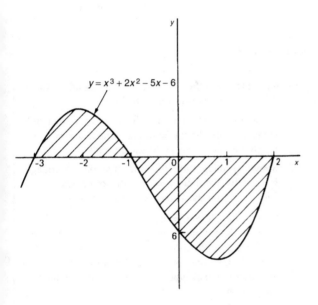

Fig. 31.5

Shaded area $= \int_{-3}^{-1} y \, dx - \int_{-1}^{2} y \, dx$, the minus sign before the second integral being necessary since the enclosed area is below the x-axis. Hence,

shaded area

$$= \int_{-3}^{-1} (x^3 + 2x^2 - 5x - 6) \, dx$$

$$- \int_{-1}^{2} (x^3 + 2x^2 - 5x - 6) \, dx$$

$$= \left[\frac{x^4}{4} + \frac{2x^3}{3} - \frac{5x^2}{2} - 6x \right]_{-3}^{-1} - \left[\frac{x^4}{4} + \frac{2x^3}{3} - \frac{5x^2}{2} - 6x \right]_{-1}^{2}$$

$$= \left[\left\{ \frac{1}{4} - \frac{2}{3} - \frac{5}{2} + 6 \right\} - \left\{ \frac{81}{4} - 18 - \frac{45}{2} + 18 \right\} \right]$$

$$- \left[\left\{ 4 + \frac{16}{3} - 10 - 12 \right\} - \left\{ \frac{1}{4} - \frac{2}{3} - \frac{5}{2} + 6 \right\} \right]$$

$$= \left[\left\{ 3\frac{1}{12} \right\} - \left\{ -2\frac{1}{4} \right\} \right] - \left[\left\{ -12\frac{2}{3} \right\} - \left\{ 3\frac{1}{12} \right\} \right]$$

$$= \left[5\frac{1}{3} \right] - \left[-15\frac{3}{4} \right]$$

$$= 21\frac{1}{12} \quad \text{or} \quad \textbf{21.08 square units}$$

Problem 19. Determine the area enclosed by the curve $y = 3x^2 + 4$, the x-axis and ordinates $x = 1$ and $x = 4$ by

(a) the trapezoidal rule,   (b) the mid-ordinate rule,

(c) Simpson's rule, and   (d) integration.

The curve $y = 3x^2 + 4$ is shown plotted in Figure 31.6.

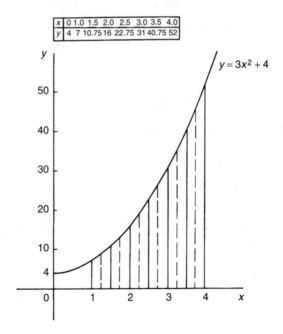

x	0	1.0	1.5	2.0	2.5	3.0	3.5	4.0
y	4	7	10.75	16	22.75	31	40.75	52

Fig. 31.6

The trapezoidal rule, the mid-ordinate rule and Simpson's rule are discussed in Chapter 23, page 177.

(a) **By the trapezoidal rule,**

$$\text{area} = \text{(width of interval)} \left[\frac{1}{2} \left(\frac{\text{first} + \text{last}}{\text{ordinate}} \right) + \text{sum of remaining ordinates} \right]$$

Selecting 6 intervals each of width 0.5 gives:

$$\text{area} = (0.5)[\frac{1}{2}(7+52)+10.75+16$$

$$+22.75+31+40.75]$$

$$= \textbf{75.375 square units}$$

(b) By the mid-ordinate rule,

area = (width of interval)(sum of mid-ordinates).

Selecting 6 intervals, each of width 0.5 gives the mid-ordinates as shown by the broken lines in Figure 31.6.

Thus, area $= (0.5)(8.7+13.2+19.2+26.7$

$$+35.7+46.2)$$

$$= \textbf{74.85 square units}$$

(c) By Simpson's rule,

$$\text{area} = \frac{1}{3}\left(\begin{array}{c}\text{width of}\\\text{interval}\end{array}\right)\left[\left(\begin{array}{c}\text{first} + \text{last}\\\text{ordinates}\end{array}\right)\right.$$

$$\left.+4\left(\begin{array}{c}\text{sum of even}\\\text{ordinates}\end{array}\right)+2\left(\begin{array}{c}\text{sum of remaining}\\\text{odd ordinates}\end{array}\right)\right]$$

Selecting 6 intervals, each of width 0.5 gives:

$$\text{area} = \frac{1}{3}(0.5)[(7+52)+4(10.75+22.75+40.75)$$

$$+2(16+31)]$$

$$= \textbf{75 square units}$$

(d) By integration, shaded area

$$= \int_1^4 y\,dx = \int_1^4 (3x^2+4)\,dx = [x^3+4x]_1^4$$

$$= (64+16)-(1+4) = \textbf{75 square units}$$

Integration gives the precise value for the area under a curve. In this case Simpson's rule is seen to be the most accurate of the three approximate methods.

Problem 20. Find the area enclosed by the curve $y = \sin 2x$, the x-axis and the ordinates $x = 0$ and $x = \dfrac{\pi}{3}$

A sketch of $y = \sin 2x$ is shown in Figure 31.7.

(Note that $y = \sin 2x$ has a period of $\dfrac{2\pi}{2}$, i.e. π radians).

$$\text{Shaded area} = \int_0^{\pi/3} y\,dx = \int_0^{\pi/3} \sin 2x\,dx = \left[-\frac{1}{2}\cos 2x\right]_0^{\pi/3}$$

$$= \left\{-\frac{1}{2}\cos\frac{2\pi}{3}\right\}-\left\{-\frac{1}{2}\cos 0\right\}$$

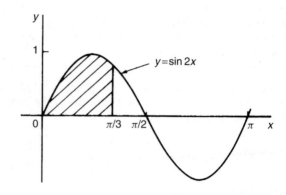

Fig. 31.7

$$= \left\{-\frac{1}{2}\left(-\frac{1}{2}\right)\right\}-\left\{-\frac{1}{2}(1)\right\}$$

$$= \frac{1}{4}+\frac{1}{2} = \frac{3}{4} \text{ or } \textbf{0.75 square units}$$

Now try the following exercise

Exercise 118 Further problems on area under curves (Answers on page 265)

1. Show by integration that the area of a rectangle formed by the line $y = 4$, the ordinates $x = 1$ and $x = 6$ and the x-axis is 20 square units

2. Show by integration that the area of the triangle formed by the line $y = 2x$, the ordinates $x = 0$ and $x = 4$ and the x-axis is 16 square units.

3. Sketch the curve $y = 3x^2+1$ between $x = -2$ and $x = 4$. Determine by integration the area enclosed by the curve, the x-axis and ordinates $x = -1$ and $x = 3$. Use an approximate method to find the area and compare your result with that obtained by integration.

4. The force F newtons acting on a body at a distance x metres from a fixed point is given by: $F = 3x+2x^2$. If work done $= \int_{x_1}^{x_2} F\,dx$, determine the work done when the body moves from the position where $x_1 = 1$ m to that when $x_2 = 3$ m.

In Problems 5 to 9, sketch graphs of the given equations, and then find the area enclosed between the curves, the horizontal axis and the given ordinates.

5. $y = 5x$; $x = 1, x = 4$

6. $y = 2x^2 - x + 1$; $x = -1, x = 2$

7. $y = 2\sin 2x$; $x = 0, x = \dfrac{\pi}{4}$

8. $y = 5\cos\ 3t;\quad t = 0, t = \dfrac{\pi}{6}$

9. $y = (x-1)(x-3);\qquad x = 0, x = 3$

10. The velocity v of a vehicle t seconds after a certain instant is given by: $v = (3t^2 + 4)$ m/s. Determine how far it moves in the interval from $t = 1$ s to $t = 5$ s.

Assignment 15

This assignment covers the material contained in Chapters 30 and 31. The marks for each question are shown in brackets at the end of each question

1. Differentiate from first principles $f(x) = 2x^2 + x - 1$
(5)

2. If $y = 4x^5 - 2\sqrt{x}$ determine $\dfrac{dy}{dx}$
(4)

3. Given $f(x) = \dfrac{3}{x^2} - \dfrac{1}{x}$ find $f'(x)$
(4)

4. Differentiate the following with respect to the variable:
   (a) $y = 4\sin\ 2x - 5\cos\ 3x$
   (b) $y = 2e^{3t} - 3\ln\ 7t$
(4)

5. Find the gradient of the curve $y = 5x^2 + 2x - 3$ at the point $(1, 4)$
(4)

6. Given $y = 3x^2 - \dfrac{2}{x^3} + \sqrt{x}$ find $\dfrac{d^2 y}{dx^2}$
(6)

7. Newtons law of cooling is given by: $\theta = \theta_0 e^{-kt}$, where the excess of temperature at zero time is $\theta_0 °C$ and at time t seconds is $\theta °C$. Determine the rate of change of temperature after 40 s, correct to 3 decimal places, given that $\theta_0 = 16°C$ and $k = -0.01$
(4)

8. Determine the following:
   (a) $\displaystyle\int (2 - 3x + 5x^2)\ dx$
   (b) $\displaystyle\int (4\cos\ 2x - 2\sqrt{x})\ dx$
(4)

9. Evaluate the following definite integrals, correct to 4 significant figures:
   (a) $\displaystyle\int_1^2 \left(2x^2 - \dfrac{1}{x}\right)\ dx$
   (b) $\displaystyle\int_0^{\pi/3} 3\sin\ 2x\ dx$
   (c) $\displaystyle\int_0^1 \left(3e^{2x} - \dfrac{1}{\sqrt{x}}\right)\ dx$
(10)

10. Sketch the curve $y = 2x^2 + 5$ between $x = -1$ and $x = 3$. Determine, by integration, the area enclosed by the curve, the x-axis and ordinates $x = -1$ and $x = 3$.
(5)

List of formulae

Laws of indices:

$$a^m \times a^n = a^{m+n} \qquad \frac{a^m}{a^n} = a^{m-n} \qquad (a^m)^n = a^{mn}$$

$$a^{m/n} = \sqrt[n]{a^m} \qquad a^{-n} = \frac{1}{a^n} \qquad a^0 = 1$$

Quadratic formula:

If $ax^2 + bx + c = 0$ then $x = \dfrac{-b \pm \sqrt{b^2 - 4ac}}{2a}$

Equation of a straight line:

$$y = mx + c$$

Definition of a logarithm:

If $y = a^x$ then $x = \log_a y$

Laws of logarithms:

$$\log(A \times B) = \log A + \log B$$

$$\log\left(\frac{A}{B}\right) = \log A - \log B$$

$$\log A^n = n \times \log A$$

Exponential series:

$$e^x = 1 + x + \frac{x^2}{2!} + \frac{x^3}{3!} + \cdots \qquad \text{(valid for all values of } x\text{)}$$

Theorem of Pythagoras:

$$b^2 = a^2 + c^2$$

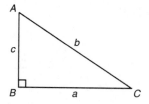

Fig. F1

Areas of plane figures:

(i) **Rectangle**   Area $= l \times b$

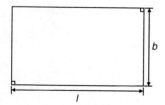

Fig. F2

(ii) **Parallelogram**   Area $= b \times h$

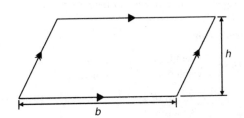

Fig. F3

(iii) **Trapezium**   Area $= \frac{1}{2}(a+b)h$

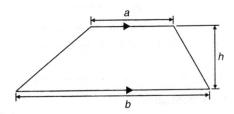

Fig. F4

(iv) **Triangle**   Area $= \frac{1}{2} \times b \times h$

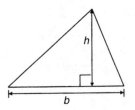

Fig. F5

(v) **Circle**   Area $= \pi r^2$ Circumference $= 2\pi r$

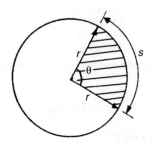

Fig. F6

Radian measure: 2π radians $= 360$ degrees

For a sector of circle:

arc length, $s = \dfrac{\theta°}{360}(2\pi r) = r\theta$ (θ in rad)

shaded area $= \dfrac{\theta°}{360}(\pi r^2) = \dfrac{1}{2}r^2\theta$ (θ in rad)

Equation of a circle, centre at origin, radius r:
$$x^2 + y^2 = r^2$$

Equation of a circle, centre at (a, b), radius r:
$$(x - a)^2 + (y - b)^2 = r^2$$

Volumes and surface areas of regular solids:

(i) **Rectangular prism (or cuboid)**

$$\text{Volume} = l \times b \times h$$

$$\text{Surface area} = 2(bh + hl + lb)$$

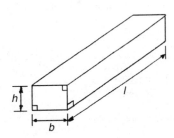

Fig. F7

(ii) **Cylinder**

$$\text{Volume} = \pi r^2 h$$

$$\text{Total surface area} = 2\pi rh + 2\pi r^2$$

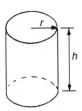

Fig. F8

(iii) **Pyramid**

If area of base $= A$ and

perpendicular height $= h$ then :

$$\text{Volume} = \frac{1}{3} \times A \times h$$

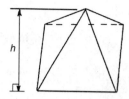

Fig. F9

Total surface area = sum of areas of triangles forming sides + area of base

(iv) **Cone**

$$\text{Volume} = \tfrac{1}{3}\pi r^2 h$$

$$\text{Curved surface area} = \pi r l$$

$$\text{Total surface area} = \pi r l + \pi r^2$$

Fig. F10

(v) **Sphere**

$$\text{Volume} = \tfrac{4}{3}\pi r^3$$

$$\text{Surface area} = 4\pi r^2$$

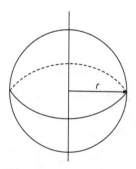

Fig. F11

Areas of irregular figures by approximate methods:

Trapezoidal rule

$$\text{Area} \approx \left(\begin{array}{c} \text{width of} \\ \text{interval} \end{array} \right) \left[\frac{1}{2} \left(\begin{array}{c} \text{first} + \text{last} \\ \text{ordinate} \end{array} \right) \right.$$

$$\left. + \text{ sum of remaining ordinates} \right]$$

Mid-ordinate rule

$$\text{Area} \approx (\text{width of interval})(\text{sum of mid-ordinates})$$

Simpson's rule

$$\text{Area} \approx \frac{1}{3} \left(\begin{array}{c} \text{width of} \\ \text{interval} \end{array} \right) \left[\left(\begin{array}{c} \text{first} + \text{last} \\ \text{ordinate} \end{array} \right) \right.$$

$$\left. + 4 \left(\begin{array}{c} \text{sum of even} \\ \text{ordinates} \end{array} \right) + 2 \left(\begin{array}{c} \text{sum of remaining} \\ \text{odd ordinates} \end{array} \right) \right]$$

Mean or average value of a waveform

$$\text{mean value}, \; y = \frac{\text{area under curve}}{\text{length of base}}$$

$$= \frac{\text{sum of mid-ordinates}}{\text{number of mid-ordinates}}$$

Triangle formulae:

Sine rule: $\quad \dfrac{a}{\sin A} = \dfrac{b}{\sin B} = \dfrac{c}{\sin C}$

Cosine rule: $\quad a^2 = b^2 + c^2 - 2bc \cos A$

Fig. F12

$$\text{Area of any triangle} = \frac{1}{2} \times \text{base} \times \text{perpendicular height}$$

$$= \frac{1}{2} ab \sin C \quad \text{or} \quad \frac{1}{2} ac \sin B \quad \text{or} \quad \frac{1}{2} bc \sin A$$

$$= \sqrt{[s(s-a)(s-b)(s-c)]} \quad \text{where} \quad s = \frac{a+b+c}{2}$$

For a **general sinusoidal function** $y = A\sin(\omega t \pm \alpha)$, then

A = amplitude

ω = angular velocity = $2\pi f$ rad/s

$\dfrac{2\pi}{\omega}$ = periodic time T seconds

$\dfrac{\omega}{2\pi}$ = frequency, f hertz

α = angle of lead or lag (compared with $y = A \sin \omega t$)

Cartesian and polar co-ordinates

If co-ordinate $(x, y) = (r, \theta)$ then

$$r = \sqrt{x^2 + y^2} \quad \text{and} \quad \theta = \tan^{-1}\frac{y}{x}$$

If co-ordinate $(r, \theta) = (x, y)$ then

$$x = r\cos\theta \quad \text{and} \quad y = r\sin\theta$$

Arithmetic progression:

If a = first term and d = common difference, then the arithmetic progression is: $a, a + d, a + 2d, \ldots$

The n'th term is : $a + (n - 1)d$

Sum of n terms, $S = \dfrac{n}{2}[2a + (n - 1)d]$

Geometric progression:

If a = first term and r = common ratio, then the geometric progression is: $a, ar, ar^2, \ldots$

The n'th term is: ar^{n-1}

Sum of n terms, $S_n = \dfrac{a(1 - r^n)}{(1 - r)}$ or $\dfrac{a(r^n - 1)}{(r - 1)}$

If $-1 < r < 1$, $S_\infty = \dfrac{a}{(1 - r)}$

Statistics:

Discrete data:

mean, $\bar{x} = \dfrac{\sum x}{n}$

standard deviation, $\sigma = \sqrt{\left[\dfrac{\sum(x - \bar{x})^2}{n}\right]}$

Grouped data:

$$\text{mean, } \bar{x} = \frac{\sum fx}{\sum f}$$

$$\text{standard deviation, } \sigma = \sqrt{\left[\frac{\sum\{f(x - \bar{x})^2\}}{\sum f}\right]}$$

Standard derivatives

y or $f(x)$	$\dfrac{dy}{dx}$ or $f'(x)$
ax^n	anx^{n-1}
$\sin ax$	$a\cos ax$
$\cos ax$	$-a\sin ax$
e^{ax}	ae^{ax}
$\ln ax$	$\dfrac{1}{x}$

Standard integrals

y	$\int y\,dx$
ax^n	$a\dfrac{x^{n+1}}{n+1} + c$ (except when $n = -1$)
$\cos ax$	$\dfrac{1}{a}\sin ax + c$
$\sin ax$	$-\dfrac{1}{a}\cos ax + c$
e^{ax}	$\dfrac{1}{a}e^{ax} + c$
$\dfrac{1}{x}$	$\ln x + c$

Answers to exercises

Exercise 1 (Page 2)

1. 19   **2.** −66   **3.** 565

4. −2136   **5.** −36 121   **6.** 107 701

7. 16   **8.** 1487   **9.** −225

10. 5914   **11.** 189   **12.** −70 872

13. −15 333   **14.** (a) 1827 (b) 4158

15. (a) 8613 (b) 1752   **16.** (a) 1026 (b) 2233

17. (a) 10 304 (b) − 4433   **18.** (a) 48 (b) 89

19. (a) 259 (b) 56   **20.** (a) 456 (b) 284

21. (a) $474\frac{100}{187}$ (b) $593\frac{10}{79}$

Eerercise 2 (Page 4)

1. (a) 2 (b) 210   **2.** (a) 3 (b) 180

3. (a) 5 (b) 210   **4.** (a) 15 (b) 6300

5. (a) 14 (b) 420 420   **6.** (a) 14 (b) 53 900

Exercise 3 (Page 5)

1. 59   **2.** 14   **3.** 88   **4.** 5   **5.** −107   **6.** 57   **7.** 68

Exercise 4 (Page 8)

1. (a) $\frac{9}{10}$ (b) $\frac{3}{16}$ **2.** (a) $\frac{43}{77}$ (b) $\frac{47}{63}$

3. (a) $8\frac{51}{52}$ (b) $1\frac{9}{40}$ **4.** (a) $1\frac{16}{21}$ (b) $\frac{17}{60}$

5. (a) $\frac{5}{12}$ (b) $\frac{3}{49}$ **6.** (a) $\frac{3}{5}$ (b) 11

7. (a) $\frac{1}{13}$ (b) $\frac{5}{12}$ **8.** (a) $\frac{8}{15}$ (b) $\frac{12}{23}$

9. $-\frac{1}{9}$ **10.** $1\frac{1}{6}$ **11.** $5\frac{4}{5}$

12. $-\frac{13}{126}$ **13.** $2\frac{28}{55}$

Exercise 5 (Page 9)

1. 91 mm to 221 mm   **2.** 81 cm to 189 cm to 351 cm

3. £1.71 and £3.23   **4.** 17 g

5. 72 kg : 27 kg   **6.** 5

7. (a) 2 h 10 min (b) 4 h 20 min

Exercise 6 (Page 11)

1. 11.989   **2.** −31.265   **3.** 24.066

4. 10.906   **5.** 2.244 6

6. (a) 24.81 (b) 24.812

7. (a) 0.006 39 (b) 0.006 4

8. (a) $\frac{13}{20}$ (b) $\frac{21}{25}$ (c) $\frac{1}{80}$ (d) $\frac{141}{500}$ (e) $\frac{3}{125}$

9. (a) $1\frac{41}{50}$ (b) $4\frac{11}{40}$ (c) $14\frac{1}{8}$ (d) $15\frac{7}{20}$ (e) $16\frac{17}{80}$

10. 0.444 44   **11.** 0.629 63   **12.** 1.563

13. 53.455   **14.** 13.84   **15.** 8.69

Exercise 7 (Page 12)

1. (a) 5.7%   (b) 37.4%   (c) 128.5%

2. (a) 21.2%   (b) 79.2%   (c) 169%

3. (a) 496.4 t   (b) 8.657 g   (c) 20.73 s

4. 2.25%   **5.** (a) 14%   (b) 15.67%   (c) 5.36%

6. 37.8 g   **7.** 7.2%   **8.** A 0.6 kg, B 0.9 kg, C 0.5 kg

9. 54%, 31%, 15%, 0.3 t

Exercise 8 (Page 15)

1. (a) 3^7 (b) 4^9 **2.** (a) 2^6 (b) 7^{10}

3. (a) 2   (b) 3^5 **4.** (a) 5^3 (b) 7^3

5. (a) 7^6 (b) 3^6 **6.** (a) 15^{15} (b) 17^8

7. (a) 2   (b) 3^6 **8.** (a) 5^2 (b) 13^2

9. (a) 3^4 (b) 1   **10.** (a) 5^2 (b) $\dfrac{1}{3^5}$

11. (a) 7^2 (b) $\dfrac{1}{2}$ **12.** (a) 1   (b) 1

Exercise 9 (Page 17)

1. (a) $\dfrac{1}{3 \times 5^2}$ (b) $\dfrac{1}{7^3 \times 3^7}$

2. (a) $\dfrac{3^2}{2^5}$ (b) $\dfrac{1}{2^{10} \times 5^2}$

3. (a) 9   (b) ± 3 (c) $\pm \dfrac{1}{2}$ (d) $\pm \dfrac{2}{3}$

4. $\dfrac{147}{148}$ **5.** $-1\dfrac{19}{56}$ **6.** $-3\dfrac{13}{45}$ **7.** $\dfrac{1}{9}$

8. $-5\dfrac{65}{72}$ **9.** 64   **10.** $4\dfrac{1}{2}$

Exercise 10 (Page 18)

1. (a) 7.39×10 (b) 2.84×10
(c) 1.9762×10^2

2. (a) 2.748×10^3 (b) 3.317×10^4
(c) 2.74218×10^5

3. (a) 2.401×10^{-1} (b) 1.74×10^{-2} (c) 9.23×10^{-3}

4. (a) 1.7023×10^3 (b) 1.004×10 (c) 1.09×10^{-2}

5. (a) 5×10^{-1} (b) 1.1875×10
(c) 1.306×10^2 (d) 3.125×10^{-2}

6. (a) 1010   (b) 932.7   (c) 54 100   (d) 7

7. (a) 0.038 9   (b) 0.674 1   (c) 0.008

Exercise 11 (Page 19)

1. (a) 1.351×10^3 (b) 8.731×10^{-1} (c) 5.734×10^{-2}

2. (a) 1.7231×10^3 (b) 3.129×10^{-3}
(c) 1.1199×10^3

3. (a) 1.35×10^2 (b) 1.1×10^{-5}

4. (a) 2×10^{-2} (b) 1.5×10^{-3}

5. (a) 2.71×10^3 kg m^{-3} (b) 4.4×10^{-1}
(c) 3.7673×10^2 Ω (d) 5.11×10^{-1} MeV
(e) 9.57897×10^7 C kg^{-1} (f) 2.241×10^{-2} m^3 mol^{-1}

Exercise 12 (Page 21)

1. order of magnitude error

2. Rounding-off error - should add 'correct to 4 significant figures' or 'correct to 1 decimal place'

3. Blunder

4. Measured values, hence $c = 55\,800$ Pa m^3

5. Order of magnitude error and rounding-off error - should be 0.022 5, correct to 3 significant figures or 0.022 5, correct to 4 decimal places

6. ≈ 30 (29.61, by calculator)

7. ≈ 2 (1.988, correct to 4 s.f., by calculator)

8. ≈ 10 (9.481, correct to 4 s.f., by calculator)

Exercise 13 (Page 23)

1. (a) 10.56   (b) 5443   (c) 96 970   (d) 0.004 083

2. (a) 2.176   (b) 5.955   (c) 270.7   (d) 0.160 0

3. (a) 0.128 7   (b) 0.020 64   (c) 12.25   (d) 0.894 5

4. (a) 558.6   (b) 31.09   **5.** (a) 109.1   (b) 3.641

6. (a) 0.248 9   (b) 500.5

7. (a) 0.057 77   (b) 28.90   (c) 0.006 761

8. (a) 2 515   (b) 146.0   (c) 0.000 029 32

9. (a) 18.63   (b) 87.36   (c) 0.164 3   (d) 0.064 56

10. (a) 0.005 559   (b) 1.900

11. (a) 6.248   (b) 0.963 0

12. (a) 1.165   (b) 2.680

13. (a) 1.605   (b) 11.74

14. (a) 6.874×10^{-3} (b) 8.731×10^{-2}

Exercise 14 (Page 24)

1. (a) 49.40 euro   (b) \$175.78   (c) £78

(d) £126   (e) \$154

2. (a) 381 mm   (b) 56.35 km/h   (c) 378.35 km

(d) 52 lb 13 oz   (e) 6.82 kg   (f) 54.55 litres

(g) 5.5 gallons

3. (a) 7.09 a.m.   (b) 51 minutes, 32 m.p.h.   (c) 7.04 a.m.

Exercise 15 (Page 27)

1. $A = 66.59\,\text{cm}^2$ **2.** $C = 52.78\,\text{mm}$

3. $R = 37.5$ **4.** 159 m/s

5. 0.407 A   **6.** 0.005 02 m or 5.02 mm

7. 0.144 J   **8.** 628.8 m^2

9. 224.5   **10.** 14 230 kg/m^3

11. 281.1 m/s   **12.** 2.526 Ω

13. £589.27   **14.** 508.1 W

15. $V = 2.61\,\text{V}$ **16.** $F = 854.5$

17. $I = 3.81\,\text{A}$ **18.** $T = 14.79\,\text{s}$

19. $E = 3.96\,\text{J}$ **20.** $I = 12.77\,\text{A}$

21. $S = 17.25\,\text{m}$ **22.** $A = 7.184\,\text{cm}^2$

23. $v = 7.327$

Exercise 16 (Page 29)

1. (a) 6_{10} (b) 11_{10} (c) 14_{10} (d) 9_{10}

2. (a) 21_{10} (b) 25_{10} (c) 45_{10} (d) 51_{10}

3. (a) $0.812\,5_{10}$ (b) $0.781\,25_{10}$

(c) $0.218\,75_{10}$ (d) $0.343\,75_{10}$

4. (a) 26.75_{10} (b) 23.375_{10}

(c) $53.437\,5_{10}$ (d) $213.718\,75_{10}$

Exercise 17 (Page 31)

1. (a) 101_2 (b) 1111_2 (c) 10011_2 (d) $11\,101_2$

2. (a) $11\,111_2$ (b) $101\,010_2$ (c) $111\,001_2$

(d) $111\,111_2$

3. (a) 0.01_2 (b) $0.001\,11_2$ (c) $0.010\,01_2$

(d) $0.10\,011_2$

4. (a) $101\,111.011\,01_2$ (b) $11\,110.110\,1_2$

(c) $110\,101.111\,01_2$ (d) $111\,101.101\,01_2$

Exercise 18 (Page 32)

1. (a) $101\,010\,111_2$ (b) $1\,000\,111\,100_2$

(c) $10\,011\,110\,001_2$

2. (a) $0.011\,11_2$ (b) $0.101\,1_2$ (c) $0.101\,11_2$

3. (a) $11\,110\,111.000\,11_2$ (b) $1\,000\,000\,010.011\,1_2$

(c) $11\,010\,110\,100.110\,01_2$

4. (a) $7.437\,5_{10}$ (b) 41.25_{10} (c) $7\,386.187\,5_{10}$

Exercise 19 (Page 35)

1. 231_{10} **2.** 44_{10} **3.** 152_{10} **4.** 753_{10}

5. 36_{16} **6.** $C8_{16}$ **7.** $5B_{16}$ **8.** EE_{16}

9. $D7_{16}$ **10.** EA_{16} **11.** $8B_{16}$ **12.** $A5_{16}$

13. $110\,111_2$ **14.** $11\,101\,101_2$ **15.** $10\,011\,111_2$

16. $101\,000\,100\,001_2$

Exercise 20 (Page 38)

1. -16 **2.** -8 **3.** $4a$

4. $2\dfrac{1}{6}c$ **5.** $-2\dfrac{1}{4}x + 1\dfrac{1}{2}y + 2z$

6. $a - 4b - c$ **7.** $9d - 2e$ **8.** $3x - 5y + 5z$

9. $-5\frac{1}{2}a + \frac{5}{6}b - 4c$ **10.** $3x^2 - xy - 2y^2$

11. $6a^2 - 13ab + 3ac - 5b^2 + bc$

12. (i) $\frac{1}{3b}$ (ii) $2ab$ **13.** $2x - y$ **14.** $3p + 2q$

Exercise 21 (Page 40)

1. $x^5 y^4 z^3$, $13\frac{1}{2}$ **2.** $a^2 b^{1/2} c^{-2}$, $4\frac{1}{2}$ **3.** $a^3 b^{-2} c$, 9

4. $x^{7/10} y^{1/6} z^{1/2}$ **5.** $\frac{1+a}{b}$ **6.** $\frac{p^2 q}{q - p}$

7. $ab^6 c^{3/2}$ **8.** $a^{-4} b^5 c^{11}$ **9.** $xy^3 \sqrt[6]{z^{13}}$

10. $\frac{1}{ef^2}$ **11.** $a^{11/6} b^{1/3} c^{-3/2}$ or $\frac{\sqrt[6]{a^{11}} \sqrt[3]{b}}{\sqrt{c^3}}$

Exercise 22 (Page 42)

1. $3x + y$ **2.** $3a + 5y$ **3.** $5(x - y)$

4. $x(3x - 3 + y)$ **5.** $-5p + 10q - 6r$

6. $a^2 + 3ab + 2b^2$ **7.** $3p^2 + pq - 2q^2$

8. (i) $x^2 - 4xy + 4y^2$ (ii) $9a^2 - 6ab + b^2$

9. $3ab + 7ac - 4bc$ **10.** 0   **11.** $4 - a$

12. $2 + 5b^2$ **13.** $11q - 2p$

14. (i) $p(b + 2c)$ (ii) $2q(q + 4n)$

15. (i) $7ab(3ab - 4)$ (ii) $2xy(y + 3x + 4x^2)$

16. (i) $(a + b)(y + 1)$ (ii) $(p + q)(x + y)$

17. (i) $(x - y)(a + b)$ (ii) $(a - 2b)(2x + 3y)$

Exercise 23 (Page 44)

1. $\frac{1}{2} + 6x$ **2.** $\frac{1}{5}$

3. $4a(1 - 2a)$ **4.** $a(4a + 1)$

5. $a(3 - 10a)$ **6.** $\frac{2}{3y} - 3y + 12$

7. $\frac{2}{3y} + 12\frac{1}{3} - 5y$ **8.** $\frac{2}{3y} + 12 - 13y$

9. $\frac{5}{y} + 1$ **10.** pq

11. $\frac{1}{2}(x - 4)$ **12.** $y\left(\frac{1}{2} + 3y\right)$

Exercise 24 (Page 45)

1. (a) 15   (b) 78

2. (a) 0.0075   (b) 3.15 litres   (c) 350 K

3. (a) 0.000 08   (b) $4.16 \times 10^{-3} A$ (c) 45 V

4. (a) 9.18   (b) 6.12   (c) 0.3375

5. (a) 300×10^3 (b) $0.375\,m^2$ (c) 24×10^3

Exercise 25 (Page 47)

1. 1   **2.** 2   **3.** 6   **4.** -4 **5.** $1\frac{2}{3}$

6. 1   **7.** 2   **8.** $\frac{1}{2}$ **9.** 0   **10.** 3

11. 2   **12.** -10 **13.** 6   **14.** -2 **15.** $2\frac{1}{2}$

16. 2   **17.** $6\frac{1}{4}$ **18.** -3

Exercise 26 (Page 49)

1. 5   **2.** -2 **3.** $-4\frac{1}{2}$ **4.** 2

5. 12   **6.** 15   **7.** -4 **8.** $5\frac{1}{3}$

9. 2   **10.** 13   **11.** -10 **12.** 2

13. 3   **14.** -11 **15.** -6 **16.** 9

17. $6\frac{1}{4}$ **18.** 3   **19.** 4   **20.** 10

21. ± 12 **22.** $-3\frac{1}{3}$ **23.** ± 3 **24.** ± 4

Exercise 27 (Page 50)

1. 10^{-7} **2.** $8\,m/s^2$ **3.** 3.472

4. (a) $1.8\,\Omega$ (b) $30\,\Omega$ **5.** 12p, 17p   **6.** $800\,\Omega$

Exercise 28 (Page 52)

1. 12 cm, $240\,cm^2$ **2.** 0.004   **3.** 30

4. $45°C$ **5.** 50   **6.** £208, £160

7. 30 kg   **8.** 12 m, 8 m

Exercise 29 (Page 54)

1. $d = c - a - b$ 2. $y = \frac{1}{3}(t - x)$ 3. $r = \frac{c}{2\pi}$

4. $x = \frac{y - c}{m}$ 5. $T = \frac{I}{PR}$ 6. $R = \frac{E}{I}$

7. $R = \frac{S - a}{S}$ or $1 - \frac{a}{S}$ 8. $C = \frac{5}{9}(F - 32)$

Exercise 30 (Page 56)

1. $x = \frac{d}{\lambda}(y + \lambda)$ or $d + \frac{yd}{\lambda}$

2. $f = \frac{3F - AL}{3}$ or $f = F - \frac{AL}{3}$ 3. $E = \frac{Ml^2}{8yI}$

4. $t = \frac{R - R_0}{R_0\alpha}$ 5. $R_2 = \frac{RR_1}{R_1 - R}$

6. $R = \frac{E - e - IR}{I}$ or $R = \frac{E - e}{I} - r$

7. $b = \sqrt{\left(\frac{y}{4ac^2}\right)}$ 8. $x = \frac{ay}{\sqrt{(y^2 - b^2)}}$ 9. $l = \frac{t^2g}{4\pi^2}$

10. $u = \sqrt{v^2 - 2as}$ 11. $R = \sqrt{\left(\frac{360A}{\pi\theta}\right)}$

12. $a = N^2y - x$ 13. $L = \frac{\sqrt{Z^2 - R^2}}{2\pi f}$

Exercise 31 (Page 57)

1. $a = \sqrt{\left(\frac{xy}{m - n}\right)}$ 2. $R = \sqrt[4]{\left(\frac{M}{\pi} + r^4\right)}$

3. $r = \frac{3(x + y)}{(1 - x - y)}$ 4. $L = \frac{mrCR}{\mu - m}$

5. $b = \frac{c}{\sqrt{1 - a^2}}$ 6. $r = \sqrt{\left(\frac{x - y}{x + y}\right)}$

7. $b = \frac{a(p^2 - q^2)}{2(p^2 + q^2)}$ 8. $v = \frac{uf}{u - f}, 30$

9. $t_2 = t_1 + \frac{Q}{mc}, 55$ 10. $v = \sqrt{\left(\frac{2dgh}{0.03L}\right)}, 0.965$

11. $l = \frac{8S^2}{3d} + d, 2.725$

12. $C = \frac{1}{\omega\{\omega L - \sqrt{Z^2 - R^2}\}}, 63.1 \times 10^{-6}$

Exercise 32 (Page 61)

1. $a = 5, b = 2$ 2. $x = 1, y = 1$

3. $s = 2, t = 3$ 4. $x = 3, y = -2$

5. $m = 2\frac{1}{2}, n = \frac{1}{2}$ 6. $a = 6, b = -1$

7. $x = 2, y = 5$ 8. $c = 2, d = -3$

Exercise 33 (Page 62)

1. $p = -1, q = -2$ 2. $x = 4, y = 6$

3. $a = 2, b = 3$ 4. $s = 4, t = -1$

5. $x = 3, y = 4$ 6. $u = 12, v = 2$

7. $x = 10, y = 15$ 8. $a = 0.30, b = 0.40$

Exercise 34 (Page 64)

1. $x = \frac{1}{2}, y = \frac{1}{4}$ 2. $\frac{1}{3}, b = -\frac{1}{2}$

3. $p = \frac{1}{4}, q = \frac{1}{5}$ 4. $x = 10, y = 5$

5. $c = 3, d = 4$ 6. $r = 3, s = \frac{1}{2}$

7. $x = 5, y = 1\frac{3}{4}$ 8. 1

Exercise 35 (Page 67)

1. $a = \frac{1}{5}, b = 4$ 2. $I_1 = 6.47, I_2 = 4.62$

3. $u = 12, a = 4, v = 26$ 4. £15 500, £12 800

5. $m = -\frac{1}{2}, c = 3$ 6. $\alpha = 0.00426, R_0 = 22.56\Omega$

7. $a = 12, b = 0.40$ 8. $a = 4, b = 10$

Exercise 36 (Page 69)

1. $4, -8$ 2. $4, -4$ 3. $2, -6$ 4. $-1, 1\frac{1}{2}$

5. $\frac{1}{2}, \frac{1}{3}$ 6. $\frac{1}{2}, -\frac{4}{5}$ 7. 2

8. $1\frac{1}{3}, -\frac{1}{7}$ 9. $\frac{3}{8}, -2$ 10. $\frac{2}{5}, -3$

11. $\frac{4}{3}, -\frac{1}{2}$ 12. $\frac{5}{4}, -\frac{3}{2}$ 13. $x^2 - 4x + 3 = 0$

14. $x^2 + 3x - 10 = 0$ 15. $x^2 + 5x + 4 = 0$

16. $4x^2 - 8x - 5 = 0$ 17. $x^2 - 36 = 0$

18. $x^2 - 1.7x - 1.68 = 0$

Exercise 37 (Page 71)

1. $-3.732, -0.268$ **2.** $-3.137, 0.637$

3. $1.468, -1.135$ **4.** $1.290, 0.310$

5. $2.443, 0.307$ **6.** $-2.851, 0.351$

Exercise 38 (Page 72)

1. $0.637, -3.137$ **2.** $0.296, -0.792$

3. $2.781, 0.719$ **4.** $0.443, -1.693$

5. $3.608, -1.108$ **6.** $4.562, 0.438$

Exercise 39 (Page 74)

1. $1.191\,s$ **2.** $0.345\,A$ or $0.905\,A$

3. $7.84\,cm$ **4.** $0.619\,m$ or $19.38\,m$

5. $0.013\,3$ **6.** $1.066\,m$

7. $86.78\,cm$ **8.** $1.835\,m$ or $18.165\,m$

9. $7\,m$ **10.** 12 ohms, 28 ohms

Exercise 40 (Page 75)

1. $x = 1, y = 3$ and $x = -3, y = 7$

2. $x = \frac{2}{5}, y = -\frac{1}{5}$ and $-1\frac{2}{3}, y = -4\frac{1}{3}$

3. $x = 0, y = 4$ and $x = 3, y = 1$

Exercise 41 (Page 81)

1. 14.5 **2.** $\frac{1}{2}$

3. (a) $4, -2$ (b) $-1, 0$ (c) $-3, -4$ (d) $0, 4$

4. (a) $2, \frac{1}{2}$ (b) $3, -2\frac{1}{2}$ (c) $\frac{1}{24}, \frac{1}{2}$

5. (a) $6, 3$ (b) $-2, 4$ (c) $3, 0$ (d) $0, 7$

6. (a) $2, -\frac{1}{2}$ (b) $-\frac{2}{3}, -1\frac{2}{3}$ (c) $\frac{1}{18}, 2$ (d) $10, -4\frac{2}{3}$

7. (a) $\frac{3}{5}$ (b) -4 (c) $-1\frac{5}{6}$

8. (a) and (c), (b) and (e)   **9.** (a) -1.1 (b) -1.4

10. $(2, 1)$ **11.** $1\frac{1}{2}, 6$

Exercise 42 (Page 85)

1. (a) $40°C$ (b) $128\ \Omega$

2. (a) 850 rev/min   (b) $77.5\,V$

3. (a) 0.25 (b) 12 (c) $F = 0.25L + 12$

   (d) $89.5\,N$ (e) $592\,N$ (f) $212\,N$

4. $-0.003, 8.73$

5. (a) 22.5 m/s   (b) $6.43s$ (c) $v = 0.7t + 15.5$

6. $m = 26.9L - 0.63$

7. (a) $1.26t$ (b) 21.68% (c) $F = -0.09w + 2.21$

8. (a) $96 \times 10^9 Pa$ (b) $0.000\,22$ (c) $28.8 \times 10^6 Pa$

9. (a) $\frac{1}{5}$ (b) 6 (c) $E = \frac{1}{5}L + 6$ (d) $12N$ (e) $65\,N$

10. $a = 0.85$, $b = 12$, 254.3 kPa, 275.5 kPa, 280 K

Exercise 43 (Page 88)

1. $x = 1, y = 1$ **2.** $x = 3\frac{1}{2}, y = 1\frac{1}{2}$

3. $x = -1, y = 2$ **4.** $x = 2.3, y = -1.2$

5. $x = -2, y = -3$ **6.** $a = 0.4, b = 1.6$

Exercise 44 (Page 92)

1. (a) Minimum $(0, 0)$ (b) Minimum $(0, -1)$

   (c) Maximum $(0, 3)$ (d) Maximum $(0, -1)$

2. -0.4 or 0.6 **3.** -3.9 or 6.9

4. -1.1 or 4.1 **5.** -1.8 or 2.2

6. $x = -1\frac{1}{2}$ or -2, Minimum at $(-1\frac{3}{4}, -\frac{1}{8})$

7. $x = -0.7$ or 1.6

8. (a) ± 1.63 (b) 1 or $-\frac{1}{3}$

9. $(-2.58, 13.31), (0.58, 0.67); x = -2.58$ or 0.58

10. $x = -1.2$ or 2.5 (a) -30

   (b) 2.75 and -1.45 (c) 2.29 or -0.79

Exercise 45 (Page 93)

1. $x = 4$, $y = 8$ and $x = -\frac{1}{2}$, $y = -5\frac{1}{2}$

2. (a) $x = -1.5$ or 3.5 (b) $x = -1.24$ or 3.24

   (c) $x = -1.5$ or 3.0

Exercise 46 (Page 94)

1. $x = -2.0, -0.5$ or 1.5

2. $x = -2, 1$ or 3, Minimum at $(2.12, -4.10)$, Maximum at $(-0.79, 8.21)$

3. $x = 1$ 4. $x = -2.0, 0.38$ or 2.6

5. $x = 0.69$ or 2.5 6. $x = -2.3, 1.0$ or 1.8

7. $x = -1.5$

Exercise 47 (Page 98)

1. 4   2. 4   3. 3   4. -3 5. $\frac{1}{3}$

6. 3   7. 2   8. -2 9. $1\frac{1}{2}$

10. $\frac{1}{3}$ 11. 2   12. $10\,000$

13. $100\,000$ 14. 9   15. $\pm\frac{1}{32}$

16. 0.01   17. $\frac{1}{16}$ 18. e^3

19. $2\log 2 + \log 3 + \log 5$

20. $2\log 2 + \frac{1}{4}\log 5 - 3\log 3$

21. $4\log 2 - 3\log 3 + 3\log 5$

22. $\log 2 - 3\log 3 + 3\log 5$

23. $5\log 3$ 24. $4\log 2$

25. $6\log 2$ 26. $\frac{1}{2}$ 27. $\frac{3}{2}$

28. $x = 2\frac{1}{2}$ 29. $t = 8$ 30. $b = 2$

Exercise 48 (Page 99)

1. 1.691   2. 3.170   3. 0.2696

4. 6.058   5. 2.251   6. 3.959

7. 2.542   8. -0.3272

Exercise 49 (Page 101)

1. (a) 81.45   (b) 0.7788   (c) 2.509

2. (a) 0.1653   (b) 0.4584   (c) 22030

3. (a) 57.556   (b) -0.26776 (c) 645.55

4. (a) 5.0988   (b) 0.064037   (c) 40.446

5. (a) 4.55848   (b) 2.40444   (c) 8.05124

6. (a) 48.04106   (b) 4.07482   (c) -0.08286

7. 2.739

Exercise 50 (Page 102)

1. 2.0601   2. (a) 7.389   (b) 0.7408

3. $1 - 2x^2 - \frac{8}{3}x^3 - 2x^4$

4. $2x^{1/2} + 2x^{5/2} + x^{9/2} + \frac{1}{3}x^{13/2} + \frac{1}{12}x^{17/2} + \frac{1}{60}x^{21/2}$

Exercise 51 (Page 104)

1. $3.97, 2.03$ 2. $1.66, -1.30$

3. (a) $27.9\,\text{cm}^3$ (b) 115.5 min

4. (a) $71.6°\text{C}$ (b) 5 minutes

Exercise 52 (Page 106)

1. (a) 0.5481   (b) 1.6888   (c) 2.2420

2. (a) 2.8507   (b) 6.2940   (c) 9.1497

3. (a) -1.7545 (b) -5.2190 (c) -2.3632

4. (a) 0.27774   (b) 0.91374   (c) 8.8941

5. (a) 3.6773   (b) -0.33154 (c) 0.13087

6. -0.4904 7. -0.5822 8. 2.197

9. 816.2   10. 0.8274

Exercise 53 (page 108)

1. $9.921 \times 10^4\,\text{Pa}$

2. (a) 29.32 volts   (b) $71.31 \times 10^{-6}\text{s}$

3. (a) 1.993 m   (b) 2.293 m

4. (a) $50°\text{C}$ (b) 55.45 s

5. 30.4 N   **6.** (a) 3.04 A   (b) 1.46 s

7. 2.45 mol/cm^3 **8.** (a) 7.07 A   (b) 0.966 s

9. £2424

Exercise 54 (page 112)

1. (a) y (b) x^2 (c) c (d) d

2. (a) y (b) $\sqrt{x}$ (c) b (d) a

3. (a) y (b) $\dfrac{1}{x}$ (c) f (d) e

4. (a) $\dfrac{y}{x}$ (b) x (c) b (d) c

5. (a) $\dfrac{y}{x}$ (b) $\dfrac{1}{x^2}$ (c) a (d) b

6. $a = 1.5$, $b = 0.4$, 11.78 mm^2

7. $y = 2x^2 + 7, 5.15$ **8.** (a) 950   (b) 317 kN

9. $a = 0.4$, $b = 8.6$ (i) 94.4   (ii) 11.2

Exercise 55 (page 116)

1. (a) $lg\ y$ (b) x (c) $lg\ a$ (d) $lg\ b$

2. (a) $lg\ y$ (b) $lg\ x$ (c) l (d) $lg\ k$

3. (a) $ln\ y$ (b) x (c) n (d) $ln\ m$

4. $I = 0.0012\ V^2$, 6.75 candelas

5. $a = 3.0, b = 0.5$

6. $a = 5.7, b = 2.6$, 38.53, 3.0

7. $R_0 = 26.0, c = 1.42$

8. $Y = 0.08e^{0.24x}$

9. $T_0 = 35.4$ N, $\mu = 0.27$, 65.0 N, 1.28 radians

Exercise 56 (page 118)

1. (a) 82°11′   (b) 150°13′   (c) 100°16′3″

    (d) 89°2′23″

2. (a) 7°11′   (b) 27°48′   (c) 18°47′49″

    (d) 66°7′23″

3. (a) 15.183°   (b) 29.883°   (c) 49.705°

    (d) 135.122°

4. (a) 25°24′0″   (b) 36°28′48″   (c) 55°43′26″

    (d) 231°1′30″

Exercise 57 (page 119)

1. (a) acute   (b) obtuse   (c) reflex

2. (a) 21°   (b) 62°23′   (c) 48°56′17″

3. (a) 102°   (b) 165°   (c) 10°18′49″

4. (a) 1 & 3, 2 & 4, 5 & 7, 6 & 8

   (b) 1 & 2, 2 & 3, 3 & 4, 4 & 1, 5 & 6, 6 & 7, 7 & 8,

      8 & 5, 3 & 8, 1 & 6, 4 & 7 or 2 & 5

   (c) 1&5, 2&6, 4&8, 3&7

   (d) 3 & 5   or   2 & 8

5. 59°20′   **6.** $a = 69°, b = 21°, c = 82°$ **7.** 51°

Exercise 58 (page 121)

1. 40°, 70°, 70°, 125°, isosceles

2. $a = 18°50′, b = 71°10′, c = 68°, d = 90°, e = 22°, f = 49°, g = 41°$

3. $a = 103°, b = 55°, c = 77°, d = 125°, e = 55°, f = 22°, g = 103°, h = 77°, i = 103°, j = 77°, k = 81°$

4. 17°   **5.** $A = 37°, B = 60°, E = 83°$

Exercise 59 (page 123)

1. (a) Congruent BAC, DAC (SAS)

   (b) Congruent FGE, JHI (SSS)

   (c) Not necessarily congruent

   (d) Congruent QRT, SRT (RHS)

   (e) Congruent UVW, XZY (ASA)

2. Proof

Exercise 60 (page 125)

1. $x = 16.54$ mm, $y = 4.18$ mm   **2.** 9 cm, 7.79 cm

3. (a) 2.25 cm   (b) 4 cm   **4.** 3 m

Exercise 61 (page 126)

1.- 5. Constructions - see similar constructions in worked problems 23 to 26 on pages 125 and 126.

Exercise 62 (page 129)

1. 11.18 cm   2. 24.11 mm   3. $8^2 + 15^2 = 17^2$

4. (a) 27.20 cm each   (b) 45°   5. 20.81 km

6. 3.35 m, 10 cm   7. 132.7 km

Exercise 63 (page 130)

1. $\sin Z = \dfrac{9}{41}$, $\cos Z = \dfrac{40}{41}$, $\tan X = \dfrac{40}{9}$, $\cos X = \dfrac{9}{41}$

2. $\sin A = \dfrac{3}{5}$, $\cos A = \dfrac{4}{5}$, $\tan A = \dfrac{3}{4}$, $\sin B = \dfrac{4}{5}$, $\cos B = \dfrac{3}{5}$, $\tan B = \dfrac{4}{3}$

3. $\sin A = \dfrac{8}{17}$, $\tan A = \dfrac{8}{15}$

4. $\sin X = \dfrac{15}{113}$, $\cos X = \dfrac{112}{113}$

5. (a) $\dfrac{15}{17}$ (b) $\dfrac{15}{17}$ (c) $\dfrac{8}{15}$

6. (a) $\sin\theta = \dfrac{7}{25}$ (b) $\cos\theta = \dfrac{24}{25}$

7. (a) 9.434   (b) -0.625 (c) $-32°$

Exercise 64 (page 132)

1. $BC = 3.50$ cm, $AB = 6.10$ cm, $\angle B = 55$

2. $FE = 5$ cm, $\angle E = 53°8'$, $\angle F = 36°52'$

3. $GH = 9.841$ mm, $GI = 11.32$ mm, $\angle H = 49°$

4. $KL = 5.43$ cm, $JL = 8.62$ cm, $\angle J = 39°$, area $= 18.19$ cm^2

5. $MN = 28.86$ mm, $NO = 13.82$ mm, $\angle O = 64°25'$, area $= 199.4$ mm^2

6. $PR = 7.934$ m, $\angle Q = 65°3'$, $\angle R = 24°57'$, area $= 14.64$ m^2

7. 6.54 m

Exercise 65 (page 134)

1. 36.15 m   2. 48 m   3. 249.5 m

4. 110.1 m   5. 53.0 m   6. 9.50 m

7. 107.8 m   8. 9.43 m, 10.56 m   9. 60 m

Exercise 66 (page 135)

1. (a) 0.4540   (b) 0.1321   (c) -0.8399

2. (a) -0.5592 (b) 0.9307   (c) 0.2447

3. (a) -0.7002 (b) -1.1671 (c) 1.1612

4. (a) 0.8660   (b) -0.1010 (c) 0.5865

5. 13.54°, 13°32′, 0.236 rad

6. 34.20°, 34°12′, 0.597 rad

7. 39.03°, 39°2′, 0.681 rad

8. 1.097   9. 5.805   10. -5.325

11. 21°42′   12. 0.07448

13. (a) -0.8192 (b) -1.8040 (c) 0.6528

Exercise 67 (page 140)

1. (a) 42.78° and 137.22°   (b) 188.53° and 351.47°

2. (a) 29.08° and 330.92°   (b) 123.86° and 236.14°

3. (a) 44.21° and 224.21°   (b) 113.12° and 293.12°

Exercise 68 (page 144)

1. 1, 120°   2. 2, 144°   3. 3, 90°   4. 3, 720°

5. $\dfrac{7}{2}$, 960°   6. 6, 360°   7. 4, 180°

Exercise 69 (page 146)

1. 40, 0.04 s, 25 Hz, 0.29 rad (or 16°37′) leading $40\sin 50\pi t$

2. 75 cm, 0.157 s, 6.37 Hz, 0.54 rad (or 30°56′) lagging $75\sin 40t$

3. 300 V, 0.01 s, 100 Hz, 0.412 rad (or 23°36′) lagging $300\sin 200\pi t$

4. (a) $v = 120\sin 100\pi t$ volts
   (b) $v = 120\sin(100\pi t + 0.43)$ volts

5. $i = 20\sin\left(80\pi t - \dfrac{\pi}{6}\right)$ amperes

6. $3.2\sin(100\pi t + 0.488)$ m

7. (a) 5 A, 20 ms, 50 Hz, 24°45′ lagging   (b) -2.093 A
   (c) 4.363 A   (d) 6.375 ms   (e) 3.423 ms

Exercise 70 (page 149)

1. (5.83, 59.04°)   or   (5.83, 1.03 rad)

2. (6.61, 20.82°)   or   (6.61, 0.36 rad)

3. (4.47, 116.57°)   or   (4.47, 2.03 rad)

4. (6.55, 145.58°)   or   (6.55, 2.54 rad)

5. (7.62, 203.20°) or (7.62, 3.55 rad)

6. (4.33, 236.31°) or (4.33, 4.12 rad)

7. (5.83, 329.04°) or (5.83, 5.74 rad)

8. (15.68, 307.75°) or (15.68, 5.37 rad)

Exercise 71 (page 151)

1. (1.294, 4.830)   **2.** (1.917, 3.960)

3. (−5.362, 4.500)   **4.** (−2.884, 2.154)

5. (−9.353, −5.400)   **6.** (−2.615, −3.207)

7. (0.750, −1.299)   **8.** (4.252, −4.233)

Exercise 72 (page 156)

1. 35.7 cm² **2.** (a) 80 m (b) 170 m

3. (a) 29 cm² (b) 650 mm² **4.** 482 m² **5.** 3.4 cm

6. $p = 105°, q = 35°, r = 142°, s = 95°, t = 146°$

7. (i) rhombus (a) 14 cm² (b) 16 cm

(ii) parallelogram (a) 180 mm² (b) 80 mm

(iii) rectangle (a) 3600 mm² (b) 300 mm

(iv) trapezium (a) 190 cm² (b) 62.91 cm

8. (a) 50.27 cm² (b) 706.9 mm² (c) 3183 mm²

9. 2513 mm²

10. (a) 20.19 mm (b) 63.41 mm

11. (a) 53.01 cm² (b) 129.9 mm² (c) 6.84 cm²

12. 5773 mm² **13.** 1.89 m² **14.** 6750 mm²

Exercise 73 (page 158)

1. 1932 mm² **2.** 1624 mm²
3. (a) 0.918 ha (b) 456 m

Exercise 74 (page 158)

1. 80 ha **2.** 80 m²

Exercise 75 (page 161)

1. 45.24 cm **2.** 259.5 mm
3. 2.629 cm **4.** 47.68 cm

Exercise 76 (page 163)

1. (a) $\dfrac{\pi}{6}$ (b) $\dfrac{5\pi}{12}$ (c) $\dfrac{5\pi}{4}$

2. (a) 0.838 (b) 1.481 (c) 4.054

3. (a) 150° (b) 80° (c) 105°

4. (a) 0°43′ (b) 154°8′ (c) 414°53′

5. 17.80 cm, 74.07 cm²

6. (a) 59.86 mm (b) 197.8 mm

7. 26.2 cm **8.** 8.67 cm, 54.48 cm

9. 82°30′ **10.** 748

11. (a) 0.698 rad (b) 804.2 m²

12. (a) 396 mm² (b) 42.24%

13. 483.6 mm **14.** 7.74 mm

Exercise 77 (page 165)

1. (a) 3 (b) (−4, 1)

2. Centre at (3, −2), radius 4

3. Circle, centre (0, 1), radius 5

4. Circle, centre (0, 0), radius 6

Exercise 78 (page 168)

1. 15 cm³, 135 g   **2.** 500 litres

3. 1.44 m³   **4.** 8796 cm³

5. 4.709 cm, 153.9 cm²   **6.** 201.1 cm³, 159.0 cm²

7. 2.99 cm   **8.** 28060 cm³, 1.099 m²

9. 7.68 cm³, 25.81 cm²   **10.** 113.1 cm³, 113.1 cm²

11. 5890 mm² or 58.90 cm²

Exercise 79 (page 171)

1. 13.57 kg   **2.** 5.131 cm

3. 29.32 cm³   **4.** 393.4 m²

5. (i) (a) 670 cm³ (b) 523 cm²

(ii) (a) 180 cm³ (b) 154 cm²

(iii) (a) 56.5 cm³ (b) 84.8 cm²

(iv) (a) 10.4 cm³ (b) 32.0 cm²

(v) (a) 96.0 cm³ (b) 146 cm²

(vi) (a) $86.5 \, \text{cm}^3$ (b) $142 \, \text{cm}^2$

(vii) (a) $805 \, \text{cm}^3$ (b) $539 \, \text{cm}^2$

6. $8.53 \, \text{cm}$ **7.** (a) $17.9 \, \text{cm}$ (b) $38.0 \, \text{cm}$

8. $125 \, \text{cm}^3$ **9.** $10.3 \, \text{m}^3$, $25.5 \, \text{m}^2$

10. 6560 litres   **11.** $657.1 \, \text{cm}^3$, $1027 \, \text{cm}^2$

Exercise 80 (page 174)

1. $147 \, \text{cm}^3$, $164 \, \text{cm}^2$ **2.** $403 \, \text{cm}^3$, $337 \, \text{cm}^2$

3. $10480 \, \text{m}^3$, $1852 \, \text{m}^2$ **4.** $1707 \, \text{cm}^2$

5. $10.69 \, \text{cm}$ **6.** $55910 \, \text{cm}^3$, $8427 \, \text{cm}^2$ **7.** $5.14 \, \text{m}$

Exercise 81 (page 175)

1. $8 : 125$ **2.** $137.2 \, \text{g}$

Exercise 82 (page 178)

1. 4.5 square units   **2.** 54.7 square units   **3.** $63 \, \text{m}$

4. $4.70 \, \text{ha}$ **5.** $143 \, \text{m}^2$

Exercise 83 (page 179)

1. $42.59 \, \text{m}^3$ **2.** $147 \, \text{m}^3$ **3.** $20.42 \, \text{m}^3$

Exercise 84 (page 182)

1. (a) $2 \, \text{A}$ (b) $50 \, \text{V}$ (c) $2.5 \, \text{A}$

2. (a) $2.5 \, \text{V}$ (b) $3 \, \text{A}$

3. $0.093 \, \text{As}$, $3.1 \, \text{A}$

4. (a) $31.83 \, \text{V}$ (b) 0

5. $49.13 \, \text{cm}^2$, $368.5 \, \text{kPa}$

Exercise 85 (page 186)

1. $C = 83°$, $a = 14.1 \, \text{mm}$, $c = 28.9 \, \text{mm}$, area $= 189 \, \text{mm}^2$

2. $A = 52°2'$, $c = 7.568 \, \text{cm}$, $a = 7.152 \, \text{cm}$,
area $= 25.65 \, \text{cm}^2$

3. $D = 19°48'$, $E = 134°12'$, $e = 36.0 \, \text{cm}$, area $= 134 \, \text{cm}^2$

4. $E = 49°0'$, $F = 26°38'$, $f = 15.08 \, \text{mm}$,
area $= 185.6 \, \text{mm}^2$

5. $J = 44°29'$, $L = 99°31'$, $l = 5.420 \, \text{cm}$, area $= 6.132 \, \text{cm}^2$
OR $J = 135°31'$, $L = 8°29'$, $l = 0.810 \, \text{cm}$,
area $= 0.916 \, \text{cm}^2$

6. $K = 47°8'$, $J = 97°52'$, $j = 62.2 \, \text{mm}$, area $= 820.2 \, \text{mm}^2$
OR $K = 132°52'$, $J = 12°8'$, $j = 13.19 \, \text{mm}$,
area $= 174.0 \, \text{mm}^2$

Exercise 86 (page 187)

1. $p = 13.2 \, \text{cm}$, $Q = 47°21'$, $R = 78°39'$, area $= 77.7 \, \text{cm}^2$

2. $p = 6.127 \, \text{m}$, $Q = 30°49'$, $R = 44°11'$, area $= 6.938 \, \text{m}^2$

3. $X = 83°20'$, $Y = 52°37'$, $Z = 44°3'$, area $= 27.8 \, \text{cm}^2$

4. $Z = 29°46'$, $Y = 53°31'$, $Z = 96°43'$, area $= 355 \, \text{mm}^2$

Exercise 87 (page 189)

1. $193 \, \text{km}$

2. (a) $122.6 \, \text{m}$ (b) $94°49'$, $40°39'$, $44°32'$

3. (a) $11.4 \, \text{m}$ (b) $17°33'$

4. $163.4 \, \text{m}$

5. $BF = 3.9 \, \text{m}$, $EB = 4.0 \, \text{m}$

6. $6.35 \, \text{m}$, $5.37 \, \text{m}$

7. $32.48 \, \text{A}$, $14°19'$

Exercise 88 (page 191)

1. $80°25'$, $59°23'$, $40°12'$ **2.** (a) $15.23 \, \text{m}$ (b) $38°4'$

3. $40.25 \, \text{cm}$, $126°3'$ **4.** $19.8 \, \text{cm}$

5. $36.2 \, \text{m}$ **6.** $x = 69.3 \, \text{mm}$, $y = 142 \, \text{mm}$

7. $130°$

Exercise 89 (page 196)

1. $47 \, \text{N}$ at $29°$ **2.** Zero

3. $7.27 \, \text{m/s}$ at $90.8°$ **4.** $6.24 \, \text{N}$ at $76.10°$

5. $2.46 \, \text{N}$, $4.12 \, \text{N}$ **6.** $5.7 \, \text{m/s}^2$ at $310°$

7. $11.85 \, \text{A}$ at $31.14°$

Exercise 90 (page 198)

1. (a) $54.0 \, \text{N}$ at $78.16°$ (b) $45.64 \, \text{N}$ at $4.66°$

2. (a) $31.71 \, \text{m/s}$ at $121.81°$ (b) $19.55 \, \text{m/s}$ at $8.63°$

Exercise 91 (page 199)

1. 83.5 km/h at 71.6° to the vertical
2. 4 minutes 55 seconds

Exercise 92 (page 200)

1. 21, 25   2. 48, 96   3. 14, 7   4. $-3, -8$

5. 50, 65   6. 0.001, 0.0001   7. 54, 79

Exercise 93 (page 201)

1. 1, 3, 5, 7, ...   2. 7, 10, 13, 16, 19, ...

3. 6, 11, 16, 21, ...   4. $5n$ 5. $6n - 2$

6. $2n + 1$ 7. $4n - 2$ 8. $3n + 6$

9. $6^3 (= 216), 7^3 (= 343)$

Exercise 94 (page 202)

1. 68   2. 6.2   3. 85.25   4. $23\frac{1}{2}$

5. 11   6. 209   7. 346.5

Exercise 95 (page 204)

1. $-\frac{1}{2}$ 2. $1\frac{1}{2}, 3, 4\frac{1}{2}$ 3. 7808   4. 25

5. $8\frac{1}{2}, 12, 15\frac{1}{2}, 19$ 6. (a) 120   (b) 26070   (c) 250.5

7. £10000, £109500   8. £8720

Exercise 96 (page 205)

1. 2560   2. 273.25   3. 512, 4096   4. 10th

5. 812.5   6. 8   7. $1\frac{2}{3}$

Exercise 97 (page 207)

1. (a) 3   (b) 2   (c) 59022   2. £1566, 11 years
3. 48.71 M   4. 71.53 g   5. (a) £599.14   (b) 19 years
6. 100, 139, 193, 268, 373, 518, 720, 1000 rev/min

Exercise 98 (page 208)

1. (a) continuous   (b) continuous   (c) discrete

    (d) continuous

2. (a) discrete   (b) continuous   (c) discrete

    (d) discrete

Exercise 99 (page 211)

1. If one symbol is used to represent 10 vehicles, working correct to the nearest 5 vehicles, gives $3\frac{1}{2}, 4\frac{1}{2}$, 6, 7, 5 and 4 symbols respectively

2. If one symbol represents 200 components, working correct to the nearest 100 components gives: Mon 8, Tues 11, Wed 9, Thurs 12 and Fri $6\frac{1}{2}$

3. 6 equally spaced horizontal rectangles, whose lengths are proportional to 35, 44, 62, 68, 49 and 41, respectively

4. 5 equally spaced horizontal rectangles, whose lengths are proportional to 1580, 2190, 1840, 2385 and 1280 units, respectively

5. 6 equally spaced vertical rectangles, whose heights are proportional to 35, 44, 62, 68, 49 and 41 units, respectively

6. 5 equally spaced vertical rectangles, whose heights are proportional to 1580, 2190, 1840, 2385 and 1280 units, respectively

7. Three rectangles of equal height, subdivided in the percentages shown in the columns of the question. P increases by 20% at the expense of Q and R

8. Four rectangles of equal height, subdivided as follows:

    week 1: 18%, 7%, 35%, 12%, 28%

    week 2: 20%, 8%, 32%, 13%, 27%

    week 3: 22%, 10%, 29%, 14%, 25%

    week 4: 20%, 9%, 27%, 19%, 25%

    Little change in centres A and B, a reduction of about 5% in C, an increase of about 7% in D and a reduction of about 3% in E

9. A circle of any radius, subdivided into sectors having angles of $7\frac{1}{2}°, 22\frac{1}{2}°, 52\frac{1}{2}°, 167\frac{1}{2}°$ and 110°, respectively

10. A circle of any radius, subdivided into sectors having angles of 107°, 156°, 29° and 68°, respectively

11. (a) £495   (b) 88

12. (a) £16450   (b) 138

Exercise 100 (page 216)

1. There is no unique solution, but one solution is:

   39.3 – 39.4   1; 39.5 – 39.6   5; 39.7 – 39.8   9;
   39.9 – 40.0   17; 40.1 – 40.2   15; 40.3 – 40.4   7;
   40.5 – 40.6   4; 40.7 – 40.8   2

2. Rectangles, touching one another, having mid-points of 39.35, 39.55, 39.75, 39.95,... and heights of 1, 5, 9, 17,...

3. There is no unique solution, but one solution is:

   20.5 – 20.9   3; 21.0 – 21.4   10; 21.5 – 21.9   11;
   22.0 – 22.4   13; 22.5 – 22.9   9; 23.0 – 23.4   2

4. There is no unique solution, but one solution is:

   1 – 10   3; 11 – 19   7; 20 – 22   12; 23 – 25   14;
   26 – 28   7; 29 – 38   5; 39 – 48   2

5. 20.95 3; 21.45 13; 21.95 24; 22.45 37; 22.95 46;
   23.45 48

6. Rectangles, touching one another, having mid-points of 5.5, 15, 21, 24, 27, 33.5 and 43.5. The heights of the rectangles (frequency per unit class range) are 0.3, 0.78, 4, 4.67, 2.33, 0.5 and 0.2

7. (20.95 3), (21.45 13), (21.95 24), (22.45 37), (22.95 46), (23.45 48)

8. A graph of cumulative frequency against upper class boundary having co-ordinates given in the answer to question 7

9. (a) There is no unique solution, but one solution is:

   2.05 – 2.09   3; 2.10 – 2.14   10; 2.15 – 2.19   11;
   2.20 – 2.24   13; 2.25 – 2.29   9; 2.30 – 2.34   2

   (b) Rectangles, touching one another, having mid-points of 2.07, 2.12, ... and heights of 3, 10, ...

   (c) Using the frequency distribution given in the solution to part (a) gives: 2.095 3; 2.145 13; 2.195 24; 2.245 37; 2.295 46; 2.345 48

   (d) A graph of cumulative frequency against upper class boundary having the co-ordinates given in part (c)

Exercise 101 (page 218)

1. Mean $7\frac{1}{3}$, median 8, mode 8

2. Mean 27.25, median 27, mode 26

3. Mean 4.7225, median 4.72, mode 4.72

4. Mean 115.2, median 126.4, no mode

Exercise 102 (page 219)

1. 171.7 cm

2. Mean 89.5, median 89, mode 89.2

3. Mean 2.02158 cm, median 2.02152 cm, mode 2.02167 cm

Exercise 103 (page 221)

1. 4.60   2. 2.83 μF

3. Mean 34.53 MPa, standard deviation 0.07474 MPa

4. 9.394 cm   5. 2.828

Exercise 104 (page 222)

1. 30, 27.5, 33.5 days   2. 27, 26, 33 faults

3. $Q = 164.5$ cm, $Q = 172.5$ cm, $Q_3 = 179$ cm, 7.25 cm

4. 37 and 38; 40 and 41   5. 40, 40, 41; 50, 51, 51

Exercise 105 (page 225)

1. (a) $\frac{2}{9}$ or 0.2222   (b) $\frac{7}{9}$ or 0.7778

2. (a) $\frac{23}{139}$ or 0.1655   (b) $\frac{47}{139}$ or 0.3381

   (c) $\frac{69}{139}$ or 0.4964

3. (a) $\frac{1}{6}$ (b) $\frac{1}{6}$ (c) $\frac{1}{36}$

4. (a) $\frac{2}{5}$ (b) $\frac{1}{5}$ (c) $\frac{4}{15}$ (d) $\frac{13}{15}$

5. (a) $\frac{1}{250}$ (b) $\frac{1}{200}$ (c) $\frac{9}{1000}$ (d) $\frac{1}{50000}$

Exercise 106 (page 227)

1. (a) 0.6   (b) 0.2   (c) 0.15

2. (a) 0.64   (b) 0.32

3. 0.0768

4. (a) 0.4912   (b) 0.4211

5. (a) 89.38%   (b) 10.25%

6. (a) 0.0227   (b) 0.0234   (c) 0.0169

Exercise 107 (page 229)

1. 1, 5, 21, 9, 61 **2.** 0, 11, −10, 21 **3.** proof

4. 8, $-a^2 - a + 8$, $-a^2 - a$, $-a - 1$

Exercise 108 (page 230)

1. 16, 8

Exercise 109 (page 232)

1. 1 **2.** 7 **3.** $8x$

4. $15x^2$ **5.** $-4x + 3$ **6.** 0

7. 9 **8.** $\dfrac{2}{3}$ **9.** $18x$

10. $-21x^2$ **11.** $2x + 15$ **12.** 0

13. $12x^2$ **14.** $6x$

Exercise 110 (page 233)

1. $28x^3$ **2.** $\dfrac{1}{2\sqrt{x}}$ **3.** $\dfrac{3}{2}\sqrt{t}$ **4.** $-\dfrac{3}{x^4}$ **5.** $3 + \dfrac{1}{2\sqrt{x^3}} - \dfrac{1}{x^2}$

6. $-\dfrac{10}{x^3} + \dfrac{7}{2\sqrt{x^9}}$ **7.** $6t - 12$ **8.** $3x^2 + 6x + 3$

9. See answers for Exercise 109

10. $12x - 3$ (a) -15 (b) 21 **11.** $6x^2 + 6x - 4$, 32

12. $-6x^2 + 4$, -9.5

Exercise 111 (page 235)

1. (a) $12\cos\ 3x$ (b) $-12\sin 6x$ **2.** $6\cos 3\theta + 10\sin 2\theta$

3. 270.2 A/s **4.** 1393.4 V/s

5. $12\cos(4t + 0.12) + 6\sin(3t - 0.72)$

Exercise 112 (page 236)

1. (a) $15e^{3x}$ (b) $-\dfrac{4}{7e^{2x}}$ **2.** $\dfrac{5}{\theta} - \dfrac{4}{\theta} = \dfrac{1}{\theta}$

3. 16 **4.** 664

Exercise 113 (page 237)

1. (a) -1 (b) 16

2. $-\dfrac{4}{x^3} + \dfrac{2}{x} + 10\sin 5x - 12\cos 2x + \dfrac{6}{e^{3x}}$

Exercise 114 (page 237)

1. (a) $36x^2 + 12x$ (b) $72x + 12$

2. (a) $\dfrac{4}{5} - \dfrac{12}{t^5} + \dfrac{6}{t^3} + \dfrac{1}{4\sqrt{t^3}}$ (b) -4.95

3. $-12\sin\ 2t - \cos\ t$

4. proof

Exercise 115 (page 238)

1. -2542 A/s **2.** (a) 0.16 cd/V (b) 312.5 V

3. (a) -1000 V/s (b) -367.9 V/s **4.** -1.635 Pa/m

Exercise 116 (page 241)

1. (a) $4x + c$ (b) $\dfrac{7}{2}x^2 + c$

2. (a) $\dfrac{2}{15}x^3 + c$ (b) $\dfrac{5}{24}x^4 + c$

3. (a) $3x + x^2 - \dfrac{4}{3}x^3 + c$ (b) $3\left(\dfrac{x^2}{2} + \dfrac{5x^3}{3}\right) + c$ or $\dfrac{3}{2}x^2 + 5x^3 + c$

4. (a) $\dfrac{3x^2}{2} - 5x + c$ (b) $4x + 2x^2 + \dfrac{x^3}{3} + c$

5. (a) $\dfrac{-4}{3x} + c$ (b) $\dfrac{-1}{4x^3} + c$

6. (a) $\dfrac{2}{3}\sqrt{x^3} + c$ (b) $4\sqrt{x} + c$

7. (a) $\dfrac{3}{2}\sin\ 2x + c$ (b) $-\dfrac{7}{3}\cos\ 3x + c$

8. (a) $\dfrac{3}{8}e^{2x} + c$ (b) $\dfrac{-2}{15e^{5x}} + c$

9. (a) $\dfrac{2}{3}\ln\ x + c$ (b) $\dfrac{x^2}{2} - \ln\ x + c$

Exercise 117 (page 243)

1. (a) 37.5 (b) 0.50 **2.** (a) 6 (b) -1.333

3. (a) 10.83 (b) -4 **4.** (a) 0 (b) 4

5. (a) 1 (b) 4.248 **6.** (a) 0.2352 (b) 2.638

7. (a) 19.09 (b) 2.457 **8.** (a) 0.2703 (b) 9.099

Exercise 118 (page 246)

1. proof **2.** proof **3.** 32 **4.** 29.33 Nm **5.** 37.5

6. 7.5 **7.** 1 **8.** 1.67 **9.** 2.67 **10.** 140 m

Index